P9-DXI-802

STANLEY M. CHERIM, Department of Chemistry, Delaware County Community College, Media, Pennsylvania

PRELIMINARY COLLEGE CHEMISTRY

W. B. SAUNDERS COMPANY • PHILADELPHIA • LONDON • TORONTO

W. B. Saunders Company: West Washington Square
Philadelphia, PA 19105

12 Dyott Street
London, WC1A 1DB

833 Oxford Street
Toronto 18, Ontario

The cover and chapter page illustration is from the cover page of a Gnostic treatise including alchemical data. (Courtesy Weidenfeld & Nicolson Ltd., Publishers, *The Art of the Alchemists.*)

Preliminary College Chemistry ISBN 0-7216-2520-7

 Made in the United States of America. Press of W. B. Saunders Company. Library of Congress catalog card number 72-86445

Print No: 9 8 7 6 5 4 3 2

to my parents

PREFACE

To Whom This Book Is Addressed

This book is primarily designed to be the basis of a thorough preparation for the higher level college chemistry courses required of science majors.

There are many excellent books available for nonscience majors. These texts strive to stimulate interest in and knowledge about chemistry. They are concerned about chemistry's relevance to life—pollution, drugs, ecology, medicines, and synthetic products. Many of these texts fulfill their purposes admirably and are to be highly recommended. However, they are not specifically written to meet the needs of many college freshmen who lack the up-to-date background necessary to cope successfully with the more advanced chemistry courses required of students who aspire to become chemists, physicists, biologists, physicians, engineers, and medical and other health-science–related professionals.

Of course, science majors are likely to be concerned about chemistry's relevance to life fully as much as, if not more than, nonscience majors. There is, at any rate, a worldwide hope that this is so. But the author is faced with a tactical problem, viz., the inclusion of these stimulating and relevant topics would bulk out the text to an unreasonable length for a one-semester course. As an alternative, the superb paperback book *Ecology, Pollution, Environment* by Turk, Turk, and Wittes (W. B. Saunders Company, 1972) is highly recommended as a companion text.

An Overview of the Content

The chapters included in this book are carefully selected, since an attempt to cover the full spectrum of subject matter found in a survey course would defeat the intended purpose. The ordering of the chapters is a matter of individual preference, and is certainly subject to modification by the instructor.

It is hoped that the format and content of this text are such that the student will develop a mastery of the fundamental skills related to problem solving in chemistry. The focus is on the systematic methods of interpreting problems, organizing relevant data, estimating the results of calculations, and understanding the importance of dimensional analysis. The use of scientific notation and the thoughtful handling of dimensional units are stressed in numerous examples and practice exercises.

The chapters dealing with chemical formulas, nomenclature, gases, types of reactions, solutions, and the techniques of balancing equations are other specific skill-oriented areas. Chapters and parts of chapters concerned with atomic structure, chemical bonding, behavior of gases, acids and bases, and chemical equilibrium are attempts to communicate some relatively difficult concepts in simplified and readable language.

It is not the author's intention that this text be completely covered in a one-semester or one-quarter course. Preferring to operate on the philosophy that an instructor's job is to *uncover* a book rather than to cover it, the author has left the depth and breadth of topic coverage to the discretion of the instructor, because he is the best judge of the requirements of his college. Furthermore, the amount of material to be covered can be tailored to the need and ability of each student. Topics that may be selectively omitted for classroom discussion will remain as permanent references for students—as need or curiosity becomes apparent. A star (★) will be used to indicate those topics that, in the author's estimation, might be classified as relatively advanced for an introductory level text. These are the topics that will be available for the more able or the more curious student.

Comments on the Chapters

Chapter 1, Measurement: The metric system is emphasized strongly. The use of conversion factors and dimensional analysis is applied to metric unit conversions. While this approach may be questioned as being tedious and a misapplication of dimensional analysis, its use is justified by experience. In 20 years of teaching this method to students from eighth grade through college, the response has been enthusiastic. When students develop a facility in handling the metric system, they find it easy to shorten the steps involved in conversion calculations. Another motive involved in the choice of the conversion-factor method for metric system conversions is simply to get students used to developing and using conversion factors—thousands of which are found in reference tables.

Chapter 2, Matter, Energy, and Change: One may wonder at the inclusion of such energy units as electron-volts, joules, and liter-atmospheres at this level. The author is convinced that students need to realize that energy can be measured in a variety of units and that these units can be interconverted by using conversion factors. The fuller understanding of the applications of energy units (in electrochemistry and thermodynamics, for example) remains a task for higher level chemistry courses. However, it is to the student's advantage to be aware of these units even though they may not be used immediately.

A certain amount of material may be described as physics rather than as chemistry. The author is firmly convinced, however, that these artificial distinctions should be eliminated to the greatest extent possible. One might raise the question, "How can the subjects of atomic structure, chemical bonding, oxidation-reduction, and basic electrochemistry be discussed meaningfully if the student has no conceptual appreciation of the basic units of electrical measurement?" Most chemists will readily agree that there are many physical facts and assumptions that are fundamental to both physics and chemistry.

Chapter 3, Elements, Compounds, and the Mole Concept: Numerous applications of the mole concept in problem solving are provided to convince the student of the importance and centrality of this concept in chemistry.

Chapter 4, The Behavior of Gases: This chapter begins with illustrations of the interrelationships among the parameters that affect the behavior of gases—volume, moles, temperature, and pressure. The kinetic molecular theory is a conceptual foundation. Special emphasis is placed on the application of the equation of state as a method of solving gas problems in terms of the mole concept.

The sequence of topics in this chapter is rather unusual. Most texts prefer a historical approach in which the laws of Boyle and Charles are clearly shown to precede the kinetic-molecular theory. The historical approach does indeed demonstrate the experimental-theoretical interplay which occurs as a scientist attempts to explore and investigate physical phenomena. The author, too, tends to be very favorably disposed toward the bold logic of this method. However, experience sometimes teaches us that the most logical approach is not *necessarily* the best way. Admittedly, this is a highly subjective statement, and, therefore, in the same breath it must be strongly reiterated that the author's choice of sequence is always subject to the individual teacher's revision.

Chapter 5, Atomic Structure and the Periodic Table: In this chapter, the topic of atomic structure is approached historically. The author demonstrates a fearless inconsistency here because he believes the historical approach in atomic structure is also the best one. The fascination of the story seems justification enough. There are so many exciting stories about the lives, experiments, and theories of scientists associated with the discovery of atomic structure that the need for selective omission tends to reduce an author to tears. But it must be so. It remains for the individual instructor to wax eloquent as he supplements the chapter with Rutherford's "gold foil experiment," Milliken's "oil drop experiment," Einstein's fixation with the doctrine of causality, or even the Rayleigh-Jeans "ultraviolet catastrophe."

The numerous problems, from the Bohr atom through the

quantum mechanical model, serve mainly to enhance the student's ability to cope with scientific notation, organization of data, use of conversion factors, dimensional analysis, and the systematic approach to problem solving.

The subsequent discussion of the periodic law and table serves to relate atomic structure to an orderly, meaningful, and useful arrangement of the elements. Emphasis is placed on how the periodic table permits useful generalizations of the physical and chemical properties of the elements.

Chapter 6, Chemical Bonds: The nature of the chemical bond and the principal types of bonds are presented in this chapter. Special attention is paid to understanding the significance of polarity in molecules. Further generalizations on the periodic table are presented.

Chapter 7, Formulas and Nomenclature: This chapter is testimony to the author's belief that careful attention to the language of chemistry is necessary. Many advanced chemistry students experience needless grief because of their uncertainty in writing and naming formulas. Both the Stock system and classical system are included. The basic arithmetic of percentage composition and empirical formulas is contained in this chapter.

Chapter 8, Chemical Equations and Stoichiometry: The importance of being able to classify, write, balance, and interpret chemical equations can hardly be overemphasized. The student is systematically introduced to the art of making meaningful predictions of reaction products by identifying the type of reaction (e.g., decomposition, simple replacement) and by referring to the "activity series" and solubility tables. In this chapter, the required calculations are based on the stoichiometry of molecular equations. The factor-label method is used throughout. Ionic equations, redox equations, and the equations for equilibrium reactions are presented in later chapters.

Chapter 9, Solutions: This fairly comprehensive chapter includes a discussion of the solution process and emphasizes the factor of polarity. Ionic equations are introduced here. Careful attention is paid to the methods of writing and balancing these equations.

Expressing concentrations of solutions is presented in considerable detail. The methods of preparing molar, normal, percentage, formal, and ppm solutions from solid reagents and by dilution of concentrated stock solutions are included.

The colligative properties of solutions, in terms of the effects of the change of vapor pressure with concentration (Raoult's law), complete the chapter.

Chapter 10, Acids and Bases: A progressive introduction to the subject of acids and bases is presented through the Arrhenius, Brønsted, and Lewis concepts. Special emphasis is given to Brønsted equations, where attention is focused on the conservation of ionic charge. Solution stoichiometry and pH are other emphasized topics.

Chapter 11, Chemical Equilibrium: Particular consideration is given to the significance of equilibrium constants. Le Chatelier's principle is central to this chapter. While the discussion of chemical equilibrium is almost entirely qualitative, it is hoped that its inclusion will give the student a sound conceptual basis for the more rigorous experience he will have in a higher level course.

Chapter 12, Oxidation and Reduction: Many examples and exercises are provided to help the student become skilful in balancing redox equations. Both the "oxidation-number" method and the "ion-electron" method are introduced. A section on redox stoichiometry completes the chapter.

Special Features

As an aid to students, *Preliminary College Chemistry* will employ **boldface** print to call attention to technical terms, and *italics* will be used for emphasis. Numerous sample problems are solved in annotated steps. Each chapter is prefaced by a list of *Behavioral Objectives*, which outline just what the student can be expected to *calculate*, *list*, *define*, *describe*, *sketch*, *write*, or *use* as a result of successfully mastering the chapter content. The appendix of the book contains a review of scientific mathematics and a section on the theory and use of the slide rule. The answers to many selected numerical problems are also provided.

A Note of Appreciation

The author is indebted to Professor Norman J. Juster of Pasadena City College for his excellent review of the text, from the initial chapters to the final manuscript. My gratitude is also extended to Professors Raymond T. O'Donnell, Alan Cunningham, John Healy, Douglas S. Russell, Tom Fredricks, Donald Slavin, and William Masterton, all of whom provided many helpful suggestions, constructive criticism, and encouragement during the writing of the manuscript. The comments of Professor Eugene Rochow of Harvard University, who reviewed the completed manuscript, are very much appreciated. For all that may be considered good in my book, these men must be given a significant portion of the credit. The blame for whatever is objectionable will sit squarely on the shoulders of the author who didn't always take their advice. Any errors or suggestions that users of my text see fit to bring to my attention will be deeply appreciated.

The fine artwork of Grant Lashbrook, Lynn Breslin, John Hackmaster, Henry McKee, and Lorraine Battista is gratefully acknowledged. The thoughtful advice and warm cordiality of the staff of W. B. Saunders Company has meant a great deal to me.

Many thanks, also, to Peg Middleton and Marge Pedrick, for their excellent job of typing parts of the manuscript. Finally, another note of thanks to my wife and children, whose patience and encouragement were so important.

Stanley M. Cherim

CONTENTS

APPENDICES

PRELIMINARY COLLEGE CHEMISTRY

CHAPTER 1

Science is valued for its practical advantages, it is valued because it gratifies disinterested curiosity, and it is valued because it provides the contemplative imagination with objects of great aesthetic charm.

J. W. N. Sullivan, *The Limitations of Science**

Behavioral Objectives

At the completion of this chapter, the student should be able to:

1. Distinguish among the metric units of length, volume, and mass.
2. Explain the difference between mass and weight.
3. List and define the metric system prefixes.
4. Convert measurement units within the metric system.
5. Perform dimensional analysis in solving problems.
6. Choose appropriate conversion factors from available tables.
7. Develop conversion factors when data are available.
8. Change one conversion factor to another.
9. Write numbers in the form of scientific notation.
10. List four types of volumetric glassware.
11. Explain what is meant by the TC and TD markings on glassware.
12. List three types of balances used for mass determinations.
13. Distinguish between temperature and heat.
14. Convert temperatures among Fahrenheit, Celsius, and Kelvin scales.
15. Define the terms *density* and *specific gravity.*
16. Calculate densities, volumes, or masses from measurement data.
17. Construct a graph and determine the slope of the line.
18. Calculate proportionality constants from data arranged in tables or graphically represented.
19. Use a hydrometer.

---

* The Viking Press, 1933.

MEASUREMENT

1.1 INTRODUCTION

When the great English scientist James Clerk Maxwell was a small child, he began to demonstrate the powerful driving force of human curiosity. For Maxwell, life in its most meaningful and exuberant sense meant asking questions about the mysteries of our environment. As the young Maxwell examined a mechanical device and asked his father, "What is the 'go' of that," he was raising the kind of question that characterizes the scientific mentality regardless of age. What is it made of? How does it work? What is it used for?

Maxwell typified the scientific approach by going first to the well-spring of recorded human experience. In other words, by learning something about the laws of nature as they were understood in the 19th century, he was better able to distinguish between the physical principles that might provide a sound basis for progress and those that needed to be questioned. It takes a prepared mind to see most clearly and objectively the cause and effect relationships in natural phenomena, and it requires that peculiarly scientific mind to evaluate subjectively the significance of the observations. Making an orderly fabric (scientific discipline) out of a chaotic tangle of threads (isolated facts) can be a beautiful and creative human endeavor. We should be glad that the aesthetic aspects of science persist. It is true that we have moved through the centuries from the art and magic of the alchemists to the very different objectives and methods of modern chemistry, but there is a debt to be acknowledged. Despite the fact that the energies of our ancestors were fruitlessly devoted to the creation of the "philosopher's stone," with which common metals might be changed into gold, and the discovery of the "elixir of life," which would secure eternal youth, science had its raw beginnings in their arcane laboratories.

When the age of charlatans passed, and with them their magical incantations and occult symbols, some light began to creep into the darkness of the dark ages. Soon men dared to question "sacred" authority, and suffered for it. (Was something really true because Aristotle said so?) However, the age of antiscience was nevertheless the time when glassblowers produced retorts, beakers, funnels, and crude volumetric glassware. The potters fashioned earthenware crocks, crucibles, mortars, pestles, and ovens. And the blacksmiths gave us support stands, forceps, tongs, and spatulas.

And so, we have moved from authoritarianism and the black arts to the modern arts, and from there to science—from magic and absolutes to tentative assumptions and systematic experimental testing. This is the scientific method. To paraphrase an old and anonymous saying: In science we have moved from cocksure ignorance to thoughtful uncertainty.

Unfortunately, science has not been a pure and beautiful monument to man's nobler instincts. Chemistry, the science concerned with the structure and behavior of matter, has provided more than an adventure in living. It has been, and still is, the basic tool for war's death and destruction. Problems of environmental pollution provide new and critical challenges and teach us something about moral responsibility. The poisoning of our waters and air brings the relationship between science and life into a new and very special focus. The young Maxwells of our age are likely to ask, "What is the 'nice' of it?" in addition to the "go" of it. The old and rather meaningless adage "Science for science's sake" is giving way to "Science for mankind's sake."

The broad physical science of chemistry is subdivided into specialty areas. *Inorganic* chemistry deals with all the known elements and combinations of elements in terms of their structures and changes. *Organic* chemistry is specially concerned with compounds of carbon, which are most familiar to us in the form of living creatures and synthetic fabrics and plastics. *Analytical* chemistry is involved with finding out what a substance is composed of and how much is there. *Physical* chemistry is most directly concerned with the complexity and mechanism of chemical change, and its relationship to energy. *Biochemistry* is naturally tied in with the nature of the chemical changes that sustain life. It is difficult to find any aspect of life upon which chemistry does not impinge. Think of ways in which the adventure of chemistry is apparent in economics, philosophy, politics, theology, business, and art.

Any serious attempt to learn something about the effect of forces and energy on the structure and behavior of matter requires an understanding of the concept of measurement. How else can the observations of "how much," "how many," "how far," and "how fast" be described? Measurement is a fundamental effort needed to increase the orderly expansion of our knowledge. Dimensional units are a must if sense is to be made of linear distance, mass, time, and temperature. And an understanding of how these basic units are interrelated is necessary, so that the more sophisticated phenomena of energy, density, force, and volume can be understood. The current international standards for measurement constitute what is known as the *metric system.** Throughout this book, the fact that all measurements include both a *number* and a *dimension* will become obvious.

1.2 THE METRIC SYSTEM

The metric system (or the *Systeme Internationale d'Unités*, SI—the International System of Weights and Measures) is the measuring system used throughout most of the world in everyday living, and by practically all chemists. Regardless of what is being measured, be it *length*,

---

* By 1975, the United States may be the only country in the world that does not use the metric system as the official standard for everyday commerce as well as for scientific work.

area, *volume*, *mass* (or *weight*, as a loosely applied synonym), or *temperature*, the system has the advantage of being directly related to the commonly used base of ten. Conversions among different units amounts to a simple change in the order of magnitude (a factor of 10). Our familiar English system of weights and measures does not enjoy this advantage. One cannot proceed from inches to feet to yards to miles simply by using a factor, or a multiple, of ten.

A marked increase in efficiency is gained when the metric system is used exclusively. The conversion of metric units to inches, quarts, and pounds is time-consuming and unproductive. It is analogous to communicating in a foreign language: until it is possible to *think* in the other tongue, instead of mentally constructing an English sentence and translating word-for-word, communication is slow and frustrating. Similarly, chemistry students have to reach the point where they can think quickly and confidently in metric units.

It takes time and effort to master the various units and prefixes that describe the "what," "how much," and "how many" of our observations. But the effort is worth it. It is helpful to remember that our American monetary system is basically a metric one.

1.3 METRIC UNITS OF LENGTH AND AREA

The basic unit of linear measure is the **meter (m).** Think of this roughly as a large stride, or the distance from your *left* ear to the fingertips of your outstretched *right* hand, as illustrated in Figure 1–1. Lengths usually are measured by a meter stick or centimeter ruler, shown in Figure 1–2.

Figure 1–1 A rough approximation of a meter.

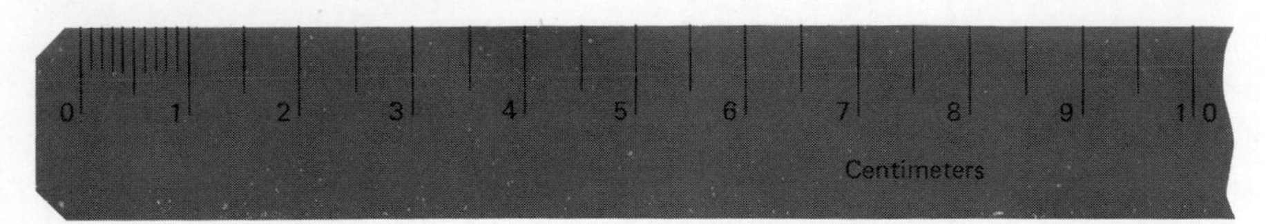

Figure 1–2 A centimeter rule.

A hundredth of a meter is a **centimeter (cm),** the prefix *centi* always meaning a hundredth part of something, just as a cent is a hundredth of a dollar:

$$1 \text{ cm} = 0.01 \text{ m} = \frac{1}{100} \text{ m} = 10^{-2} \text{ m*}$$

The centimeter is about equal to the diameter of a stick of chalk, as illustrated in Figure 1–3.

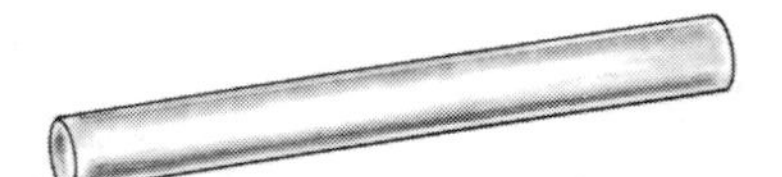

Figure 1–3

The diameter of a stick of chalk $\cong$ 1 cm.

The large practical unit of length is equivalent to 1000 meters and is called a **kilometer (km).** The prefix *kilo* always means a multiple of 1000. For example, a "kiloglop" would mean 1000 "glops." The 60 mile distance between Philadelphia and the Atlantic Ocean might be visualized (by easterners, at any rate) as being about 100 km.

Examining the smaller units of linear measure, we find the **millimeter (mm).** This is one-thousandth of a meter, or one-tenth of a centimeter. The millimeter is approximately equal to the thickness of a dime (Fig. 1–4). The prefix *milli* always means a thousandth part of the fundamental unit:

$$1 \text{ mm} = 0.001 \text{ m} = \frac{1}{1000} \text{ m} = 10^{-3} \text{ m}$$

$$1 \text{ mm} = 0.1 \text{ cm} = \frac{1}{10} \text{ cm} = 10^{-1} \text{ cm}$$

The word "millimeter" is familiar to almost everyone because photo-

thickness of a dime

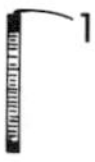

1 mm

Figure 1–4 The thickness of a dime $\cong$1 mm.

* See Appendix A for an explanation of scientific notation. Also included in Appendix A are some basic rules for working mathematically with numbers having exponents—i.e., addition, subtraction, multiplication, and division.

graphic film is commonly described this way. One can take 8 mm home movies or use 35 mm film for color slides.

Further subdivisions are more difficult to visualize. The next one is the **micrometer, μm** (μ, the Greek letter mu). The micrometer is a thousandth of a millimeter and a millionth of a meter. The dimensions of bacteria might be measured in micrometers. The prefix *micro* always means a millionth part of the basic unit:

$$1\ \mu\text{m} = 0.000001\ \text{m} = \frac{1}{1{,}000{,}000}\ \text{m} = 10^{-6}\ \text{m}$$

Subdividing the micrometer into a thousand parts produces the **nanometer (nm).*** This tiny unit of measure is most appropriate for measuring large molecules and lengths of light waves. The wavelength of green light is illustrated in Figure 1–5. The nanometer will be

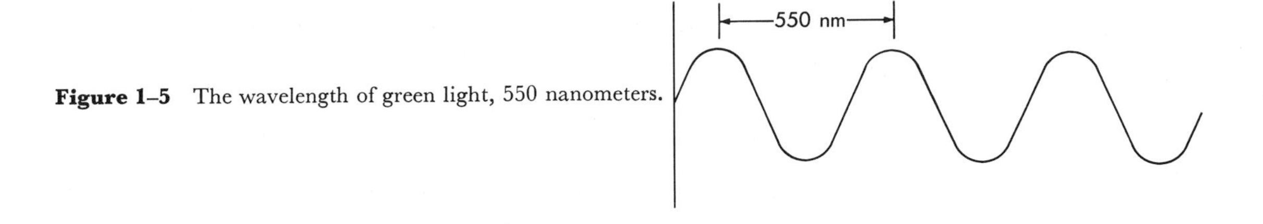

Figure 1–5 The wavelength of green light, 550 nanometers.

observed to have special application in solving the problems in Chapter 5, Atomic Structure and The Periodic Table.

$$1\ \text{nm} = 0.000000001\ \text{m} = \frac{1}{1{,}000{,}000{,}000}\ \text{m} = 10^{-9}\ \text{m}$$

The smallest useful unit of linear measure is the **angstrom (Å),** and it is a tenth of a nanometer. Angstroms are used to measure the dimensions of small molecules and individual atoms as well as very short wavelengths of light. A typical atom has a diameter of one to two angstroms.

$$1\ \text{Å} = 0.1\ \text{nm} = \frac{1}{10}\ \text{nm} = 10^{-1}\ \text{nm} = 10^{-8}\ \text{cm} = 10^{-10}\ \text{m}$$

Three smaller units that should be mentioned briefly are the *picometer* (pm), the *femtometer* (fm), and the *attometer* (am). These incredibly small units represent 10^{-12} m, 10^{-15} m, and 10^{-18} m, respectively. Although they will not be included in further discussions

---

* It should be noted that some people have not shifted completely to the term nanometer. Many older books still use the term **millimicron** (mμ) to indicate a billionth part (10^{-9}) of a meter.

because their practical application in chemistry is limited, the student should be aware of the existence of such units.

Before continuing with other metric units, the linear system will be examined with regard to the method of performing conversions between units.

1.4 CONVERSIONS BETWEEN UNITS

A practically foolproof way of interconverting metric units is found in the use of **dimensional analysis** and **conversion factors.** Dimensional analysis involves the logical and very desirable method of assigning units, such as centimeters, grams, seconds, and calories, to the appropriate numbers when performing calculations. Dimensional analysis is sometimes described as the **unit factoring** method, or the **factor-label** method. Regardless of what it is called, this method is used throughout this text and is strongly recommended. A conversion factor is a numerical value that relates different units of measure and permits easy changes between the units by multiplication or division.

A conversion factor is obtained by the following steps:

1. Take a fundamental equivalence, such as 1 cm = 0.01 m.
2. Express the equivalence by scientific notation:

$$1 \text{ cm} = 10^{-2} \text{ m}$$

3. For uniformity and convenience, restate the equivalence in terms of positive exponents:

$$10^2 \text{ cm} = 1 \text{ m}$$

4. If 100 centimeters equal 1 meter, the relationship can be stated as:

$$10^2 \text{ cm } \textbf{PER} \text{ m}$$

5. The term **per** signifies a fraction:

$$10^2 \text{ cm/m} \qquad \text{or} \qquad \frac{10^2 \text{ cm}}{\text{m}}$$

6. The dimensions cm and m are *first power* expressions, although the notations cm^{+1} and m^{+1} are not normally used. However, the reciprocal of $\frac{1}{\text{m}^{+1}}$ is m^{-1}, and the expression $\frac{10^2 \text{ cm}}{\text{m}}$ can be written 10^2 cm m^{-1}.
7. This technique of assigning a negative exponent to a dimension may be translated as **per**; therefore:

10^2 cm m^{-1} means 100 centimeters per meter

8. Most important, 10^2 cm m^{-1} is a *conversion factor.*

EXAMPLE 1.1

How many meters are there in 12,500 centimeters?

1. In scientific notation, 12,500 cm $= 1.25 \times 10^4$ cm.

2. Visualizing the change, it would seem that many centimeters (small units) are equal to few meters (comparatively large units). The mathematical operation most likely to produce a small numerical value is *division.* Divide 1.25×10^4 cm by the conversion factor:

$$\frac{1.25 \times 10^4 \not{c}\text{m}}{10^2 \not{c}\text{m m}^{-1}} = \underset{\text{answer}}{1.25 \times 10^2 \text{ m}}$$

Note: This simple calculation shows dimensional analysis in action. The centimeter units cancel out and the reciprocal of 1/m^{-1} is meters—exactly the units required:

$$\frac{1}{\text{m}^{-1}} = \frac{\text{m}}{1}$$

The final value for the order of magnitude was found according to the rule for division of exponents (see Appendix A):

$$\frac{10^4}{10^2} = 10^2$$

subtract 2 from 4

3. Assume that the 1.25×10^4 cm was mistakenly multiplied by the conversion factor:

$$1.25 \times 10^4 \text{ cm } (10^2 \text{ cm m}^{-1}) = 1.25 \times 10^6 \text{ cm}^2 \text{ m}^{-1}$$

The answer 1.25×10^6 *centimeters squared per meter* is nonsense, and therefore demonstrates the foolproof nature of dimensional analysis.

EXAMPLE 1.2

Convert 5.50×10^{-5} cm to nm.

1. Obtain an equivalence between centimeters and nanometers from Table 1–1:

TABLE 1–1 CONVERSION FACTORS (LINEAR MEASUREMENT)

Unit	Symbol	Conversion Factor
kilometer	km	10^3 m km^{-1}
meter	m	basic unit
centimeter	cm	10^2 cm m^{-1} 10^5 cm km^{-1}
millimeter	mm	10^1 mm cm^{-1} 10^3 mm m^{-1}
micrometer	μm	10^6 μm m^{-1} 10^4 μm cm^{-1} 10^3 μm mm^{-1}
nanometer (millimicron)	nm (mμ)	10^9 nm m^{-1} 10^7 nm cm^{-1} 10^6 nm mm^{-1} 10^3 nm μm^{-1}
angstrom	Å	10^{10} Å m^{-1} 10^8 Å cm^{-1} 10^7 Å mm^{-1} 10^4 Å μm^{-1} 10^1 Å nm^{-1}

10^7 nm = 1 cm, which means:

10^7 nm *per* cm, written as:

10^7 nm cm^{-1}, the conversion factor

2. Try multiplication, since one centimeter equals millions of nanometers:

$$5.50 \times 10^{-5}\ \cancel{\text{cm}}\ (10^7\ \text{nm}\ \cancel{\text{cm}}^{-1}) = 5.50 \times 10^2\ \text{nm}$$

a. cm and cm^{-1} cancel out because cm^{+1} (cm^{-1}) = cm^0 = 1.
b. The product of 10^{-5} (10^7) = 10^2.

3. Again, observe the foolproof aspect of dimensional analysis. Assume that division by the conversion factor was used:

$$\frac{5.50 \times 10^{-5}\ \text{cm}}{10^7\ \text{nm cm}^{-1}} = 5.50 \times 10^{-12}\ \text{cm}^2\ \text{nm}^{-1}$$

a. The reciprocals of $\frac{1}{10^7\ \text{nm cm}^{-1}}$ are $\frac{10^{-7}\ \text{nm}^{-1}\ \text{cm}^{+1}}{1}$.

b. Simplifying the expression

$$\frac{5.50 \times 10^{-5}\ \text{cm}^{+1}\ (10^{-7}\ \text{nm}^{-1}\ \text{cm}^{+1})}{1} = 5.50 \times 10^{-12}\ \text{cm}^2\ \text{nm}^{-1}$$

c. The answer, 5.50×10^{-12} *centimeters squared per nanometer,* is nonsense, as expected.

Exercise 1.1

Perform the following linear measure conversions:

a. 1.53 cm = ____ m

b. 0.027 km = ____ m

c. 153 nm = ____ cm

d. 2.41×10^3 Å = ____ cm

e. 1.2 μm ____ m

f. 8.1×10^{-5} m = ____ nm

g. 2.3×10^7 nm = ____ km

h. 1.5×10^{-6} cm = ____ nm

i. 0.44 Å = ____ μm

j. 3.5×10^{-3} m = ____ mm

1.5 METRIC UNITS OF VOLUME

The basic unit of volume is the **liter (ℓ).** This is close to the size of the familiar quart of milk. Today, by international agreement,* the liter is equal to 1000 cubic centimeters (cm^3). Therefore, cubic centimeters and milliliters (a thousandth of a liter) can be used interchangeably. This double system of volume measure is similar to the choice in the English system between the units of cubic inches, cubic feet, pecks, and bushels, and the units of ounces, pints, quarts, and gallons. In short, the liter and its subdivisions are commonly applied to liquid and gas volumes, while solids are likely to be measured in cm^3.

Since 1000 cm^3 equals a liter, a single cm^3 equals a **milliliter (ml),** as illustrated in Figure 1–6.*

* In 1963, the International Bureau of Standards modified the *near* equality of a liter and 1000 cm^3 so that they have become *exactly* equal.

Figure 1–6

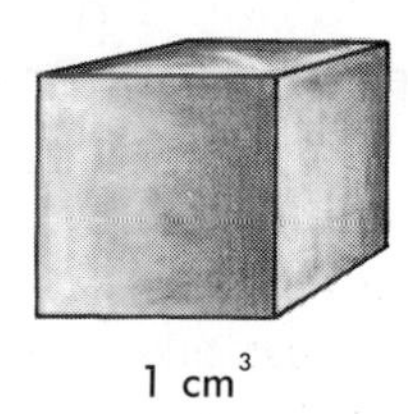

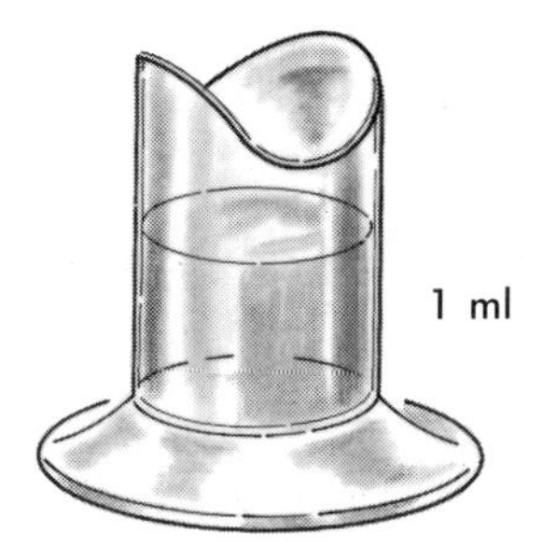

$$1 \text{ ml} = 0.001 \text{ liter} = \frac{1}{1000} \text{ liter} = 10^{-3} \text{ liter}$$

It is extremely useful to note that one ml of cold water weighs about a gram. In fact, the units of mass were based on the weight of a milliliter of water at 4°C. The temperature is important because the volume occupied by a sample of water varies with the temperature.

The one other useful unit of volume is the **microliter (μl).** The microliter volume is symbolized also by the Greek letter *lambda* (λ). The microliter is extremely small. It is a millionth of a liter and a thousandth of a milliliter. Microliter volumes are used in the laboratory, however, and there are small (and expensive) types of **pipets** and syringes used for very small measurements (Fig. 1–7).

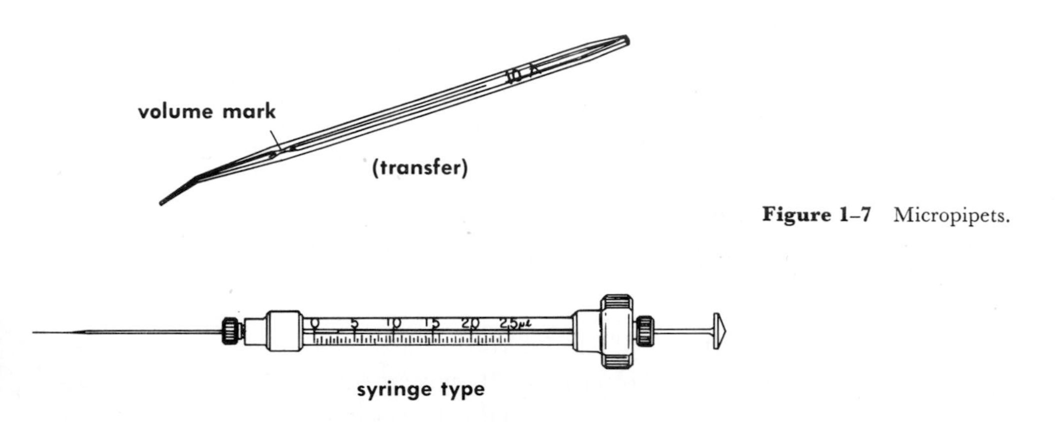

Figure 1–7 Micropipets.

A variety of types of pipets, cylinders, burets, and flasks are used to measure volumes. These items vary in capacity, purpose, and precision. The *volumetric flasks* and *graduated cylinders*, for example, are designed to measure what they *contain* more accurately than what they *deliver* when poured. The amount of liquid adhering to the walls of such vessels prevents accurate measurements of poured volumes. A piece of glassware designed for contained volume measurement is usually marked *TC* (to contain). See Figure 1–8 for illustrations of volumetric flasks and graduated cylinders.

A *buret* and a *pipet* (really a simplified buret) are more likely to be made to measure accurately the volume delivered, and they are appropriately marked *TD* (to deliver). Pipets are designed to be used much like a straw. When poisonous or otherwise dangerous solutions are to be drawn, a rubber bulb or a more sophisticated controlled-volume plunger attachment should be used rather than sucking by mouth. Many people prefer the more cautious approach in every case. Some examples of burets are illustrated in Figure 1–9.

When reading the volume of liquid in a graduated piece of glassware, one must take into account the tendency of liquids to adhere to the walls of the vessels. The standard procedure is to hold the vessel at

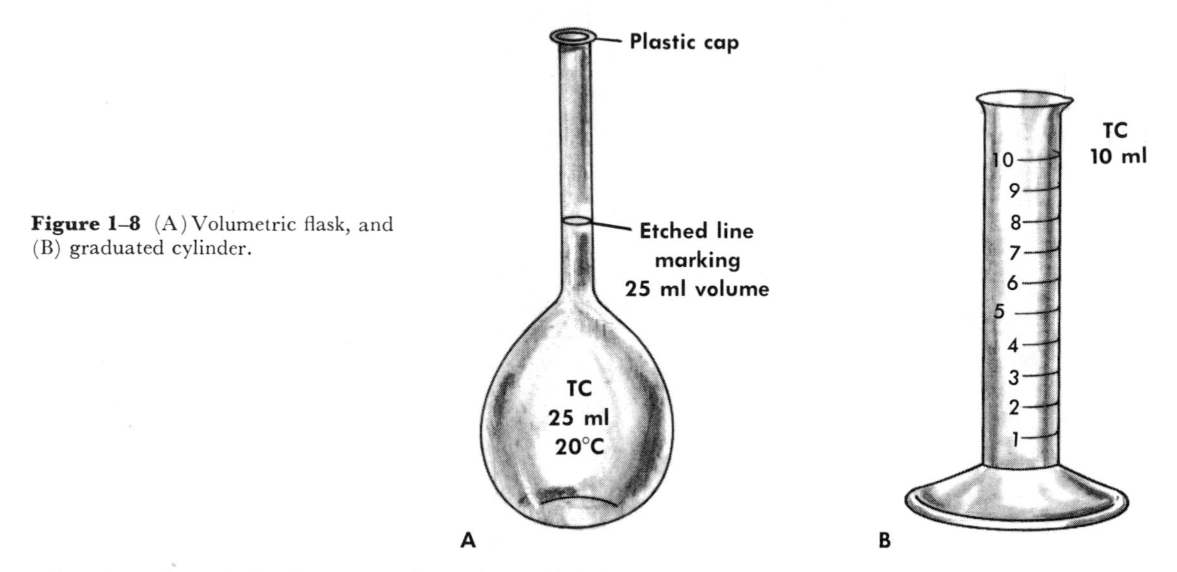

Figure 1–8 (A) Volumetric flask, and (B) graduated cylinder.

eye level and read the lowest point of the liquid when it sags, or the highest point when it bulges upward. The name given to the surface of the liquid is the **meniscus** (Fig. 1–10). Some plastic measuring vessels eliminate this bulge, because water or water solutions do not adhere to such material.

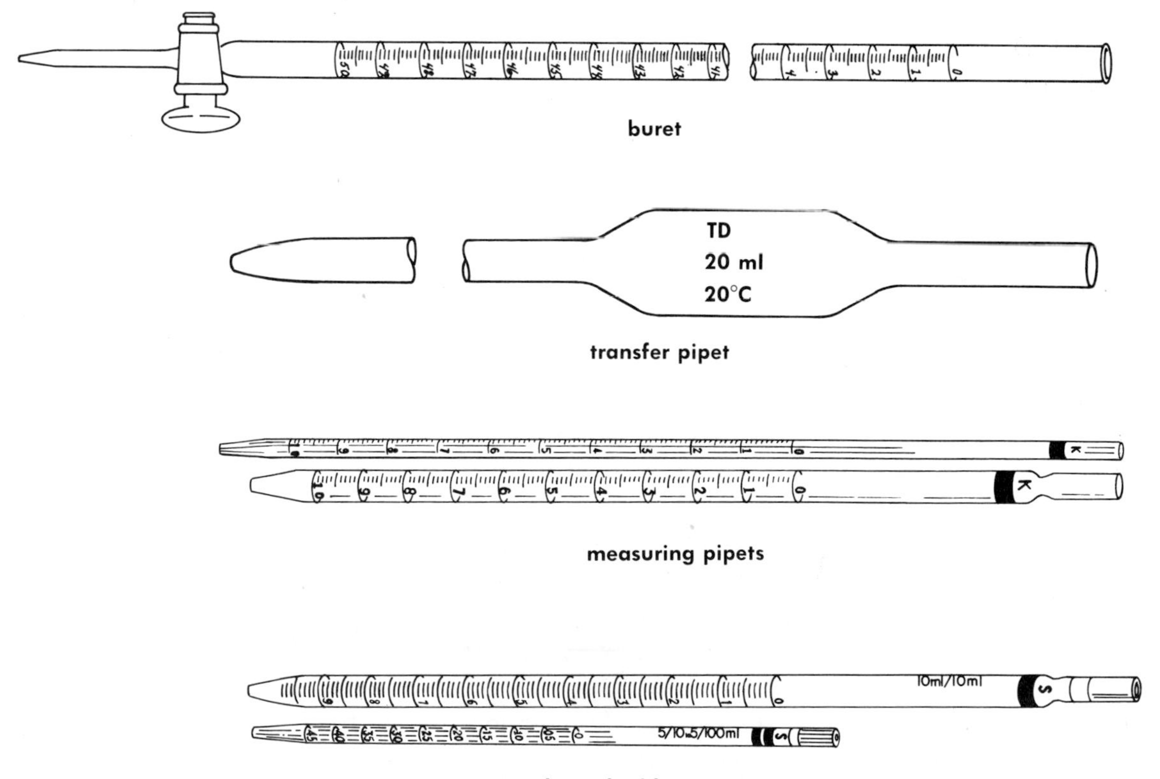

Figure 1–9 Variety of pipets.

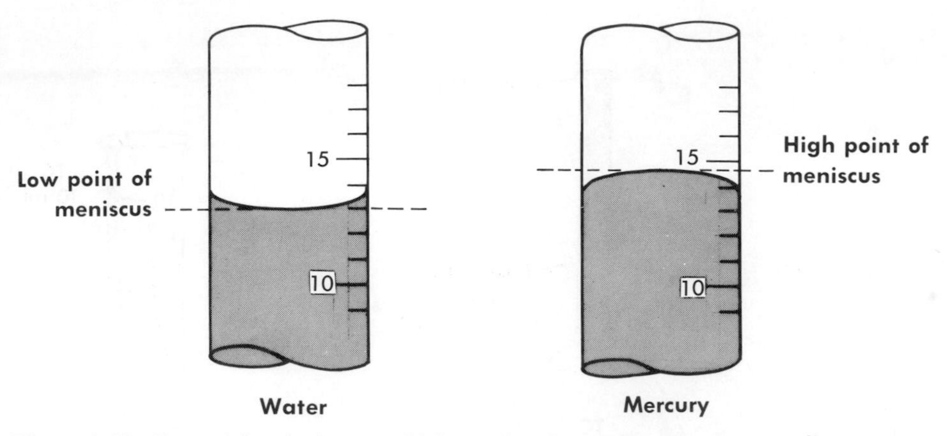

Figure 1–10 Determining the lowest or highest point of a meniscus in volume reading.

TABLE 1–2 CONVERSION FACTORS (VOLUME)

Unit	Symbol	Conversion Factor
liter	ℓ	basic unit
milliliter	ml	10^3 ml ℓ^{-1}
microliter (lambda)	μl (λ)	10^3 μl ml^{-1} 10^6 μl ℓ^{-1}
cubic meter	m^3	basic unit
cubic centimeter	cm^3 (cc)	10^6 cm^3 m^{-3}

EXAMPLE 1.3

Convert 300 ml to liters.

1. Change 300 ml to scientific notation form:

$$300 \text{ ml} = 3.00 \times 10^2 \text{ ml}$$

2. The conversion factor is 10^3 ml ℓ^{-1}.
3. Considering that the liter is a much larger unit of volume than the milliliter, it would seem correct to predict that many ml equal few liters. Try division as the operation that will produce a smaller numerical value:

$$\frac{3.00 \times 10^2 \not{m}l}{10^3 \not{m}l\ \ell^{-1}} = 3.00 \times 10^{-1}\ \ell = 0.3\ \ell$$

4. Once again, to support the foolproof aspect of dimensional analysis, multiply

$$3.00 \times 10^2 \text{ ml } (10^3 \text{ ml } \ell^{-1}) = 3.00 \times 10^5 \text{ ml}^2\ \ell^{-1}$$

The resulting units, milliliters squared per liter, are totally meaningless.

EXAMPLE 1.4

Express 0.4 μl of water in ml units.

1. Change 0.4 μl to scientific notation form:

$$0.4\ \mu\text{l} = 4 \times 10^{-1}\ \mu\text{l}$$

2. The conversion factor is $10^3\ \mu\text{l ml}^{-1}$.
3. Since a microliter is such a small fraction of a milliliter, a very small numerical value is expected as an answer. Division seems to be indicated:

$$\frac{4 \times 10^{-1}\ \cancel{\mu\text{l}}}{10^3\ \cancel{\mu\text{l}}\ \text{ml}^{-1}} = 4 \times 10^{-4}\ \text{ml}$$

Exercise 2

Perform the following volume conversions:

a. 27.2 ml = ____ ℓ

b. 0.084 ℓ = ____ ml

c. 176 μl = ____ ml

d. 20 ml = ____ μl

e. $5.6 \times 10^7\ \mu$l = ____ ℓ

f. 0.002 ℓ = ____ λ

g. 5.4 cm^3 = ____ ml

h. 260 cc = ____ ℓ

i. $3.1 \times 10^{-5}\ \text{m}^3$ = ____ cm^3

j. 0.049 ml = ____ μl

1.6 METRIC UNITS OF MASS (WEIGHT)

The terms **mass** and **weight** are loosely equated in practice. However, there is a difference that should be pointed out for the sake of accuracy. Mass is a description of the quantity of matter in an object, while weight describes the pull of gravity on the object. Consider a block of copper, for example. Regardless of where this block of copper is measured or what the conditions of temperature or pressure may be, the *amount* of it is constant. Under the sea, in outer space, or on the moon, this block of copper represents a precise amount of matter, measured in units of *mass*. The *weight* is something else. In outer space the block may be weightless. On the moon, it weighs only $\frac{1}{6}$ of its weight on earth. The reasons are that in outer space the effects of gravity may be cancelled by the movement of the block away from the earth, and on the moon the pull of gravity is only about $\frac{1}{6}$ that of the earth's.

Why, then, are mass and weight commonly used as synonymous terms? Very simply, since the standards for mass were established on the surface of the earth, and since they were originally derived from weight measurements in surface-of-the-earth laboratories, the distinction has become negligible. Until scientists find themselves in extraterrestrial laboratories, this distinction need not be a critical one.

The most frequently used unit of mass is the **gram(g).** On the earth's surface, a gram might be thought of as being almost as heavy as a dime. The units of mass are definitely preferred for the accurate designation of quantity of a solid. It is pointless to talk about so many milliliters of salt, for example, because such a volume measurement unavoidably includes the air spaces among the particles.

The unit of mass that is a multiple of a gram is the **kilogram (kg):**

$$1\ \text{kg} = 1000\ \text{g} = 10^3\ \text{g}$$

An example that may help in the visualization of a kilogram is the five pound bag of potatoes, which is roughly equivalent to two kilograms.

High-precision work often involves the use of the **milligram (mg),** which is three orders of magnitude lighter than the gram:

$$1\ \text{mg} = 0.001\ \text{g} = \frac{1}{1000}\ \text{g} = 10^{-3}\ \text{g}$$

A small "pinch" of an aspirin tablet would probably have a mass (or, "weigh") in the milligram range. The weight of ink used in a person's signature would be about three or four milligrams.

When extremely potent or expensive materials are required in an experiment, it may be necessary to use an even smaller unit of mass called the **microgram (µg).** The microgram is sometimes symbolized by the Greek letter gamma (γ). The prefix *micro* indicates one-millionth, as usual:

$$1\ \mu\text{g} = 0.000001\ \text{g} = \frac{1}{1{,}000{,}000}\ \text{g} = 10^{-6}\ \text{g}$$

TABLE 1–3 CONVERSION FACTORS (MASS)

Unit	Symbol	Conversion Factor
kilogram	kg	10^3 g kg^{-1}
gram	g	basic unit
milligram	mg	10^3 mg g^{-1}
microgram (gamma)	µg (γ)	10^3 µg mg^{-1} 10^6 µg g^{-1}

The actual type of balance that should be used to determine mass depends on the amount of material to be weighed and the required precision. It would not make sense to attempt to weigh several hundred grams of a substance on an analytical balance that is designed to weigh small samples in the milligram range. It would be equally poor procedure to use a coarse triple-beam or double-pan balance when directions call for weighing out 0.055 g of a solid. The guidelines are based on common sense. When a precision (reproducibility) of 0.1 g is called for, a relatively coarse triple-beam or double-pan balance is satisfactory. Examples of such balances are illustrated in Figure 1–11.

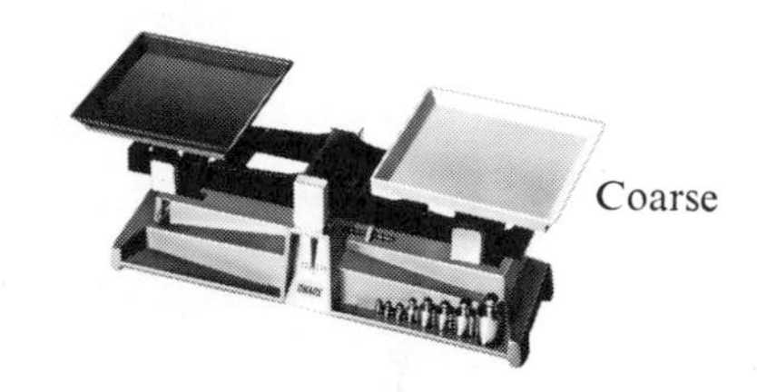

Figure 1–11 Typical coarse balances.

If the directions call for a precision of 0.01 g, 0.001 g, or 0.0001 g, higher precision analytical balances are used. Some examples of modern analytical balances are illustrated in Figure 1–12.

Mettler

Figure 1–12 Analytical balances.

EXAMPLE 1.5

Convert 4.2 mg to grams.

1. Select the conversion factor:

$$10^3 \text{ mg g}^{-1}$$

2. Since milligrams are small compared to grams, many mg will equal few g. The need for a smaller number as an answer suggests division:

$$\frac{4.2 \text{ mg}}{10^3 \text{ mg g}^{-1}} = 4.2 \times 10^{-3} \text{ g}$$

3. Assume an error in reasoning in order to support the foolproof aspect of dimensional analysis:

$$4.2 \text{ mg } (10^3 \text{ mg g}^{-1}) = 4.2 \times 10^3 \text{ mg}^2 \text{ g}^{-1}$$

The units, milligrams squared per gram, are nonsense, as expected.

EXAMPLE 1.6

Change 0.0067 mg to micrograms.

1. Express the value in scientific notation:

$$0.0067 \text{ mg} = 6.7 \times 10^{-3} \text{ mg}$$

2. Find the conversion factor:

$$10^3 \ \mu\text{g mg}^{-1}$$

3. Reasoning that micrograms are very small units, multiplication seems to be indicated:

$$6.7 \times 10^{-3} \text{ mg } (10^3 \ \mu\text{g mg}^{-1}) = 6.7 \ \mu\text{g}$$

Exercise 1.3

Perform the following mass conversions:

a. 325 mg = ____ g

b. 1.4 g = ____ kg

c. 0.0033 kg = ____ mg

d. 3×10^{-2} mg = ____ μg

e. 1.5×10^{-7} g = ____ μg

f. 4280 γ = ____ g

g. 0.09 mg = ____ g

h. 6×10^{-4} kg = ____ g

i. 0.023 γ = ____ mg

j. 72 mg = ____ μg

1.7 TEMPERATURE

Temperature is a measure of how vigorously molecules of a given mass of matter are moving in straight lines in all directions.

In other words, temperature is a measure of the degree of random motion of molecules. Temperature is *not* heat, nor is it a measure of heat directly. Heat is a form of energy and its unit of measure is a calorie rather than a degree. Temperature differences can be used to find calories. The topic of heat will be taken up in the next chapter.

Fahrenheit and Celsius (Centigrade) Scales

A feature common to both the **Fahrenheit** and **Celsius** temperature scales is the arbitrary zero point. An arbitrary zero signifies that it is a starting point that has been selected for some convenience. All temperatures above the starting point would have positive values, while those temperatures below the starting point would be negative. The difference between the zero points on the Fahrenheit and Celsius scales is 32 Fahrenheit units because 0°F is set 32 Fahrenheit degrees lower than the freezing point of water (which is the starting point for the Celsius scale). See Figure 1–13.

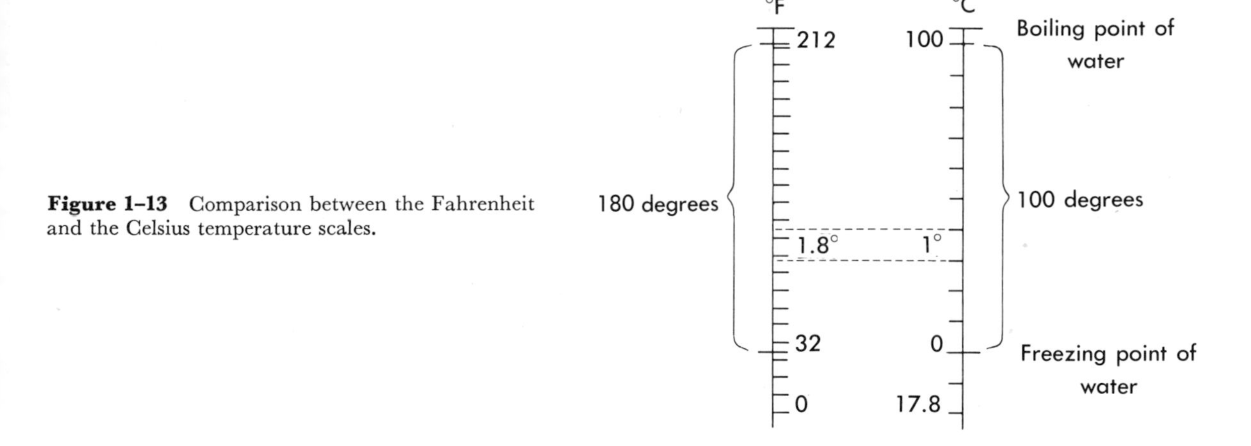

Figure 1–13 Comparison between the Fahrenheit and the Celsius temperature scales.

Since both scales have definite degree intervals, and they measure the same phenomenon in the same way, they may be compared graphically. (See Appendix A for the basic rules of graphing.) The slope of the line will yield the proportionality constant—a conversion factor in this case—and the y intercept then becomes the **correction factor** (Fig. 1–14).

The mathematical relationship that produces the straight line observed in Figure 1–14 (or any equation that expresses such a relationship) has the form: $y = mx + b$. The value for y represents a Fahrenheit reading, which is equivalent to the conversion factor, m (1.8 Fahrenheit degrees per Celsius degree), times the x value for the

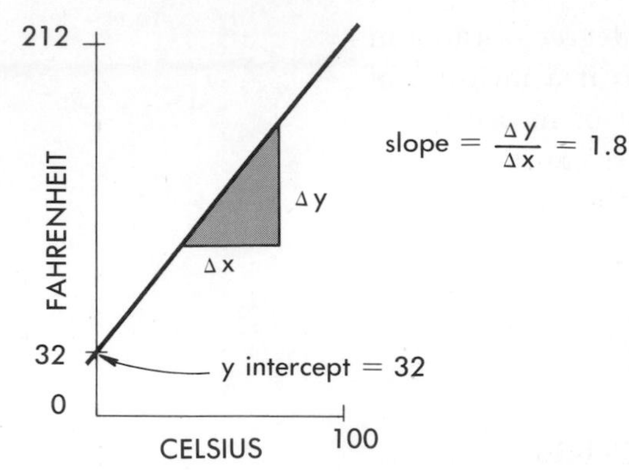

Figure 1–14

Celsius reading, plus b, which represents the correction factor, 32. Substituting:

$$y = mx + b$$

$$F = 1.8\ C + 32°$$

If the temperature conversion involves changing a known Fahrenheit temperature to degrees Celsius, the equation can be rearranged algebraically. (See Appendix A for some basic rules of algebra.)

(1) $y = mx + b$ (subtract b from the equation)

(2) $-b + y = mx$ (divide the equation by m)

(3) $\frac{y - b}{m} = x$

Substituting:

$$C = \frac{F - 32°}{1.8}$$

EXAMPLE 1.7

If the temperature in the laboratory is 72°F, what is this in degrees Celsius?

1. Select the appropriate form of the equation:

$$C = \frac{F - 32°}{1.8}$$

2. Substitute the data:

$$C = \frac{72° - 32°}{1.8}$$

3. Simplify the numerator, estimate the answer, and divide by slide rule. (See Appendix B for the technique of using the slide rule.)

$$C = \frac{40^\circ}{1.8} = \text{approximately } 40^\circ/2 = 20^\circ$$

$$\text{slide rule} = 22.2^\circ \text{ C}$$

EXAMPLE 1.8

The temperature of a dry ice and alcohol mixture is about $-70°C$. Convert this to Fahrenheit.

1. Select the appropriate form of the equation:

$$F = 1.8\ C + 32^\circ$$

2. Substitute the data:

$$F = 1.8\ (-70^\circ) + 32^\circ$$

3. Estimating the answer to be about -100, multiply by slide rule and add 32:

$$F = -126^\circ + 32^\circ$$

$$\text{answer} = -94^\circ \text{ F}$$

The Kelvin (Absolute) Scale

The striking difference between the Kelvin scale and other temperature scales is the fact that the zero point is not an arbitrary starting point. It is the point that indicates a theoretical absence of temperature, i.e., where molecular motion theoretically stops. Basing the definition on the behavior of "perfect" gases, it is the point at which a volume of a perfect gas would be reduced to zero. It should be mentioned that a gas volume reduction to zero is a mathematical concept and not a reality. There is no known substance that exists in the gaseous state below 4°K. The relationship between gas volume and temperature will be discussed in Chapter 4, The Behavior of Gases.

Using the Celsius scale units to measure absolute zero, it can be shown from the behavior of gases that 273 degrees is the difference between 0°C and 0°K (Fig. 1–15). The equation, easily extracted from a comparison of the two scales, incorporates a correction factor of 273°. It is expressed as:

$$K = C + 273^\circ$$

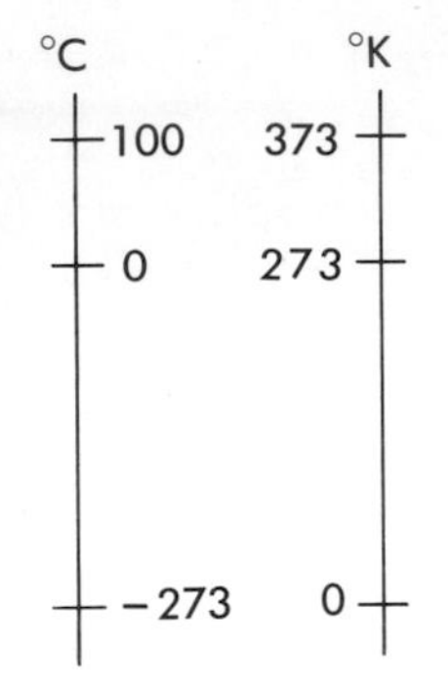

Figure 1–15 Comparison between the Celsius and the Kelvin scales.

EXAMPLE 1.9

If the temperature of a water bath is 27°C, convert this to Kelvin degrees.

1. Write the appropriate equation:

$$K = C + 273°$$

2. Substitute to obtain the answer:

$$K = 27° + 273°$$

$$\text{answer} = 300\ K$$

EXAMPLE 1.10

Liquid oxygen boils at 90°K. What is this in degrees Celsius?

1. Write the appropriate equation:

$$C = K - 273°$$

2. Substitute to obtain the answer:

$$C = 90° - 273°$$

$$\text{answer} = -183°\ C$$

Exercise 1.4

Perform the following conversions:

a. 17° F = ____ °C

b. 200° C = ____ °F

c. 6.82° F = ____ °C

d. −220° C = ____ °F

e. 10° C = ____ °K

f. 115° K = ____ °C

g. −30.5° C = ____ °K

h. 40° F = ____ °K

1.8 DENSITY AND SPECIFIC GRAVITY

A very important physical property of matter is **density.** Densities of materials are used as a fundamental basis for comparison and they have extensive application to the solutions of many practical problems in chemistry. Density is a precise type of description that takes into account both mass and volume.

If a person were asked to compare a room full of cork to a small cube of gold, there might be a nagging suspicion that this is an absurd basis for a meaningful comparison. This suspicion would be well founded. If any comparison is to be meaningful, and perhaps useful, pieces of cork and gold of equal size or equal mass are required. This is what density is describing; it is the mass of a definite volume of some specific substance. The definite volume is usually taken to be one milliliter (or cubic centimeter) for liquids and solids, and one liter for gases. These are known as unit volumes. Therefore, by definition, *the density of a substance is its mass per unit of volume.*

In order to develop this relationship mathematically so that a useful equation may be derived, start with the observation that the volume (size) of an object is directly proportional to its mass (weight). A shorthand expression of this statement is

M	$\propto$	V
mass	sign of proportionality	volume

The proportionality constant, which represents volume change as a function of mass change, can be obtained from the slope of the line when the data are represented graphically. (See Appendix A for an explanation of proportionality.) Consider silver, for example, where the following measurements have been recorded:

2 cm^3 block weighs 21.0 g

3 cm^3 block weighs 31.5 g

4 cm^3 block weighs 42.0 g

5 cm^3 block weighs 52.5 g

The graph, with the calculated value for the slope of the line, is illustrated in Figure 1–16.

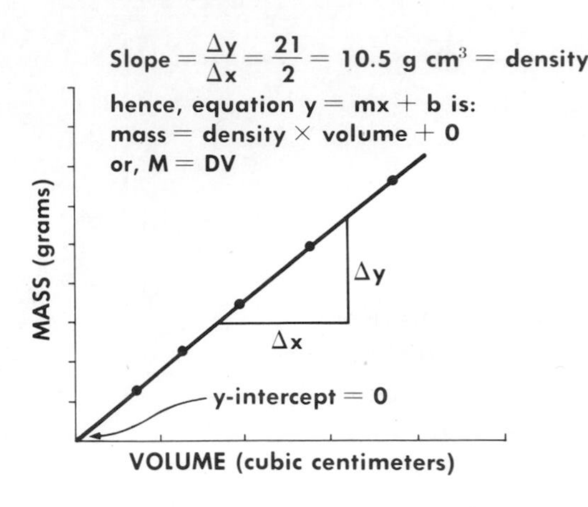

Figure 1–16

The slope, which by definition is also the proportionality constant, is read as 10.5 grams per cubic centimeter. This constant is known as the *density of silver*. The units of density will always be g cm^{-3} (or, g ml^{-1}) for solids such as silver, as a result of dimensional analysis. Since the constant, density, equals the mass of an object divided by its volume, the equation that emerges can be used to solve an impressive variety of problems throughout chemistry:

$$D = \frac{M}{V}$$

EXAMPLE 1.11

Find the density of a plastic substance if a 57 cm^3 block weighs 52 g.

1. Organize the data:

$$D = ?$$
$$M = 52 \text{ g}$$
$$V = 57 \text{ cm}^3$$

2. Use the equation that relates the data mathematically:

$$D = \frac{M}{V}$$

3. Substitute in the equation and estimate the answer:

$$D = \frac{52 \text{ g}}{57 \text{ cm}^3} = \text{slightly less than 1 g per cm}^3$$

4. Find the answer on the slide rule and include dimensional analysis:

$$D = 0.91 \text{ g cm}^{-3}$$

EXAMPLE 1.12

Find the density of benzene (liquid) if 214 ml have a mass of 184 g at 16°C.*

1. Organize the data:

$$D = ?$$
$$M = 184 \text{ g}$$
$$V = 214 \text{ ml}$$

2. Substitute in the equation as in Example 1.11:

$$D = \frac{184 \text{ g}}{214 \text{ ml}} = 0.860 \text{ g ml}^{-1}$$

EXAMPLE 1.13

Find the weight of 30.0 ml of a liquid having a density of 1.84 g ml^{-1}.

1. Organize the data:

$$D = 1.84 \text{ g ml}^{-1}$$
$$M = ?$$
$$V = 30.0 \text{ ml}$$

2. Develop the appropriate equation by simple algebra:

$$D = \frac{M}{V} \quad \text{(multiply the equation by V)}$$

$$VD = \frac{M}{\cancel{V}}\cancel{V} \quad \text{or,}$$

$$M = DV$$

3. Substitute the data and estimate the answer:

$$M = (1.84 \text{ g } \cancel{\text{ml}}^{-1})(30.0 \ \cancel{\text{ml}})$$
$$M = \text{approximately } 2 \times 30 = \sim 60$$

---

* It is often important to note the temperature at which the density is measured, since the volume of an object varies with the temperature.

4. Multiply on the slide rule and complete the dimensional analysis:

$$M = 55.2 \text{ g}$$

EXAMPLE 1.14

If a metal has a density of 9.71 g cm^{-3} and a mass of 205.6 g, what volume will it occupy?

1. Organize the data:

$$D = 9.71 \text{ g cm}^{-3}$$
$$M = 205.6 \text{ g}$$
$$V = ?$$

2. Develop the appropriate equation by simple algebra:

$$D = \frac{M}{V} \qquad \text{(multiply equation by V)}$$

$$M = DV \qquad \text{(divide the equation by D)}$$

$$\frac{M}{D} = \frac{\not{D}V}{\not{D}} \qquad \text{or,}$$

$$V = \frac{M}{D}$$

3. Substitute the data and estimate the answer:

$$V = \frac{205.6 \not{g}}{9.71 \not{g} \text{ cm}^{-3}} = \text{approximately } \frac{200}{10} = 20$$

4. Divide on the slide rule and complete the dimensional analysis:

$$V = 21.2 \text{ cm}^3 \text{ (three significant figures*)}$$

The fact that density has the dimensions of mass and volume is what distinguishes it, as a physical property of matter, from **specific gravity (sp. gr.)**. Specific gravity is a comparison in the form of a fraction, or ratio, between the density of some substance and the density of a standard. Water, very conveniently, has a density of 1.0 g per milliliter at 4°C ($D_{water} = 1.000$ g ml^{-1}) and is the standard for

* See Appendix A for an explanation of significant figures.

solids and liquids at specified temperatures. For gases, the standard is dry *air* at **STP.** STP are the initials for *standard temperature* and *pressure.* In relation to gases, these conditions are 0°C and 1 atmosphere of pressure at sea level. The equation for specific gravity is

$$\text{Sp. Gr.} = \frac{D_{\text{substance}}}{D_{\text{standard}}}$$

EXAMPLE 1.15

What is the specific gravity of sulfuric acid if its density is 1.84 g ml^{-1}?

$$\text{Sp. Gr.} = \frac{D_{\text{sulfuric acid}}}{D_{\text{water}}}$$

$$\text{Sp. Gr.} = \frac{1.84\ \cancel{\text{g ml}^{-1}}}{1.00\ \cancel{\text{g ml}^{-1}}} = 1.84$$

With this answer it can be seen that the numerical values of density and specific gravity for solids and liquids are the same; only the units of mass and volume are missing from the specific gravity value. This is true *only* in the metric system.

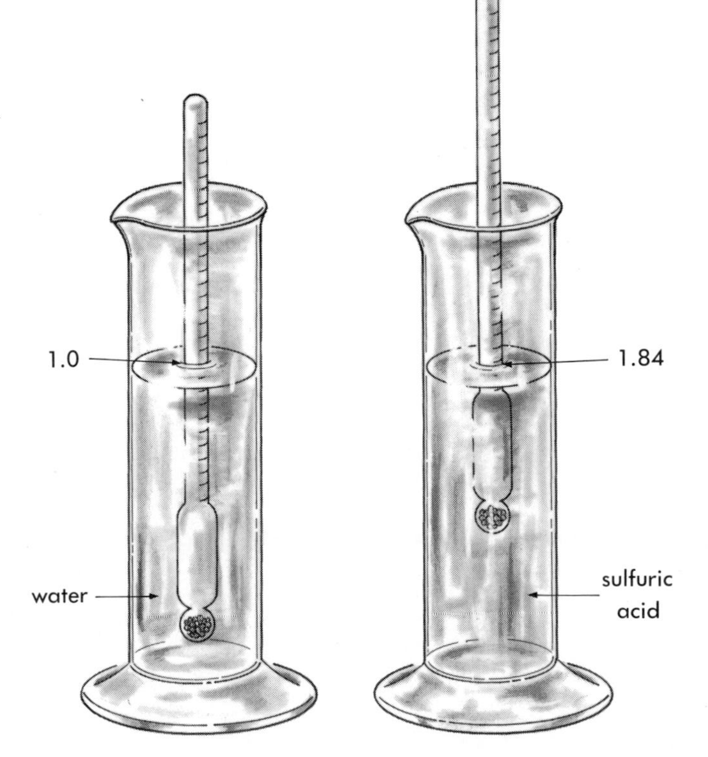

Figure 1–17 Hydrometers are used to measure specific gravity.

A **hydrometer** is an instrument used to measure the specific gravity of liquids. A weighted bulb at the bottom causes the hydrometer to sink until it displaces a volume of liquid that has a weight equal to that of the hydrometer. The specific gravity is read from a calibrated scale that extends above the surface of the liquid, as illustrated in Figure 1–17.

Exercise 1.5

1. What is the density of a liquid if 35.0 ml has a mass of 22.5 g?
2. If the density of mercury is 13.6 g ml^{-1}, what volume of it must be added to 3.2 g of tin to give a total mass of 5.8 g?
3. What volume of brass (D = 8.0 g cm^{-3}) is equal in mass to a 2 cm^3 block of gold (D = 19.3 g cm^{-3})?
4. What is the specific gravity of sulphuric acid if 0.100 liter has a mass of 184.0 grams?
5. What is the density of oxygen if 22.4 liters have a mass of 32.0 grams?

PROBLEMS AND QUESTIONS

1.1 Perform the following metric system conversions for linear measurement:

a. 351 m = ______ km
b. 14 mm = ______ cm
c. 650 nm = ______ cm
d. 52 nm = ______ Å
e. 0.02 mm = ______ μm
f. 1.3×10^{-6} cm = ______ nm

1.2 Metric system mass conversions:

a. 0.4 kg = ______ g
b. 512 mg = ______ g
c. 7×10^4 μg = ______ g
d. 4.5 mg = ______ μg
e. 2×10^{-3} g = ______ mg
f. 6.7×10^5 γ = ______ mg

1.3 Metric system volume conversions:

a. 0.12 ℓ = ______ ml
b. 3.7×10^{-5} ml = ______ μl
c. 185 ml = ______ ℓ
d. 28 ml = ______ cm^3
e. 0.8 μl = ______ ml
f. 5×10^3 λ = ______ ml

1.4 What is the significance of the notations TD and TC on volumetric glassware?

1.5 Compare heat and temperature?

1.6 Perform the following temperature scale conversions;

a. 5 C = ______ F
b. −70 C = ______ F
c. 160 F = ______ C
d. −80 C = ______ K
e. 230 K = ______ C
f. 15 F = ______ K

1.7 If one calorie equals 4.2×10^7 ergs, write a conversion factor for converting ergs to calories.

1.8 What volume is occupied by 454 g of mercury, which has a density of 13.6 grams per milliliter?

1.9 What is the density of a cylindrical rod that is 15.0 cm long and 2.0 cm in diameter, and has a mass of 45.0 g?

1.10 What is the mass of 600 ml of sulfuric acid, which has a specific gravity of 1.54?

1.11 What total volume results when 160.0 g of silver are added to 35.0 ml of mercury? The density of silver is 10.5 g ml^{-1}.

1.12 If a ring weighs 7.3 g in air and 6.9 g when suspended in water, find its density.

CHAPTER 2

It has been said that science has no ethical basis, that it is no more than a cold, impersonal way of arriving at the objective truth about natural phenomena. This view I wish to challenge, since it is my belief that by examining critically the nature, origins, and methods of science we may logically arrive at a conclusion that science is ineluctably involved in questions of values, is inescapably committed to standards of right and wrong, and unavoidably moves in the large toward social aims.

Bentley Glass, *Science and Ethical Values**

Behavioral Objectives

At the completion of this chapter, the student should be able to:

1. Define matter.
2. Classify the types of matter and give examples.
3. Define *elements* and *compounds.*
4. List the states of matter.
5. List four fundamental forces of attraction.
6. Define *entropy* in terms of a driving force in nature.
7. Distinguish between physical and chemical change.
8. List four observations that serve as evidence for a chemical change.
9. List four conditions which may be needed to permit a chemical change to occur.
10. Define the term *catalyst.*
11. Explain the laws of conservation of matter and energy.
12. Distinguish between the terms *energy, force,* and *work.*
13. Define *kinetic energy* in words and by a mathematical equation.
14. Distinguish between *kinetic* and *potential* energy.
15. List and define the fundamental units of energy.★
16. Define the term *specific heat.*
17. Calculate energy values in calories from given data.
18. Use energy conversion factors to change energy units.★
19. Define electromagnetic radiation.★
20. Describe waves in terms of length, frequency, and amplitude.★
21. Calculate solutions to problems involving wavelengths, frequencies, and the speed of light.★
22. List the colors composing the visible range of the electromagnetic spectrum.★
23. Define the term *photon* (or *quantum*).★
24. Calculate the energies of photons at various wavelengths or frequencies.★
25. List and define the principal units of electrical measurement.★
26. Describe, in terms of Coulomb's law, the factors that affect the force of attraction between oppositely charged objects.★
27. Define Ohm's law.★
28. Solve problems involving the relationships between volts, amps, and ohms.★

---

* University of North Carolina Press, 1965.

MATTER, ENERGY, AND CHANGE

Chemistry is often defined as the systematic investigation of the properties, structure, and behavior of matter, and of the changes that matter undergoes. Such a generalized definition raises a number of questions. What is matter? What is meant by change? What are the driving forces behind changes? What is it about the structure of matter that lets it respond to these "driving forces"? What are the results of change? These questions, and many others of a similar fundamental nature, are what chemistry is about.

The development of new alloys, plastics, and medications comes about as a result of the practical application of the whole network of assumptions, hypotheses, theories, and laws that chemists have formulated from their pure research into the nature of matter, energy, and change.

2.1 MATTER

Matter is usually described as anything having mass and occupying space (volume). In other words, matter is supposed to have a measurable density. Minerals, air, milk, plants, and animals are a few of the countless examples of matter. While the examples are obvious, the definition is limited to that aspect of the world familiar to man. For example, "things" such as electrons and photons of light blur the definitions and distinctions that scientists have constructed for their convenience. Familiar types of matter may be subdivided into pure substances and mixtures:

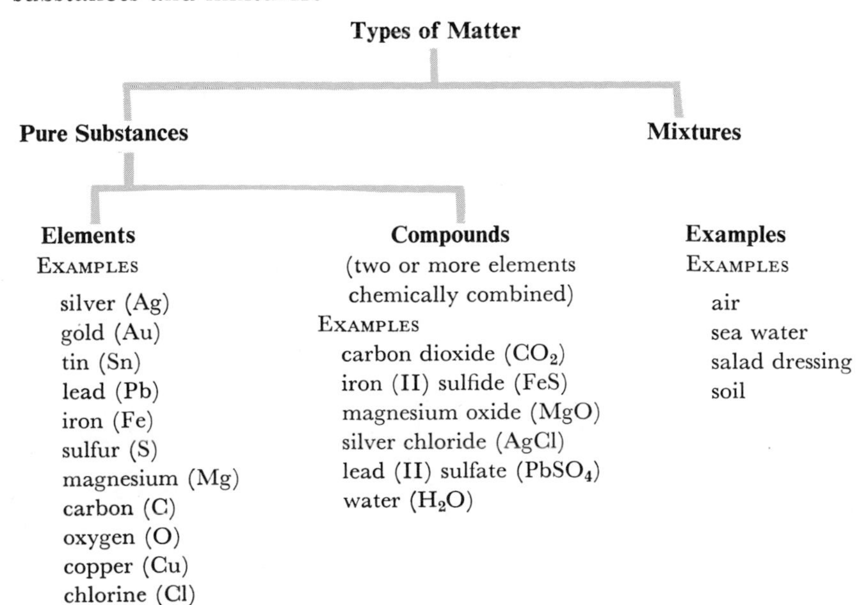

2.2 VARIABILITY AND STABILITY OF MATTER

Familiar types of matter are observed to have variable states. A *solid* can usually be changed to a *liquid* by increasing the temperature to a value described as the *melting point* of that particular substance. A further increase in temperature brings the liquid to its *boiling point*, where the liquid is changed to the *vapor* (gas) state. This changing of state is necessarily linked to energy because it is energy that enables molecules to move more vigorously and overcome attractive forces. Change in state is a story of forces in conflict. There is a tendency for matter to achieve maximum stability; at the same time, it seems rather mysteriously to strive for maximum randomness in arrangement.

The forces of attraction between unit particles of matter (atoms or molecules) are universal phenomena. Without fully understanding the nature of these forces, they are classified nevertheless as *nuclear*, *electrical*, *magnetic*, and *gravitational*. The nuclear and gravitational forces provide an interesting contrast. While the nuclear force is extremely powerful, it operates effectively over a distance of one angstrom or less. On the other hand, the gravitational force is effective over distances of millions of miles, yet it is comparatively weak. Regardless of how the forces of attraction are classified, they all tend to pull particles of matter together in an orderly geometric arrangement that characterizes stability (Fig. 2–1). There are forces of repulsion that must be reckoned with, also. Such forces will be investigated in the chapters on atomic structure and chemical bonding (Chapters 5 and 6).

This tendency of matter to sacrifice energy in its striving toward stability is one of the fundamental driving forces involved in the changes of matter. There are many common examples to illustrate this observation. A ball on top of a hill rolls down and finally stops. A machine wears out and is finally deposited on a scrap pile. Living organisms grow old, break down, and finally die. However, nature

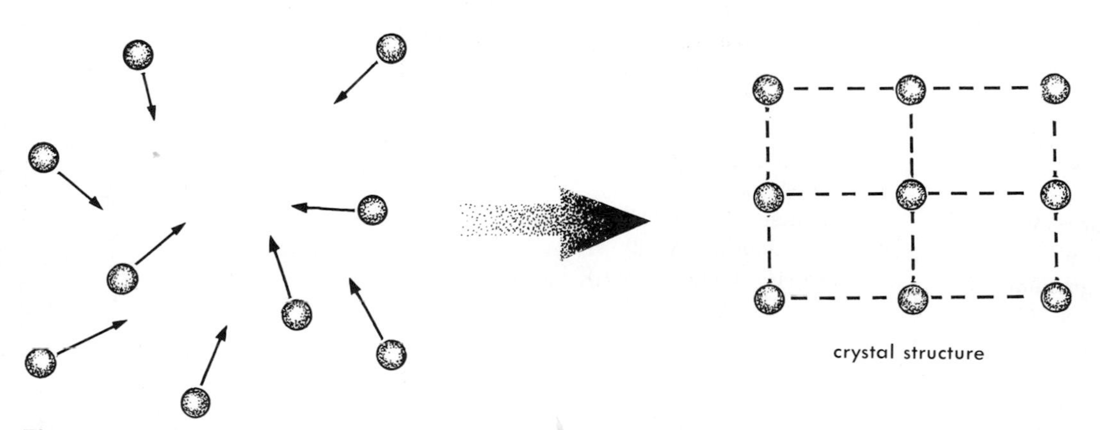

Figure 2–1 Attractive forces among unit particles, resulting in a stable crystal-like structure.

seems to provide a counterforce for every one-directional tendency, just as each skeletal muscle has another muscle operating in opposition to it. In this condition of balanced tension, there is smoothness and control.

When the unit particles have some external source of energy that they can absorb, there is a movement toward disorder. With the absorption of energy, the unit particles in the solid state can overcome the forces of attraction because of their increased ability to shift through the liquid state to the gas state in which there is maximum disorder. This tendency toward randomness (a probability in any highly ordered system) is another fundamental driving force behind changes in matter. The degree of randomness is often loosely described as a measure of the **entropy** of a system (Fig. 2–2).

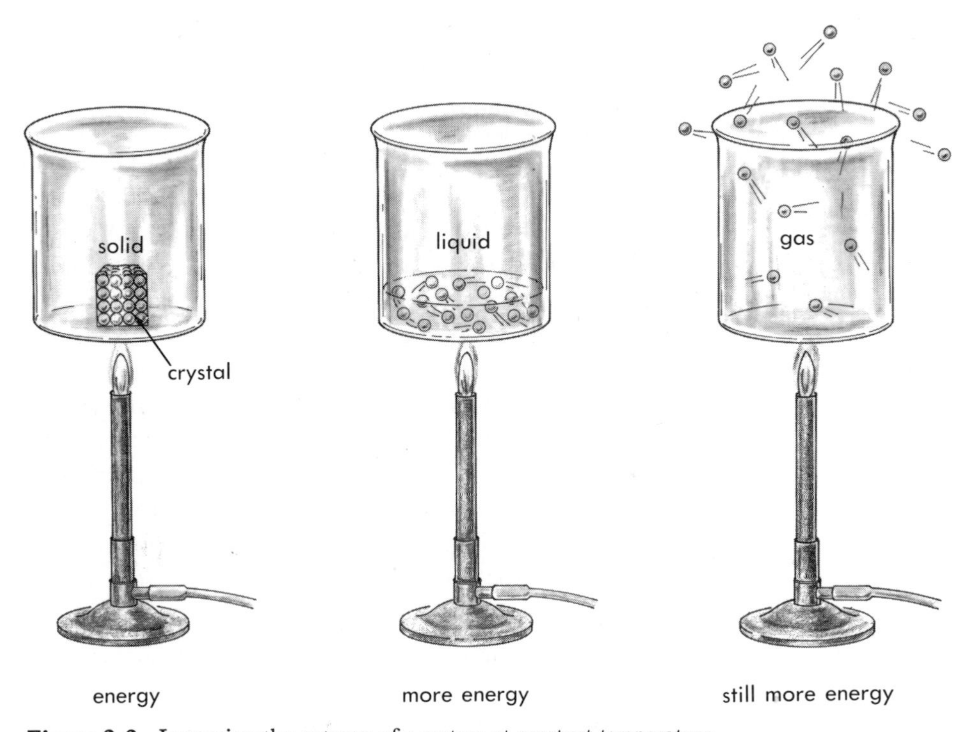

Figure 2–2 Increasing the entropy of a system at constant temperature.

Occasionally this drive toward an increase in entropy is so powerful that a highly ordered solid may spontaneously absorb energy in order to achieve greater randomness. An example would be the remarkable drop in the temperature of water when solid, crystalline ammonium chloride is dissolved in it, as illustrated in Figure 2–3. The dissolved particles of ammonium chloride are much more random in their arrangement than they were in the original solid because they have absorbed heat energy from the water.

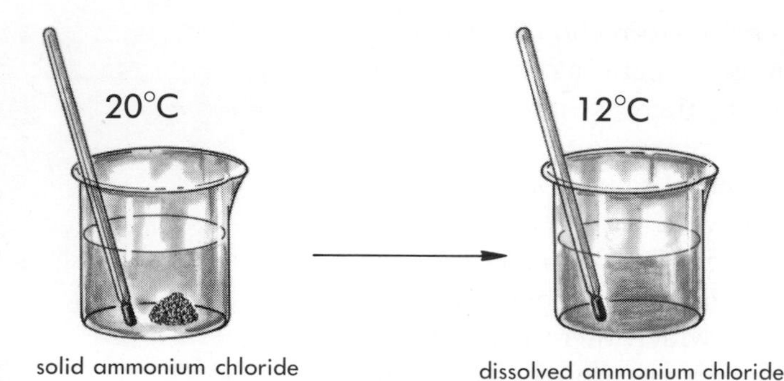

Figure 2–3 The drive toward randomness, resulting in the absorption of energy.

2.3 CHANGES OF MATTER

The changes of matter are conveniently divided into *physical* and *chemical* changes.

Physical Changes

Changes of state serve as examples of physical change. Water, ice, and steam are the same substance; all are composed of exactly the same unit particles (water molecules). While the arrangement of unit particles may vary from orderly to random, it is only when the original unit particles are broken and new types of particles are formed that a chemical change occurs.

If a pane of glass is ground into powder, it is still glass. When blackish iron powder is blended with yellow sulfur to form a homogeneous gray mixture, the particles of iron can be removed by a magnet or the sulfur can be dissolved in carbon disulfide, a liquid which can easily be evaporated after separation from the iron by filtration. Mashing, melting, or blending are all physical changes. When no new substances are formed, the change is said to be physical.

Chemical Change

A chemical change is one which begins with **reactants** having a set of unique characteristics (properties) quite different from those of the **products** that result from the change. When a **reaction** (a chemical change) goes to completion, i.e., when essentially *all* of the reactants are converted to products at the end, the reaction is said to be **stoichiometric.** Stoichiometry is concerned with very specific weights or volumes of reactants and products. The subject of stoichiometry will be discussed in detail in Chapter 8.

$$\underset{\text{reactants}}{A + B} \xrightarrow{\text{stoichiometric}} \underset{\text{products}}{C + D}$$

Reactions that have products which interact to re-form the original reactants until a dynamic balance is established between the rates of two opposing reactions are called **equilibrium** reactions:

$$A + B \xrightleftharpoons{\text{equilibrium}} C + D$$

The details of equilibrium reactions are usually more difficult to observe visually, and the quantitative (mathematical) considerations are more complicated. For the sake of expediency, the following discussion on the evidence of and the conditions for chemical change will apply mostly to stoichiometric reactions.

Evidence for the possible occurrence of a chemical change is most obvious when one or more of the following occur:

1. A *gas* is produced.
2. A *precipitate* (visible solid particles) is formed as insoluble solid or liquid separates from the bulk of the mixture.
3. A *large energy change* is observed in the form of light or heat or both.
4. A distinct *color change* occurs.

Conditions necessary for a chemical change to take place may be:

1. Simple *contact* between two reacting substances at room temperature:
   a. yellow phosphorus + iodine crystals → red phosphorus triiodide
   b. silvery sodium + pale green chlorine gas → white crystals of table salt
2. Supplying enough energy (called the *energy of activation*) to jostle sluggish reactants into an active condition which allows the breaking of the original bonds of attraction and the formation of new ones. For example, Figure 2–4 diagrammatically illustrates the reaction between silvery gray iron and yellow sulfur in the formation of dull gray iron (II) sulfide:

$$Fe + S \rightarrow FeS$$

3. Getting the reactants into water **solution** often is necessary. Figure 2–5 illustrates the absence of noticeable change when white crystals of silver nitrate are mixed with white crystals of table salt. The addition of water provides a medium through which the ions can move and finally form the white precipitate, silver chloride.
4. The use of a catalyst frequently is necessary if a reaction is to occur in a reasonable length of time (Fig. 2–6). A catalyst is anything that speeds up a chemical reaction without becoming permanently altered itself. Large complex substances called **enzymes** often function as catalysts in living organisms. For example, the breakdown of our food into simple substances (the process of digestion) involves complicated chemical changes that must be catalyzed by enzyme action. If it were not for enzymes, the first meal ever eaten by a person would probably

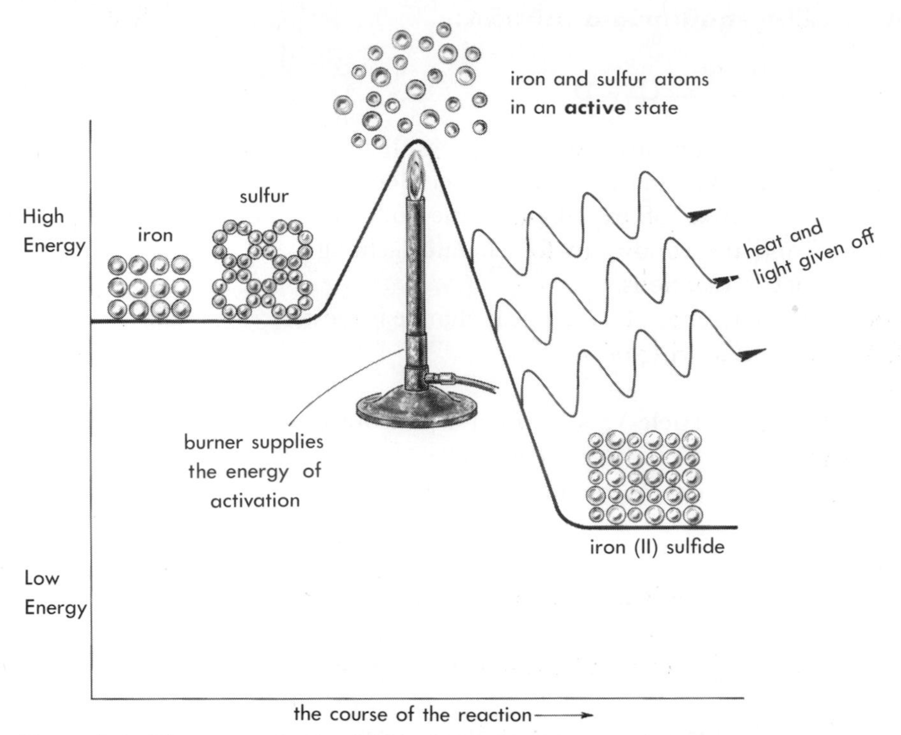

Figure 2–4 The energy of activation breaks the bonds between iron atoms, and also the bonds between sulfur atoms, so that the new iron-sulfur bonds can form.

not be fully digested after a period of many years. A catalyst works by acting as a gathering point for normally spread-out, randomly distributed reactants, or it may actually enter into a reaction as a temporary intermediate product which lowers the energy of activation and results in more molecules possessing that required activation energy at a given temperature.

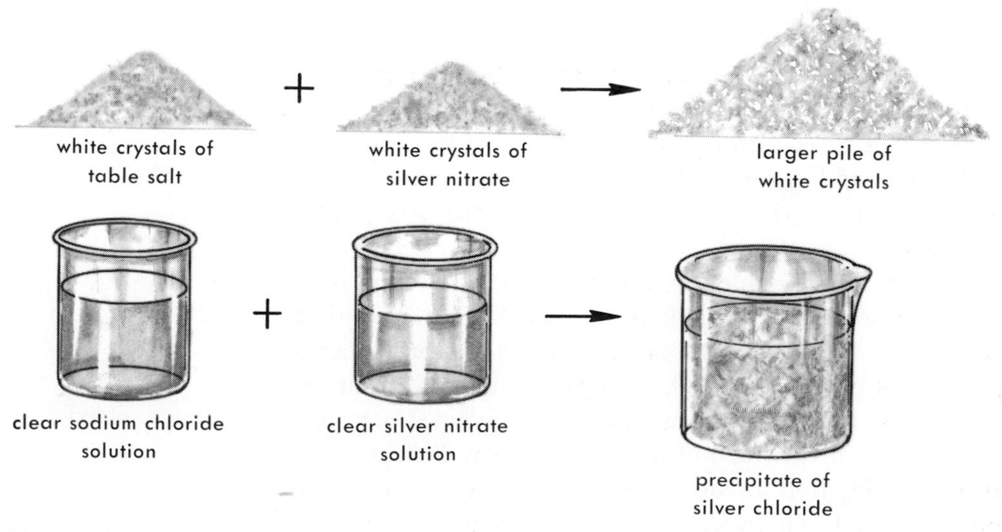

Figure 2–5

Figure 2–6 Catalysts speed up the rate of a reaction.

2.4 CONSERVATION OF MATTER

One other consideration, central to the process of stoichiometry, is the fact that *under ordinary conditions* matter cannot be created or destroyed. When a reaction begins with 10 grams of reactants, it must end with 10 grams of products—or an explanation is necessary to account for any increase or decrease in mass. The increase may be something like the invisible addition of oxygen from the air. A decrease may be due to the loss of some invisible gas product. Ordinary laboratory apparatus is usually incapable of detecting the loss of matter because of its conversion to energy. Only in *extra*ordinary changes, such as a nuclear bomb explosion, is the "annihilation" of matter a measurable quantity. (The term "annihilation" is somewhat misleading, because what happens is that matter is converted into an equivalent amount of energy.) The popularized Einstein equation states that in

Albert Einstein

reality (although not obviously) matter and energy are the two sides of the same mysterious coin:

$E = mc^2$ E = energy

m = mass

c^2 = the proportionality constant, which is the speed of light, squared:

$(3 \times 10^{10}\ \text{cm s}^{-1})^2 = 9 \times 10^{20}\ \text{cm}^2\ \text{s}^{-2}$

2.5 ENERGY

The previous discussion of matter often involved the concept of energy. Indeed, the law of conservation most directly underscores the fundamental unity of matter and energy. This raises a number of important questions regarding the definition, uses, types, and transformations of energy.

Energy is difficult to define adequately in one concise statement. The usual attempt, "Energy is the ability to do work," is, at best, a starting point. It remains for examples and analogies to give substance to a poor skeleton. **Work,** rigidly defined as a physical action, means to *move* an object. A push or pull applied to an object is described as exerting a **force** and forces can only come about if there is energy.

Energy has many familiar faces: heat, light, magnetic, mechanical, electrical, and atomic, to mention a few. These forms of energy are often easily interconverted. For example, the mechanical energy of falling water at a hydroelectric plant is converted to electrical energy, which in turn can be converted to heat and light energy in the home or factory. The scientist's need to deal quantitatively with energy leads to a classification of types of energy, the units of measurement, and the methods of conversion between units.

2.6 TYPES OF ENERGY

Potential Energy

A sharpshooting basketball player is described as a scorer. When actually playing in a game he *is* scoring. Before a game, the coach plans his strategy on the basis of his star's *potential* scoring ability. Energy, as the *ability* to do work, is described in much the same way. A book perched on a table edge has the potential to burn and so liberate heat energy (which has the theoretical possibility of doing work), or it has the potential to fall and move a hapless insect closer to the ground (Fig. 2–7).

Figure 2–7 Potential energy.

Kinetic Energy

When the book actually does fall or burn, it is effectively converting *potential* energy into moving energy, which is known as *kinetic* energy. The symbols for these forms of energy are E_p for potential energy and E_k for kinetic energy. The interrelationship can be illustrated by a boy on a swing (Fig. 2–8). Kinetic energy is a property of

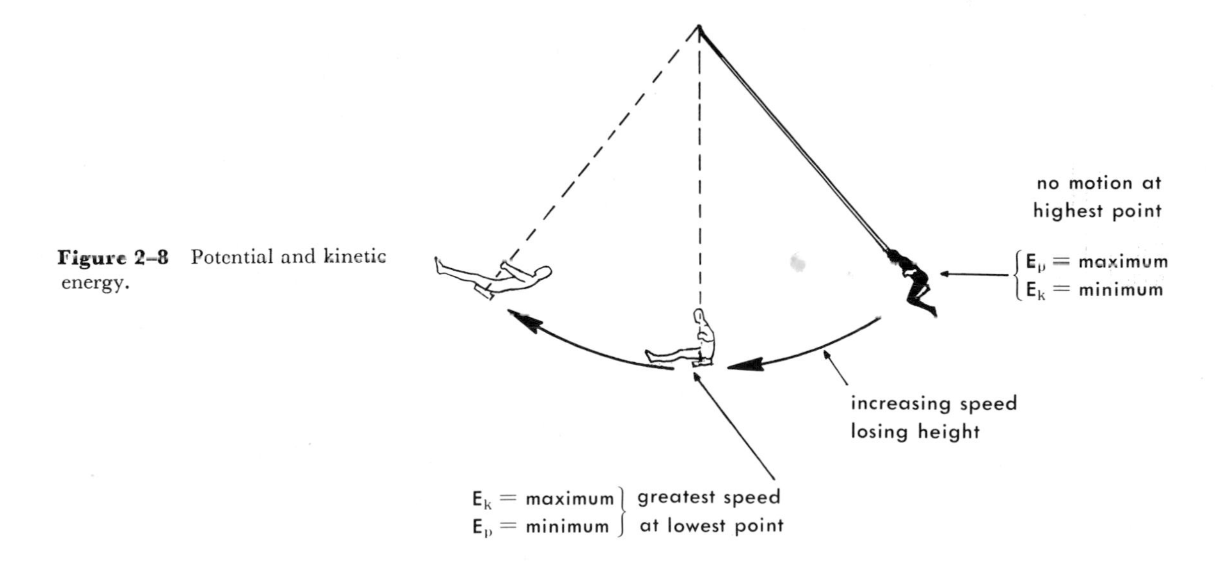

Figure 2–8 Potential and kinetic energy.

matter. In fact, one of the criteria for classifying anything as a particle of matter is its ability to transfer kinetic energy. This can be illustrated by a line of billiard balls that rapidly perform this transfer (Fig. 2–9).

Figure 2–9 The transfer of kinetic energy.

2.7 ENERGY UNITS AND CONVERSIONS*

Kinetic energy is described mathematically as being equivalent to one-half of the product of the mass of the particle and its velocity squared:

$$E_k = \frac{mv^2}{2}$$

If the units used are mass in grams and velocity in centimeters per second, the product of these units is gram-centimeters squared per second squared, or

$$E_k = g\ cm^2\ s^{-2}$$

This cumbersome group of units is more commonly known as an **erg**. One erg is roughly the amount of energy a mosquito requires to do a push-up.

Just as a line can be measured in different kinds of units (centimeters, inches, miles, rods), energy can be measured in different kinds of units. There are times when it may not be appropriate to use ergs. The energy required to lift a load of ship's cargo would be so great that ergs would be as awkward as measuring the distance between New York and Paris in centimeters. In other words, energy values can be expressed in a number of equivalent units. If a conversion factor can be found or developed, mechanical energy can be expressed as an *equivalent* amount of heat or electrical energy.

An extremely useful energy unit is the **joule (J)**. As a measure of the kinetic energy of particles, a joule is equal to half the product of a kilogram weight moving at a velocity of one meter squared per second squared:

$$E_k = \frac{kg\ (m^2\ s^{-2})}{2} = \text{joules}$$

The great value of the joule is its direct relationship to electrical energy. A joule is the product of one volt and one coulomb. A **coulomb** is an amount of electric charge carried by a current of 1 ampere in one second. Therefore

$$1\ \text{joule} = \text{volt-coulomb}$$

The volt, amp, and other basic units of electrical measurement will be discussed later in this chapter. When the gram and centimeter units of the erg are converted to the kilogram and meter units of the joule, a conversion factor emerges:

$$10^7\ \text{erg}\ J^{-1}\ \text{(ten million ergs per joule)}$$

EXAMPLE 2.1

If the kinetic energy of a moving ball is 5×10^7 ergs, what is the energy value in joules?

1. Predict a smaller numerical value for the answer, because one joule equals so many ergs.

2. Try division by the conversion factor as the operation most likely to yield a smaller number:

$$\frac{5 \times 10^7 \cancel{\text{erg}}}{10^7 \cancel{\text{erg}}\ \text{J}^{-1}} = 5 \text{ joules}$$

EXAMPLE 2.2

Electrical measurements on a motor indicate 2.5×10^4 joules of energy being used. What is the mechanical equivalent of this in ergs?

1. Predict a larger value for the answer by the same reasoning used in Example 2.1.
2. Try multiplication by the conversion factor:

$$(2.5 \times 10^4 \cancel{\text{J}})(10^7 \text{ erg } \cancel{\text{J}^{-1}}) = 2.5 \times 10^{11} \text{ erg}$$

Calories

The single most useful unit for energy measurement in the laboratory is the **calorie.** This is the unit of heat energy. Chemical reactions, while fundamentally a matter of electrical interactions among atoms, are most often described in terms of the amount of heat absorbed (endothermic reactions) or the amount of heat liberated (exothermic reactions). On the other hand, there are numerous reactions, especially in the investigations of biological chemistry, where electrical measurements are more readily obtained than direct heat measurements. There is no question about the great practicality of being able to interconvert the heat and electrical units of energy measurement.

Considering the calorie first, it is defined as the amount of heat needed to raise the temperature of one gram of water through one degree Celsius.* Calories, then, could be related to a number of grams of water undergoing a significant temperature change:

$$\text{cal} = \underset{\substack{\text{(grams}\\ \text{of}\\ \text{water)}}}{\text{g}} \times \underset{\substack{\text{(temperature}\\ \text{change)}}}{(\Delta t)}$$

EXAMPLE 2.3

How many calories are required to raise 5 g of water from 10°C to 25°C?

1. Organize the data:

$$\text{cal} = x$$
$$\text{g} = 5$$
$$\Delta t = 25 - 10 = 15 \text{ C}^\circ$$

*A more up-to-date definition of the calorie is a joule equivalence value: 1 cal = 4.18J. However, for practical purposes the old definition suffices.

2. Develop the equation:

$$x = g(\Delta t) = 5\text{ g }(15\text{ C}^\circ)$$

$$x = 75\text{ cal}$$

Notice that the answer in Example 2.3 is 75 calories, rather than 75 gram-degrees as dimensional analysis would indicate. The explanation is supplied in the form of the **specific heat**, which is commonly omitted in the case of water. The specific heat, by definition, is the number of calories needed to raise 1 gram of a specified substance through 1 degree Celsius. It should be mentioned that specific heat is more accurately defined as the ratio of the heat needed to raise 1 g of substance through 1 degree, to the heat needed to raise 1 g of water through 1 degree. What chemists call "specific heat" is in reality *heat capacity*. Since the heat capacity of water is 1.0 cal g^{-1} $°C^{-1}$, it does not affect the numerical value of the answer. However, it would solve the dimensional analysis problem in Example 2.3:

$$\text{cal} = \text{g}(\Delta t)(\text{sp. ht.})$$

$$\text{cal} = 5\,\cancel{\text{g}}\,(15\,°\cancel{\text{C}})(1\text{ cal }\cancel{\text{g}^{-1}}\,°\cancel{\text{C}^{-1}})$$

answer is 75 cal

The specific heat values for other substances, however, cannot be ignored. Specific heats of substances may be obtained experimentally or found in reference tables.

EXAMPLE 2.4

What temperature rise can be expected if 15.0 grams of glycerol absorbs 210.0 cal? The specific heat of glycerol is 0.314 cal g^{-1} $°C^{-1}$.

1. Organize the data:

$$\begin{aligned} \text{cal} &= 210.0 \\ \text{g} &= 15.0 \\ \text{sp. ht.} &= 0.314\text{ cal g}^{-1}\,°\text{C}^{-1} \\ \Delta t &= \text{? degrees} \end{aligned}$$

2. Develop the equation:

$$\text{cal} = \text{g}(\Delta t)(\text{sp. ht.})$$

$$\frac{\text{cal}}{\text{g(sp. ht.)}} = \frac{\cancel{\text{g}}(\Delta t)(\cancel{\text{sp. ht.}})}{\cancel{\text{g}}(\cancel{\text{sp. ht.}})} \qquad \text{(divide the equation by g and sp. ht.)}$$

$$\Delta t = \frac{\text{cal}}{\text{g(sp. ht.)}}$$

3. Substitute the data and solve:

$$\Delta t = \frac{210.0\ \text{cal}}{15.0\ \text{g}(0.314\ \text{cal}\ \text{g}^{-1}\,^{\circ}\text{C}^{-1})}$$

$$\Delta t = 44.6 \text{ degree rise}$$

If calories are to be measured directly, the device often used is called a **calorimeter.** This is a well-insulated container that works on the principle of conservation of heat. The principle of the conservation of heat refers to the conservation of energy, and it is stated in the same way as the law of the conservation of matter. The energy absorbed by water surrounding a chemical reaction chamber must be equal to the energy produced by the reaction. In other words,

$$\text{cal lost in reaction} = \text{cal gained by water}$$

A diagrammatic picture of a calorimeter is illustrated in Figure 2–10.

Exercise 2.1

1. What is the kinetic energy in ergs of a 10.0 mg particle moving with a velocity of 50.0 meters per second? (Hint: convert units to grams and centimeters.)
2. Express the answer to the above problem in joules.
3. Calculate the velocity, in cm s^{-1}, of a particle having a kinetic energy of 3.0×10^{-12} erg and a mass of 6.0×10^{-26} g.
4. How many calories are required to raise the temperature of 45.0 g of water from 3.2°C to 17.4°C?
5. What is the specific heat of a substance if 2.1 kcal raises 50.0 g of the substance 72.0 degrees?

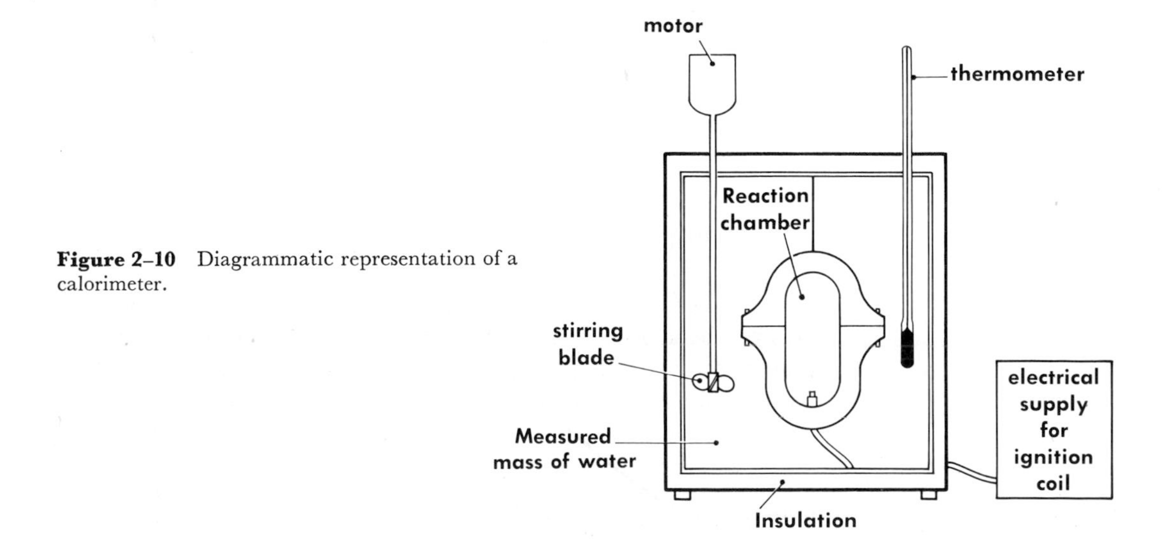

Figure 2–10 Diagrammatic representation of a calorimeter.

Electron-Volts

An energy unit that is especially applicable to the measurement of small amounts of energy is the electron volt (*eV*). An electron volt is the amount of energy acquired by an electron as it passes between two points having a potential difference of one volt. This rather technical definition of an electron volt is probably not a very satisfying one. In addition to the possibility that "potential difference" may be meaningless to a student at this stage, the amount of energy equal to 1 eV is so fantastically small that it doesn't readily lend itself to description by analogy—not even a feeble analogy. But try this: if a mosquito uses an erg to do a push-up, he will use a few electron volts in just mentally preparing himself for the exertion.

Laboratories involved in the study of nuclear reactions frequently describe energies of moving atomic particles in MeV (million electron volts) or BeV (billion electron volts) because of the small amount of energy represented by one electron volt. It takes about ten trillion trillion (10^{19}) electron volts to equal one joule, or close to ten thousand trillion trillion (10^{22}) electron volts to equal one kilocalorie.

Chemists sometimes use eV to describe the energy required to remove an electron from an isolated atom. This energy is known as the *ionization energy.*

EXAMPLE 2.5

Calculate the ionization energy of one atom of an element in kilocalories if the measured value is 5.1 eV per atom. The conversion factor is 2.6×10^{19} eV cal^{-1}.

1. Change the conversion factor so that its units are eV and kcal:

$$2.6 \times 10^{19}\ \text{eV}\ \cancel{\text{cal}}^{-1}\ (10^{3}\ \cancel{\text{cal}}\ \text{kcal}^{-1}) = 2.6 \times 10^{22}\ \text{eV kcal}^{-1}$$

2. Dimensional analysis indicates that the 5.1 eV should be divided by the conversion factor:

$$\frac{5.1\ \cancel{\text{eV}}}{2.6 \times 10^{22}\ \cancel{\text{eV}}\ \text{kcal}^{-1}} = 2.0 \times 10^{-22}\ \text{kcal}$$

Liter-Atmospheres

Another energy unit most directly encountered in problems dealing with the behavior of gases is the liter-atmosphere (ℓ-atm). A ℓ-atm is the amount of energy required to permit a gas to expand one liter against an external pressure of one atmosphere. One atmosphere of pressure may be defined as the force per unit area (dynes* per square centimeter) needed to

* Dynes × centimeters are ergs. In other words, a dyne is the amount of force needed to accelerate a mass of one gram by one centimeter per second per second. Dyne = g cm s^{-2}.

support the weight of a column of mercury that is 76.0 cm high. While familiar English units describe an atmosphere as a pressure of 14.7 lb in^{-2}, the metric equivalent is about 1×10^6 dynes cm^{-2}, or about 1 kg cm^{-2}.

Many problems in chemistry must deal with the interrelationship between heat and mechanical energy. It is often convenient to express the ℓ-atm units of mechanical energy as equivalent calories of heat energy. The conversion factor has been found to be 0.041 ℓ-atm cal^{-1}.

EXAMPLE 2.6

An important constant related to gases is 0.082 ℓ-atm per mole per degree Kelvin. Convert this value to kilocalories mole^{-1} °K^{-1}.

1. Change the conversion factor from calorie units to kilocalories.

   a. $0.041\ \ell\text{-atm cal}^{-1} = 4.1 \times 10^{-2}\ \ell\text{-atm cal}^{-1}$

   b. $4.1 \times 10^{-2}\ \ell\text{-atm}\ \cancel{\text{cal}}^{-1}\ (10^3\ \cancel{\text{cal}}\ \text{kcal}^{-1})$

   $= 4.1 \times 10^1$, or, $41\ \ell\text{-atm kcal}^{-1}$

2. Dimensional analysis indicates that the ℓ-atm value should be divided by the conversion factor:

$$0.082\ \ell\text{-atm mol}^{-1}\ \text{K}^{-1} = 8.2 \times 10^{-2}\ \ell\text{-atm mol}^{-1}\ \text{K}^{-1}$$

$$\frac{8.2 \times 10^{-2}\ \cancel{\ell\text{-atm}}\ \text{mol}^{-1}\ \text{K}^{-1}}{41\ \cancel{\ell\text{-atm}}\ \text{kcal}^{-1}}$$

$$= \frac{82 \times 10^{-3}\ \text{mol}^{-1}\ \text{K}^{-1}}{41\ \text{kcal}^{-1}} = 2 \times 10^{-3}\ \text{kcal mol}^{-1}\ \text{K}^{-1}$$

TABLE 2–1 TABLE OF ENERGY CONVERSION FACTORS

4.2×10^7 erg cal^{-1}
2.6×10^{19} eV cal^{-1}
4.2 J cal^{-1}
1.6×10^{-12} erg eV^{-1}
1×10^7 erg J^{-1}
4.1×10^{-2} ℓ-atm cal^{-1}

Exercise 2.2

Use Table 2–1 to perform the following energy conversions:

1. 2.0×10^4 ergs to cal.
2. 4.5×10^{-5} joules to ergs.
3. 5.2×10^{16} eV to kcal.

4. 8.3×10^4 cal to ℓ-atm.
5. 2.4×10^{-11} ergs to eV.
6. 0.05 ℓ-atm to cal.
7. 1.6×10^{-18} kcal to eV.
8. 8.7×10^6 joules to cal.

2.8 LIGHT AND THE QUANTUM MODEL*

A very important aspect of energy relations is the production and measurement of light. The importance of light will become very apparent when the subject of atomic structure is investigated. Visible light is a small part of a huge range of types of **radiant energy**, or **electromagnetic radiation.**

Electromagnetic radiation encompasses a range, or **spectrum,** from alternating current (low energy) to the cosmic radiation of the stars (high energy). Electromagnetic radiation may be interpreted as the outward radiation of energy waves because of the vibrations (oscillations) of electrically charged particles. The part of the electromagnetic spectrum being radiated depends on the energy involved. Since radiant energy is classically considered to be a wave phenomenon, it is necessary to relate the characteristics of waves to the amounts of energy associated with these characteristics.

The Wavelength

It helps to think of waves in terms of what results when a stone is dropped into a still pond. This old but durable analogy presents a picture of crests and troughs radiating outward from the point of disturbance. The essential difference between the water wave and the electromagnetic wave is that the electromagnetic wave apparently does not require matter in order to move. It can proceed through a vacuum. In fact, precise measurements indicate the speed of electromagnetic radiation in a vacuum to be 3×10^{10} cm s^{-1} (186,000 miles per second). This universal constant is designated by the letter *c* (remember, $E = mc^2$).

The distance from the peak of one crest to the peak of the next is called the **wavelength,** and it is symbolized by the Greek letter lambda (λ). The other symbol in Figure 2–11 (**a**) represents the height of the wave, called the **amplitude**.

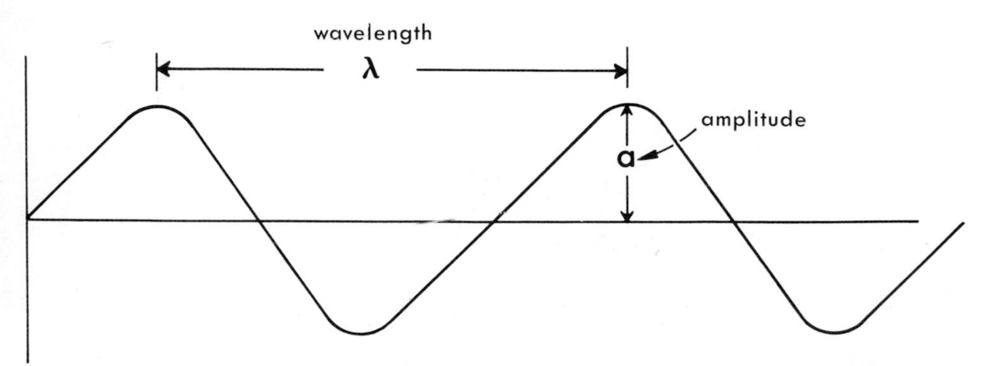

Figure 2–11 The wavelength and amplitude.

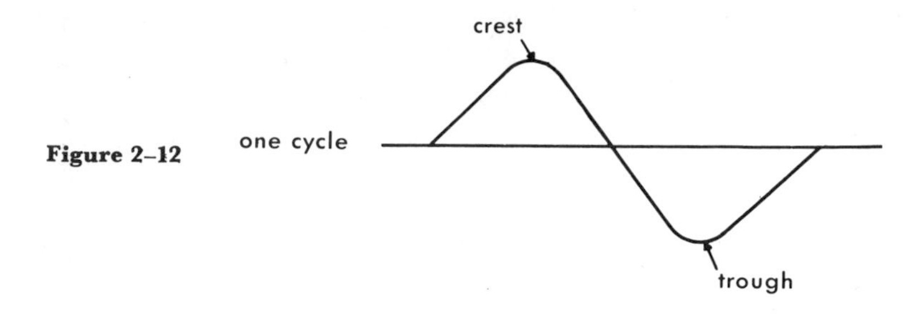

Figure 2–12

The Frequency

The other essential characteristic of a wave is its **frequency**, symbolized by another Greek letter, *nu* (ν). Frequency means the number of waves, or crest–trough cycles, that will pass an observation point in one second (Fig. 2–12).

The unit of measure for frequency is reciprocal seconds (per second, or s^{-1}) since time is the only unit of measure. "Cycles per second" has been replaced by a modern label **hertz** (Hz), named after Heinrich Hertz, who discovered electromagnetic waves.

Relating Wavelength, Frequency, and Energy

Comparing two "bits" of electromagnetic radiation as they pass an observation point, moving at the same speed (the speed of light), it can be seen that more cycles will pass the observation point if the wavelength is shorter (Fig. 2–13).

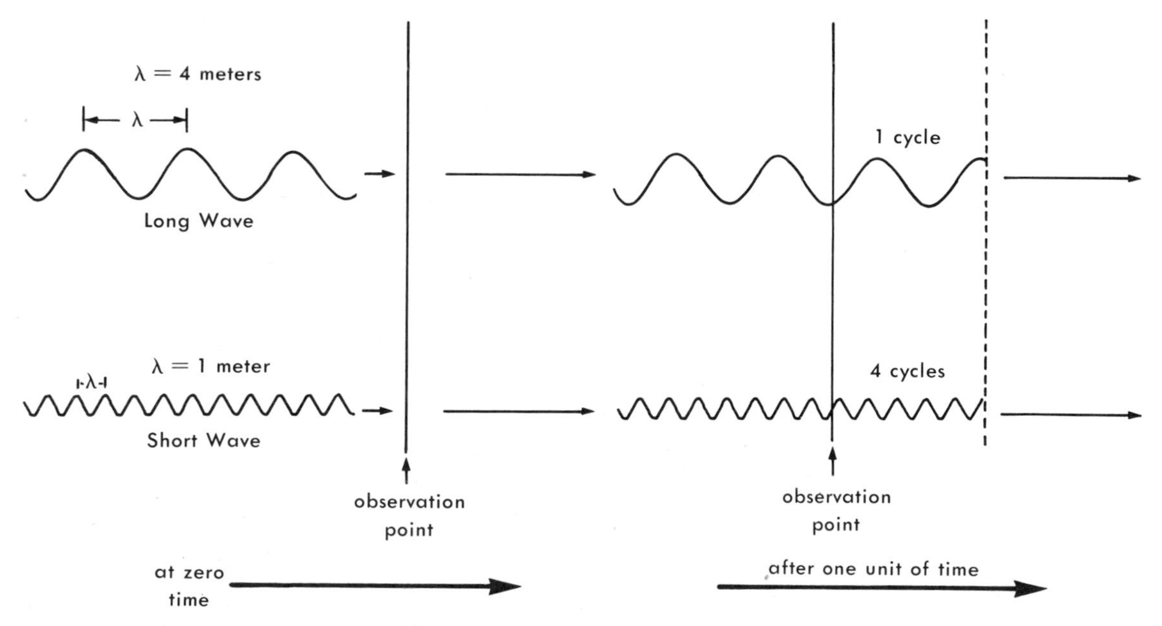

Figure 2–13 Wavelength and frequency.

From the idealized diagram shown in Figure 2–13, it is also seen that the wavelength is inversely proportional to the frequency. The equation for this relationship may be developed:

$$\lambda \propto \frac{1}{\nu}$$

$$\lambda = k\frac{1}{\nu}$$

The proportionality constant (k) relating the frequency and wavelength is the speed of light. Therefore, the equation may be rewritten:

$$\boxed{\lambda = \frac{c}{\nu}}$$

The other relationships derived from this equation are

$c = \lambda\nu$ (speed of light = the product of frequency and wavelength)

and

$$\nu = \frac{c}{\lambda}\left(\text{frequency} = \frac{\text{speed of light}}{\text{wavelength}}\right)$$

EXAMPLE 2.7

What is the frequency of radiation having a wavelength of 500 nm?

1. Organize the data and be sure the units are compatible:

$c = 3 \times 10^{10}$ cm s^{-1} —— compatibility

$\lambda = 500$ nm $= 5 \times 10^{-5}$ cm

$\nu = ?$

$$\frac{500 \text{ nm}}{10^7 \text{ nm cm}^{-1}} = 500 \times 10^{-7} \text{ cm} = 5.00 \times 10^{-5} \text{ cm}$$

2. Substitute and solve the equation:

$$\nu = \frac{c}{\lambda} = \frac{3 \times 10^{10} \text{ cm s}^{-1}}{5 \times 10^{-5} \text{ cm}} = 0.6 \times 10^{15} \text{ s}^{-1}, \text{ or } 0.6 \times 10^{15} \text{ Hz}$$

$\nu = 6 \times 10^{14}$ Hz

Many observations over the years clearly indicate that higher energy radiation has a shorter wavelength and thus a higher frequency. Radio waves having wavelengths in the meter or kilometer range are much less energetic than x-rays of the same amplitude. The wavelengths of x-rays are in the angstrom region. This difference is related to the source of the radiation (see Fig. 2–14).

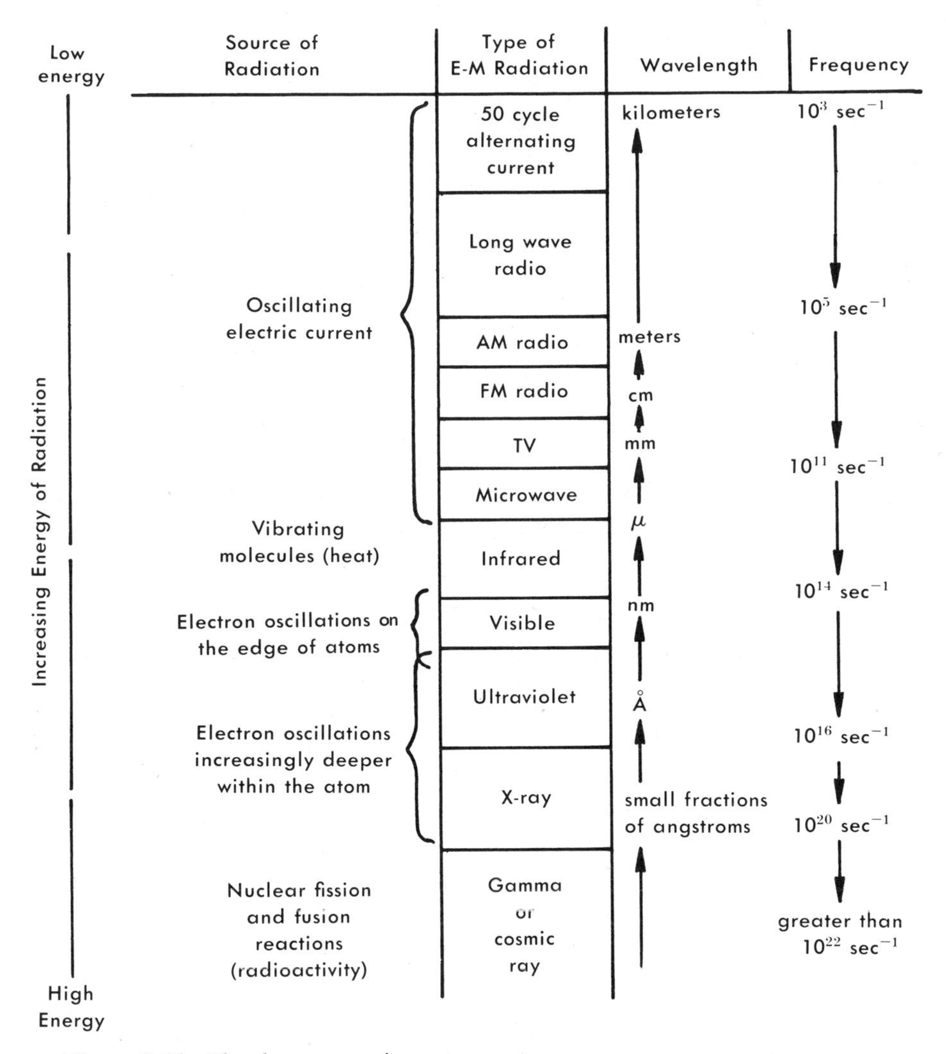

Figure 2–14 The electromagnetic spectrum.

Exercise 2.3

1. What is the frequency of electromagnetic radiation having a wavelength of 200 nm?
2. Calculate the wavelength of light if the frequency is 4.0×10^{12} Hz.
3. What is the speed of sound if a tuning fork vibrating at 120 cycles per second produces a wavelength of 2.76 meters?
4. What frequency of light has a wavelength of 210 Å?

A closer look at the visible light region of the electromagnetic spectrum will be helpful before considering the structure of the atom in Chapter 5.

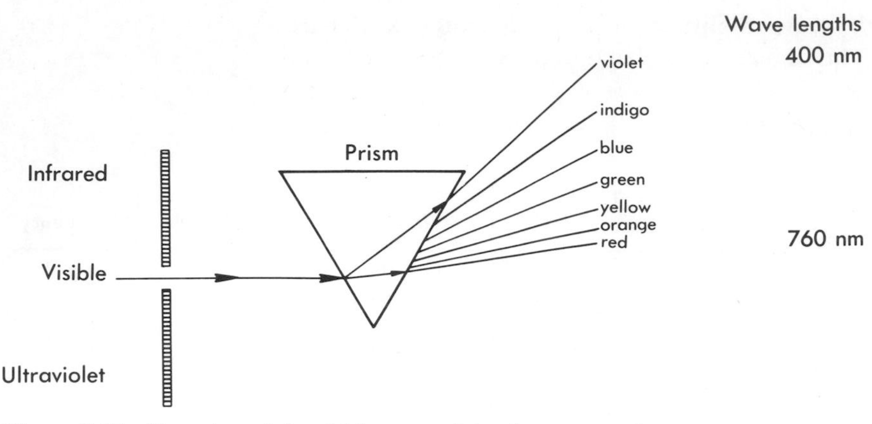

Figure 2–15 The colors of the visible range of the electromagnetic spectrum.

Figure 2–15 illustrates the components of the visible part of the spectrum. The violet part of the spectrum has the shortest wavelength and, therefore, the highest energy. All wavelengths of light traveling through a vacuum have the same speed (c), but through a **quartz prism,** the violet is slowed more than the red and as a result is bent more sharply. The speed increases with the color as the colors move toward red.

Optical instruments for measurement and analysis make use of prisms or **diffraction gratings** (a film etched with fine lines, producing the same spreading-out of a spectrum as does the prism) in selecting specific wavelengths with their associated energy values.

The method of relating energy and wavelength was puzzling and not too useful for many years. In classical physics, light was considered to be only a continuous wave phenomenon. There was little serious consideration given to the fact that light might have a particle nature until the German physicist Max Planck suggested it early in this century. Today, light is thought of as being made up of distinct "pulses" of energy that are called **photons.** A widely used term that means a "definite quantity of energy" (just described as "pulse" or "photon") is the **quantum.** Planck's proposal is called the **quantum theory.** There is considerable doubt as to the reality of a photon as a "particle," but in some ways these pulses of energy may be thought of as having a few characteristics of material particles. The great difficulty in properly describing the photon and the electron in absolute physically descriptive terms is that they are unlike anything in our common experience. There is no basis for comparison. When photons and electrons are described sometimes as waves and at other times as particles, one is inclined to accept the "Principle of Complementarity" proposed by the great Danish scientist, Niels Bohr. The principle says, in effect, that both views together—the wave and particle models—probably give a more accurate picture of what is actually true than does either view alone.

Planck expressed the relationship between energy and frequency of radiation by showing that the *energy* of a photon is directly related to the *frequency*:

$$E \propto \nu$$

The proportionality constant that emerges is known as **Planck's constant,** or the *quantum of action*. Curiously, it has the dimensions of energy and time, the product of which is called *action*, and the symbol is h:

$$E = h\nu$$

$$\text{Energy of photon} = \text{Planck's constant} \times \text{frequency}$$

The value of Planck's constant can be determined from observed data:

$$E = 3.6 \times 10^{-12} \text{ erg}$$

$$h = ?$$

$$\nu = 5.5 \times 10^{14} \text{ s}^{-1} \text{ (green light)}$$

$$h = \frac{E}{\nu} = \frac{3.6 \times 10^{-12} \text{ erg}}{5.5 \times 10^{14} \text{ s}^{-1}}$$

$$h = 6.6 \times 10^{-27} \text{ erg-s}$$

EXAMPLE 2.8

What is the energy of a photon of violet light, $\lambda = 420$ nm?

1. Express the equation relating the data:

$$E = h\nu$$

2. Organize the data:

$$E = ?$$

$$h = 6.6 \times 10^{27} \text{ erg-s}$$

$$\nu = \frac{c}{\lambda} = \frac{3 \times 10^{10} \text{ cm s}^{-1}}{420 \text{ nm}}$$

3. Change 420 nm to cm so that the units are compatible:

$$\frac{420 \cancel{\text{nm}}}{10^7 \cancel{\text{nm}} \text{ cm}^{-1}} = 4.2 \times 10^{-5} \text{ cm}$$

4. Substitute in the modified equation:

$$E = \frac{hc}{\lambda}$$

$$E = \frac{6.6 \times 10^{-27} \text{ erg-}\cancel{\text{s}} (3 \times 10^{10} \cancel{\text{cm}} \cancel{\text{s}}^{-1})}{4.2 \times 10^{-5} \cancel{\text{cm}}}$$

5. Estimate the answer:

$$\frac{7 \times 10^{-27}(3 \times 10^{10})}{4 \times 10^{-5}} = \sim 5 \times 10^{-12}$$

6. Solve the equation:

$$E = 4.7 \times 10^{-12} \text{ erg}$$

EXAMPLE 2.9

What is the energy in calories of a photon of light having a frequency of 2.0×10^{15} Hz?

1. Express the equation:

$$E = h\nu$$

2. Organize the data and adjust the units so they are compatible:

$$E = ?$$

$$h = \frac{6.6 \times 10^{-27} \text{ erg-s}}{4.2 \times 10^{7} \text{ erg cal}^{-1}}$$

$$h = 1.6 \times 10^{-34} \text{ cal-s}$$

3. Substitute in the equation:

$$E = 1.66 \times 10^{-34} \text{ cal-s } (2.0 \times 10^{15} \text{ s}^{-1})$$

$$E = 3.2 \times 10^{-19} \text{ cal}$$

Exercise 2.4

1. Calculate the energy of a photon, in ergs, of light having a wavelength of 2.0×10^{3} Å.
2. What is the wavelength of radiation where the energy of a photon is 4.5×10^{-22} kcal?
3. Find the frequency of electromagnetic radiation which has an energy 8.1×10^{-12} ergs per photon.
4. What is the total energy, in kcal, of 6.0×10^{23} photons having a frequency of 2.4×10^{15} Hz?

2.9 PRINCIPAL UNITS OF ELECTRICAL MEASUREMENT

The introduction of the joule as a unit of energy has been related to the use of such terms as *volt*, *ampere* (*amp*), and *coulomb*. Furthermore, the extensive use of the concept of electrical charge in chemistry makes it very important for the chemistry student to understand the interrelationships among units of electrical measurement. It is helpful to attempt to visualize some of the electrical fundamentals by the use of such models and analogies that describe abstract terminology in terms of man's everyday experience. As the topics of atomic structure, electrical interactions between ions, charge distribution in molecules, and oxidation-reduction are discussed, the questions about the nature of electricity and electrical charge are raised.

Electrical measurement is concerned with the description of electron behavior in terms of direction of movement, the quantity of electrons, the resistance to flow, and the effects. While the electron is an essentially mysterious entity, it is usefully described as a unit of negative charge. The designation "negative" is an arbitrary term used as a contrast to the opposite charge of "positive." Although electrons constitute an integral part of an atom, they are found abundantly outside of atoms. Electrons can be harvested by running a comb through hair, scraping shoes across a carpet, or pulling a wool sweater over a nylon blouse. Electrons can be "boiled out" of a glowing wire or lamp filament. They can be "kicked out" of metal by the energy of sunlight, and electrons can be induced to flow in a conducting wire by moving the wire so as to pass through the invisible lines of force between two magnetic poles. There is also a tendency for electrons to flow between two different metals in contact.

Some of the above examples of electron sources suggest the common definition of electricity, which is the "flow" of electrons through a conductor (usually a wire). The electrons flow from a region of high density (many negative charges) to a region of low density (comparatively few electrons present and therefore called positive), as illustrated in Figure 2–16.

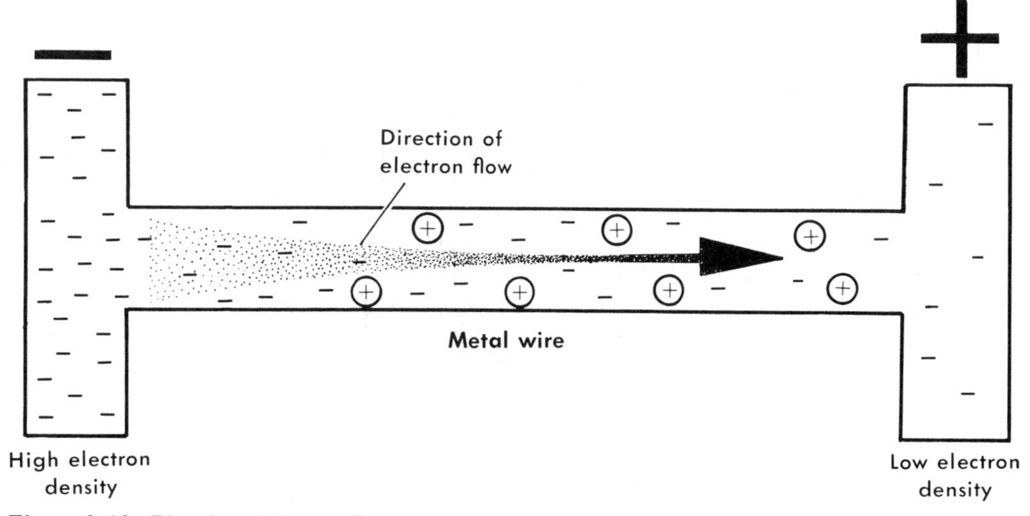

Figure 2–16 Direction of electron flow.

A well-known axiom pertaining to electricity is that opposite charges attract and like charges repel. This statement is largely explained by observing nature, which illustrates a tendency of unequal forces to strive toward a balance, or equilibrium. In the case of electrons, this drive toward balance is achieved by the electrons flowing from an area of high population to an area of low population. For example, if opposite charges (i.e., excess electrons and few electrons) are placed on separate metal spheres by a static electricity generator, a force of attraction can be seen operating. Furthermore, measurements made on the force of attraction between the spheres indicate that the strength of the attractive force depends on three factors: *the magnitude of the electrical charges*, *the distance between the charged spheres*, and *the material separating the spheres* (air, vacuum, water, or any one of a variety of substances). The force of attraction is found to be directly proportional to the product of charges, and inversely proportional to the square of the distance. This latter factor is commonly described as an inverse square function. It means that a doubling of the distance between oppositely charged particles (in this case, spheres) reduces the force of attraction to one-fourth the original value. The stated relationship,

$$F \propto \frac{q_1 q_2}{r^2}$$

mathematically expressed, is known as **Coulomb's law:**

$$F = \frac{q_1 q_2}{\epsilon r^2}$$

F = force of attraction between oppositely charged particles q_1 and q_2 = the charges, positive and negative

ϵ = a constant depending on the nature of the material separating the charges

r^2 = the distance squared

The electrical interaction between the charged particles is often described as *electrostatic* or *coulombic* interaction.

A more vivid example of the tendency of electricity to flow from a region of high electron density to one of low electron density is observed when a person scrapes his shoes over a carpet as a means of gathering free electrons from the fibers. As a radiator is approached, the load of electrons from the person's body surface leaps to the radiator and produces a visible spark.

The question that now presents itself is "What are the fundamental units of electrical measurement?"

The Volt

The volt may be described as a unit of electrical pressure. A high voltage is not necessarily dangerous. It simply means that there is a large difference in the electron density between two separate points, so that the point having the greater electron population possesses a kind of potential energy, i.e. stored up electrical energy that will move electrons to the region of low density if a pathway (such as a wire) is provided. **Potential, IR drop, EMF** (electromotive force), and **potential difference** are common synonyms for voltage.

People are familiar with 1.5 volt dry cells (flashlight batteries), with the 110 volt power lines in homes and buildings, and with the many-thousand volt power lines from the electric company's generators. A useful analogy in visualizing electric pressure is to think of the water behind a dam. The water pressure (grams per square centimeter or pounds per square foot) is something like the electrical pressure in this model.

The Ampere (amp)

The amp is a measure of the *rate* at which electricity is actually being drawn from a supply line (amount per unit of time). There is a relationship between amps and volts, but it is not as simple as one might suppose. It is as possible to draw a great deal of water rapidly from a small dam as it is to draw a trickle from a huge dam. The larger dam, however, would be more efficient for obtaining water flow at very fast rates with the least strain.

Familiarity with amps comes about through the use of household appliances and fuse boxes or circuit breakers. It is an annoying experience when too many appliances on the same line draw more current than the safety fuse will permit, and the fuse is "blown." The water analogy here is like that of a faucet set in the dam, and the amount of water that can be drawn depends on how far the valve is opened to supply the existing need. If the valve is opened too far, it may blow apart ("blows its fuse") owing to the energy of the rushing water. Electronic instruments, with their complex circuitry, have parts requiring very small quantities of current, measured in milliamps or microamps.

The Ohm

The ohm is a measure of the degree of **resistance** to the flow of electrons. Resistance to current flow is provided by using materials that are poor conductors of electricity, or else by using restricted pathways (such as a wire of very small diameter). The water analogy would have the dam itself as a fixed resistance to the flow of water. Consider the diameter of the hose from the dam outlet as related to resistance to water flow through it. Also consider obstructions to flow.

The relationship of these three units—volt, amp, and ohm—is mathematically stated in **Ohm's law**:

$$\underset{\text{(rate)}}{\text{amount of current}} = \frac{\text{voltage}}{\text{resistance}}$$

Symbolically,

$$I = \frac{V}{R}$$

In summary, Figure 2–17 illustrates the principal electrical units in terms of the water analogy.

EXAMPLE 2.10

What is the amount of current that can be drawn from a 1.5 volt dry cell when a 3000 ohm (3 kΩ) resistance is in the circuit?

1. Organize the data:

$$V = 1.5 \text{ volts}$$
$$R = 3 \times 10^3 \text{ ohms}$$
$$I = ?$$

2. Develop the equation:

$$I = \frac{V}{R} = \frac{1.5 \text{ volts}}{3 \times 10^3 \text{ ohms}} = 0.5 \times 10^{-3} \text{ amp}$$

$$I = 5 \times 10^{-4} \text{ amp}$$

or

0.5 mA (milliamps)

Electrical unit (and synonyms)	What It Measures	Symbols	Water analogy	Electrical diagram
VOLT (potential) (potential difference) (IR drop)	Electrical "pressure"	V or E 	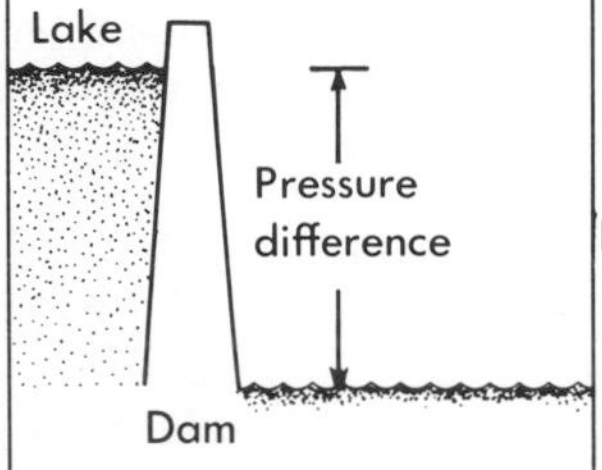 Lake; Pressure difference; Dam	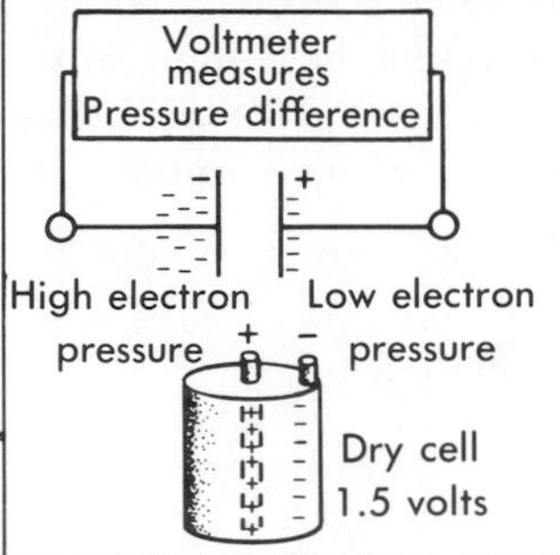 Voltmeter measures Pressure difference; High electron pressure; Low electron pressure; Dry cell 1.5 volts
AMP (current)	Amount of current being drawn per unit of time (rate)	I or A	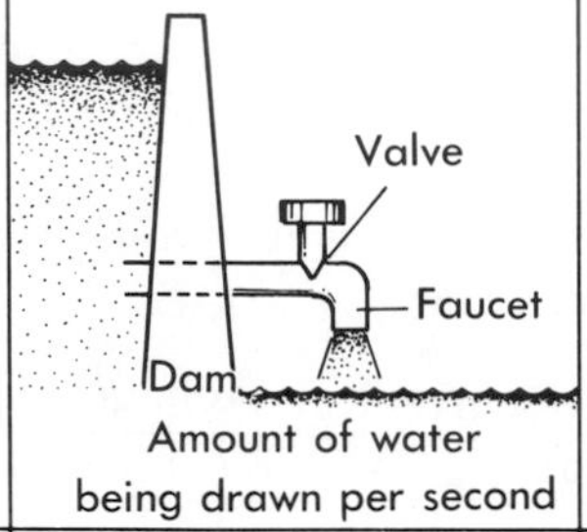 Valve; Faucet; Dam; Amount of water being drawn per second	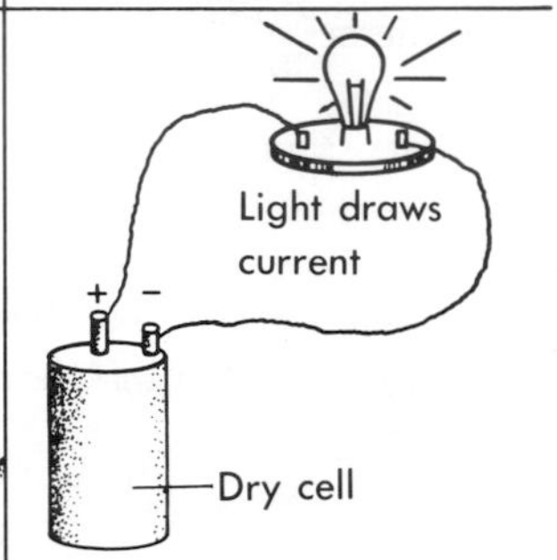 Light draws current; Dry cell
OHM	Amount of resistance to electrical flow	R or Ω (omega) 	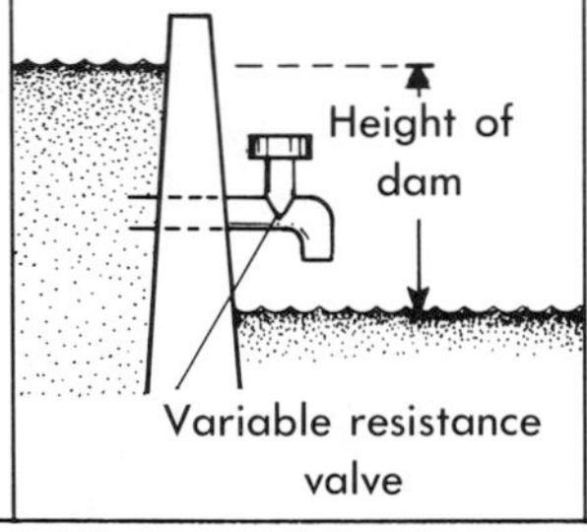 Height of dam; Variable resistance valve	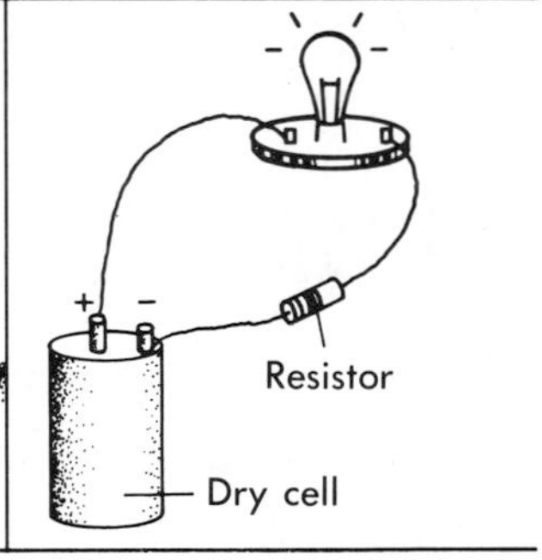 Resistor; Dry cell

Figure 2–17 The water analogy for electrical units.

EXAMPLE 2.11

What resistance value must be in a circuit in order to limit the current from a 50.0 volt battery to 20 mA?

1. Organize the data:

$$V = 50 \text{ volts}$$

$$R = ?$$

$$I = 20 \text{ mA} = 0.02 \text{ A} = 2 \times 10^{-2} \text{ amp}$$

2. Develop the equation:

$$I = \frac{V}{R} \text{ is rewritten as}$$

$$R = \frac{V}{I} = \frac{50 \text{ volts}}{2 \times 10^{-2} \text{ amp}} = 25 \times 10^2 \text{ ohms}$$

$$R = 2500 \text{ ohms}$$

or

$$2.5 \text{ k}\Omega \text{ (kilo ohms)}$$

Exercise 2.5

1. What resistance is required to draw only 15 μA (microamps) from a 6.0 volt dry cell?
2. If an electrochemical cell develops a 0.96 volt potential, what is its internal resistance if a maximum of 0.04 amps of current can be used?
3. What is the EMF of a battery when 50.0 mA can be drawn through a resistance 12.0 kΩ?

PROBLEMS AND QUESTIONS

2.1 What is matter? What are types, states, and changes of matter?

2.2 What are the two fundamental driving forces behind the changes of matter?

2.3 Name three observations that identify a chemical reaction as being stoichiometric.

2.4 What is a catalyst?

2.5 Label the following changes as physical or chemical: rusting iron, souring milk, dissolving sugar, baking a cake, changing carbon dioxide gas to dry ice.

2.6 Identify the following as elements, mixtures, or compounds: silver, table salt, air, milk, sulfur dioxide, sulfur, tea, brass, nitric acid, ice.

2.7 How fast must a 2.0 g object move in order to have a kinetic energy of 8×10^6 ergs?★

2.8 If the energy absorbed in a chemical change is 3×10^{-5} joules, what is the equivalent energy in ergs?★

2.9 How many calories are required to raise the temperature of 140 g of a substance by 22 degrees if the specific heat is 0.232 cal g^{-1} $°C^{-1}$?★

2.10 Convert 150 calories to liter-atmospheres and electron volts.★

2.11 What is the frequency of radiation having a wavelength of 320 nm?★

2.12 What is the energy (in ergs) of a photon if its frequency is 5.0×10^{15} Hz?★

2.13 What is Coulomb's law? What is Ohm's law?★

2.14 How many ohms are needed in a circuit having a 60 volt battery in order to protect an instrument requiring 30 milliamps?★

CHAPTER 3

Alice laughed. "There's no use trying," she said, "one can't believe impossible things."
"I daresay, you haven't had much practice," said the Queen. "When I was your age I always did it for half-an-hour a day. Why, sometimes I've believed as many as six impossible things before breakfast."

Lewis Carroll, *Through the Looking-Glass*

Behavioral Objectives

At the completion of this chapter, the student should be able to:

1. Define elements and compounds and give examples of each.
2. List five distinguishing characteristics of metallic and nonmetallic elements.
3. Write symbols correctly for elements.
4. Define the law of definite composition.
5. Define the law of multiple proportions.
6. List the three principal subatomic particles by name, symbol, and electrical charge.
7. Define an atomic mass unit.
8. Calculate atomic numbers, atomic weights, and the number of each type of subatomic particle from given data.
9. Define the term *isotope*.
10. Calculate average atomic weights from natural isotope distribution.
11. Define the term *mole*.
12. Calculate the mole weights of elements and compounds.
13. Find the number of moles of a substance when the mass is given.
14. Calculate the number of particles in a given number of moles.
15. Express moles of solid and liquid elements and compounds in grams.
16. Explain the significance of Avogadro's number.
17. Relate moles of vapor to liters occupied at STP.
18. Define, operationally, the symbols STP.
19. Express the law of Dulong and Petit mathematically.*
20. Calculate approximate atomic weights from specific heat data.*
21. Define the term equivalent weight, and calculate equivalent weights of elements from experimental data.*

ELEMENTS, COMPOUNDS, AND THE MOLE CONCEPT

In Chapter 2, matter was classified as being either a pure substance or a mixture. Pure substances may be subdivided into *elements* and *compounds*. Since the chemist is concerned with the properties and behavior of pure substances, some fundamental questions must be answered: What are elements and compounds? What are they composed of? How are the descriptions of their properties, composition, and changes quantitatively handled? Is there some central and unifying concept that can provide the chemist with a solid basis as he attempts to interrelate his quantitative observations? The elements of matter provide the logical starting point for such inquiries.

3.1 ELEMENTS: TYPES, SYMBOLS, AND NAMES

Elements, the simplest form of pure substances, are commonly described as **metallic** or **nonmetallic,** while recognizing that a number of them fall into the gray region that indistinctly separates the two. Those elements that exhibit the properties of both metals and nonmetals are called **metalloids.** Further reference to metalloids will be made in Chapter 5, which deals with the periodic chart.

Of the 103* recorded elements, only about 20 percent are nonmetals. The general properties that differentiate metals from nonmetals are summarized in Table 3–1.

The elements have distinct names and symbols. Many of the symbols bear little resemblance to the common English name because they have been derived from the Latin or Greek name of the element. This is especially true of metals such as copper, tin, lead, gold, and others that were known to ancient alchemists.

The rules for writing elemental symbols are as follows:

1. A single-letter symbol representing an element is always capitalized.

Examples

uranium	U
sulfur	S
oxygen	O

* Actually, 105 elements have been reported and named! However, Kurchatovium (Ku) or Rutherfordium (Rf), 104, and Hahnium (Ha), 105, have not yet been officially recognized.

TABLE 3–1 GENERAL CHARACTERISTICS OF METALLIC AND NONMETALLIC ELEMENTS

Metals	Nonmetals
relatively dense solids	gas or light powder
silvery	variety of colors
lustrous (except copper and gold)	
malleable	not malleable or ductile; fragile
good conductors of heat and electricity	poor conductors of heat and electricity
COMMON EXAMPLES	COMMON EXAMPLES
gold	oxygen
silver	sulfur
zinc	chlorine
iron	carbon

2. A two-letter symbol for an element has only the first letter capitalized.

Examples

calcium	Ca
iron	Fe
zinc	Zn

Table 3–2 lists a sampling of elements with their symbols and meanings.

A familiarity with the names and symbols of the elements is very important if one is to work efficiently in chemistry. Constant reference to a periodic chart or a table of the elements helps make the recommended memorization of the most commonly used elements a rather painless experience.

TABLE 3–2

bromine, Br (from the Greek *bromos*, meaning stench)
cobalt, Co (from the German *kobold*, meaning goblin)
copper, Cu (from the Latin name of the island of Cyprus, *cuprum*)
gold, Au (from the Latin *aurum*, meaning shining dawn)
iron, Fe (from the Latin word for iron, *ferrum*)
lead, Pb (from the Latin word for lead, *plumbum*)
mercury, Hg (from the Latin word for mercury, *hydrargyrum*)
phosphorus, P (from the Greek word for light-bearing, *phosphoros*)
potassium, K (from the Latin, *kalium*)
sodium, Na (from the Latin, *natrium*)
silver, Ag (from the Latin, *argentum*)
tin, Sn (from the Latin, *stannum*)

John Dalton

3.2 THE UNIT STRUCTURE OF ELEMENTS

While the concept of the atom as a microscopic unit structure of matter dates from the ancient Greeks, it remained for John Dalton at the beginning of the 19th century to suggest a remarkable theory of the structure of matter. It was with very limited experimental evidence that Dalton's logic suggested, or eventually led to, the following points:

1. Elements are composed of indivisible particles called atoms (from the Greek word *atomos*, meaning indivisible).
2. All atoms of a single element have the same mass and size.
3. Different elements have atoms that differ in mass and size.
4. Molecules of a compound are formed by the union of two or more atoms of different elements.
5. The atoms that combine to form molecules do so in simple whole-number ratios. For example

   2 H atoms and 1 O atom = one molecule of water, H_2O

   1 N atom and 3 H atoms = one molecule of ammonia, NH_3

   2 C atoms and 6 H atoms and 1 O atom = one molecule of ethyl alcohol, C_2H_5OH

6. Atoms of some elements may combine in different ratios to form different molecules. For example

1 C atom and 1 O atom = carbon monoxide, CO

1 C atom and 2 O atoms = carbon dioxide, CO_2

1 N atom and 1 O atom = nitrogen monoxide, NO

2 N atoms and 3 O atoms = dinitrogen trioxide, N_2O_3

1 N atom and 2 O atoms = nitrogen dioxide, NO_2

2 N atoms and 5 O atoms = dinitrogen pentoxide, N_2O_5

Today, of course, the model of the atom is vastly more sophisticated than Dalton's model. The atom is more complex than his solid-sphere concept, since he had no knowledge of the subatomic particles called protons, neutrons, and electrons. Furthermore, atoms can be decomposed by the extraordinary methods used in making atomic energy available, and not all the atoms of a particular element have the same mass (see *Isotopes*, p. 67).

Dalton's efforts, however, did give rise to two fundamental laws of chemistry that still hold true. They are the **law of definite composition** and the **law of multiple proportions.**

3.3 THE LAW OF DEFINITE COMPOSITION

This law says, in effect, that the weight proportions of elements in a given compound are always the same. If a sample of pure water is analyzed and found to be 88.8 percent oxygen by weight and 11.2 percent hydrogen, this is now and forever the weight ratio of oxygen to hydrogen. Any 100 g mass of water, be it from the laboratory, the North Pole, or Mars, must be composed of 88.8 g oxygen and 11.2 g hydrogen. If the mass of what is supposed to be water does not have the specified weight ratio, then it is, in fact, not water.

3.4 THE LAW OF MULTIPLE PROPORTIONS

This law says that when two elements can combine to form more than one compound, the weight ratios of one of the elements will vary by small whole-number multiples when compared to a fixed weight of the other element. For example, nitrogen combines with oxygen to form several compounds. Table 3–3 shows the weights of oxygen in those compounds that combine with 100.0 g of nitrogen.

The data in Table 3–3 illustrate the law of multiple proportions, insofar as the weights of oxygen that combine with 100.0 g of nitrogen are 1×57.0 g, 2×57.0 g, 3×57.0 g, 4×57.0 g, and 5×57.0 g.

3.5 THE SUBATOMIC PARTICLES

One of the first significant advances from the Dalton atom was made by J. J. Thomson in 1897 when he discovered the units of negative

TABLE 3–3

Compound	Nitrogen	Oxygen	Whole-Number Multiples of Oxygen Weight
N_2O	100.0 g	57.0 g	$\frac{57.0}{57.0} = 1$
NO	100.0 g	114.0 g	$\frac{114.0}{57.0} = 2$
N_2O_3	100.0 g	171.0 g	$\frac{171.0}{57.0} = 3$
NO_2	100.0 g	228.0 g	$\frac{228.0}{57.0} = 4$
N_2O_5	100.0 g	285.0 g	$\frac{285.0}{57.0} = 5$

Sir Joseph John Thomson

electrical charge called **electrons.** In 1907, R. A. Millikan found the charge on any one of these electrons to be 1.6×10^{-19} coulombs. This value, used in connection with Thomson's discovery of the charge-to-mass ratio for the electron (e/m), 1.8×10^{8} coul g^{-1}, indicates that the apparent mass of the electron is 9.1×10^{-28} g. This is about as close to nothing as one can get.

This information led Thomson to postulate that an atom was a sphere of positive charge in which electrons were distributed much like raisins in a pudding. This "plum pudding" model did not last very long. It is interesting to note that the nature of the electron is less clearly understood by today's scientists than by scientists at the turn of the century, who supposed their theories fairly accurate. Although the electron is commonly pictured as a particle having a specific charge and mass, its mass is not a very meaningful value, and its size (after all, a particle of matter should have physical dimensions) is practically nil. That is to say, the electron is described as an approximate *point* of charge, and a point has no dimensions! But the electron sometimes behaves as though it had size. Depending on the conditions under which electrons are observed, the size of a single electron may *seem* to be as large as or greater than a whole atom. This recitation of the mysterious aspects of the electron is meant to emphasize the need to distinguish between models and reality. It is very useful to think of the electron as a particle when it *behaves* like a particle. However, an analogy may be found in the fact that while a man can swim and sometimes behave like a fish, this does not make him a fish in reality.

In 1911, Ernest Rutherford proved experimentally that the atom

Ernest Rutherford

TABLE 3–4

Particle	Symbol	Electrical Charge	Mass in amu	Mass in grams
proton	p^+	1+	1.0073	1.6×10^{-24}
neutron	n^0	0	1.0087	1.6×10^{-24}
electron	e^-	1−	0.0005	9.1×10^{-28}

has an extremely small and dense core. This discovery led to the acceptance of the nuclear model of the atom, in which a dense nucleus is surrounded by a complement of electrons. While the electrons consitutute most of the volume of the atom (very close to 100 percent), the remaining subatomic particles compose the nucleus and contain almost all the mass.

Wilhelm Wein and J. J. Thomson identified the subatomic particle of positive charge, which was given the name **proton.** It remained for Rutherford to establish the proton as a nuclear particle. The charge on the proton is the same as that of the electron, *but opposite in sign*, and its mass is nearly 2000 times as great as that of the electron. The proton is assigned the arbitrary mass of 1 **atomic mass unit,** 1 amu. Today, an atomic mass unit is alternately defined as 1/12 the mass of a carbon-12 atom. The electron, by comparison, is 0.00055 amu.

To account for the difference between the number of protons and the larger value for the mass of the entire atom, Rutherford, in 1920, predicted the existence of an electrically neutral particle in the nucleus which would have about the same mass as the proton. In 1932, James Chadwick discovered the *neutron*, and thus validated Rutherford's prediction. Table 3–4 summarizes the characteristics of the principal subatomic particles.

3.6 ATOMIC WEIGHTS

In the periodic chart of the elements, carbon is represented as shown in Figure 3–1. The **atomic number,** 6, indicates the number of protons in the nucleus of the atom. This is also the number of electrons in a neutral atom. The rounded-off atomic weight, 12,

Figure 3–1 Carbon, as shown on the periodic chart.

is the mass of the most abundant carbon atom. The difference between the atomic weight and the atomic number is the number of neutrons. The conventional way of indicating the atomic weight and the atomic number of an atom is to write the atomic weight as a *superscript* to the left of the symbol, and the atomic number as a *subscript* to the left of the symbol:

$^{12}_{6}C$ 6 protons
6 electrons
$12 - 6 = 6$ neutrons

$^{56}_{26}Fe$ 26 protons
26 electrons
$56 - 26 = 30$ neutrons

$^{238}_{92}U$ 92 protons
92 electrons
$238 - 92 = 146$ neutrons

The rounded-off atomic weight (often called the *mass number*) is always the sum of the number of protons and neutrons, because both nuclear particles have the approximate value of 1 atomic mass unit. The mass of the electron is ignored because it is so very small by comparison. Remember, it would take nearly 2000 electrons to have a mass of 1 amu, while the largest atoms have only about 100 electrons.

Exercise 3.1

1. Write the symbols, with atomic numbers as subscripts and atomic weights (rounded off) as superscripts, for the following:
   a. sodium
   b. phosphorus
   c. nickel
   d. molybdenum
   e. lead

2. Complete the following chart:

Element	Atomic Number	Number of Neutrons	Number of Electrons	Number of Protons	Atomic Weight
$^{27}_{13}Al$					
Chromium	24				52
$_{35}Br$		45			
			79		197
^{32}S					

It was noted previously that the mass of a proton is about 1.6×10^{-24} g. In other words, if atomic mass units were converted to grams, 1 amu would equal 1.6×10^{-24} g. This system would lead to the expression of a carbon atom mass as $12 \times 1.6 \times 10^{-24}$ g $= 19.2 \times 10^{-24} = 1.92 \times 10^{-23}$ g. To describe this method of expressing atomic weights as cumbersome is something of an understatement. The alternative that chemists have adopted is to establish a system of *relative atomic weights.* In other words, atomic weights are based on a comparison to the carbon-12 **isotope.** Magnesium, for example, has an atomic weight of 24 because a magnesium atom is about twice as massive as a carbon atom. What isotopes are, and their significance, will be discussed below.

A method for establishing a comparative system of weights when it is impractical or impossible to compare single units, is to compare *equally large numbers* of the units. For example, the weights of a penny, nickel, and quarter could be compared as shown in Figure 3–2.

In the same manner in which 1000 of each coin produces the same weight ratio as single coins, very large but equal numbers of atoms may be compared. The specific large number used, 6.02×10^{23}, known as **Avogadro's number,** provides the basis for the *mole concept,* which is one of the most important quantitative tools in chemistry.

3.7 ISOTOPES

The isotope, carbon-12, and the question of its definition and significance were mentioned previously. Isotopes of an element are

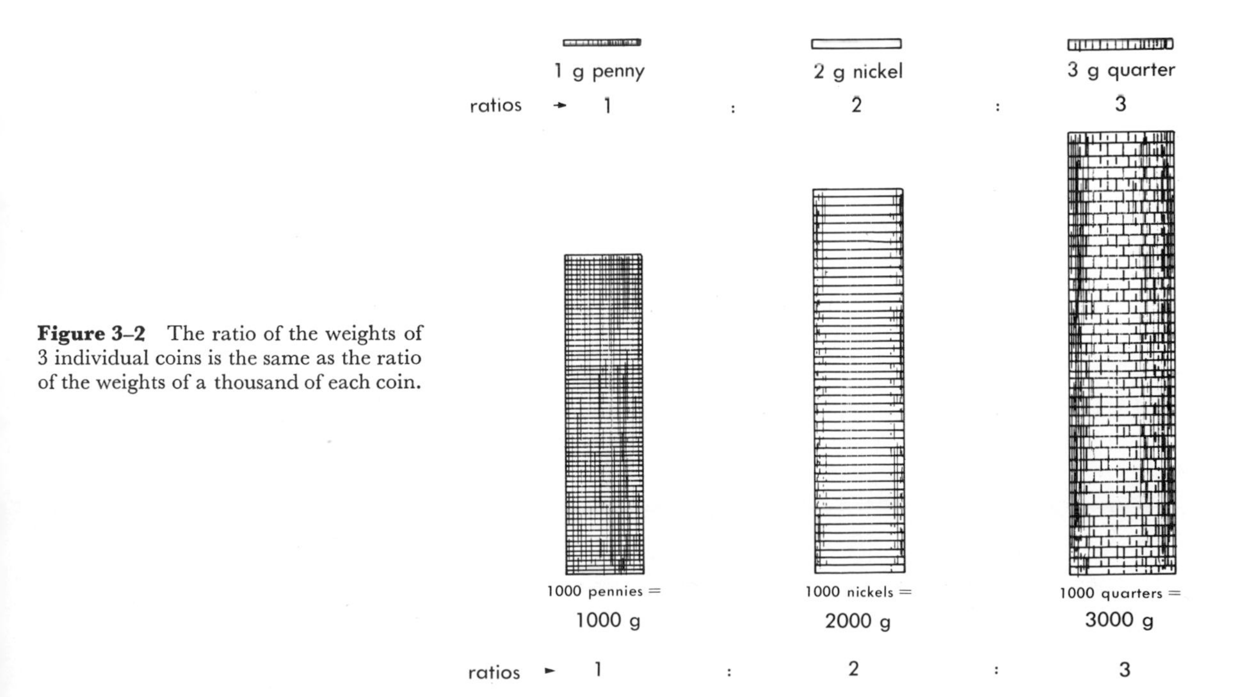

Figure 3–2 The ratio of the weights of 3 individual coins is the same as the ratio of the weights of a thousand of each coin.

TABLE 3–5

Isotope	A Reason for Fame or Infamy
carbon-14, ^{14}C	dating of objects belonging to ancient cultures
cobalt-60, ^{60}Co	cancer therapy
strontium-90, ^{90}Sr	radioactive fallout from atomic bombs
iodine-121, ^{121}I	thyroid therapy
uranium-235, ^{235}U	atomic energy
hydrogen-3, ^{3}H (tritium)	active substance of hydrogen bombs
phosphorus-32, ^{32}P	biologic tracer

atoms that have the same number of protons (atomic number), but differing numbers of neutrons. The varying numbers of neutrons do not make different elements. The atoms have only different masses and, occasionally, different degrees of stability, but they have the same number of electrons and the same chemical behavior. Some isotopes are notoriously unstable. They tend to undergo nuclear decomposition at rates that vary from microseconds to centuries, emitting radioactive particles in the process. Table 3–5 illustrates some of the isotopes that may be familiar because their uses or dangers have often been discussed in newspapers and magazines.

The reason that most elements have atomic weights which are not perfectly whole numbers is because they have two or more isotopes, and the naturally occurring element is a mixture. Any natural sample of chlorine gas would be found to be 75.53 percent $^{35}_{17}Cl$ and 24.47 percent $^{37}_{17}Cl$, and have a resultant atomic weight of 35.45. The average atomic weight has to be weighted toward the $^{35}_{17}Cl$, since three-fourths of the naturally occurring chlorine has that mass. It would not make sense to add the two masses and divide by 2. The weighted average is obtained by multiplying each value by its percent abundance and calculating the sum.

EXAMPLE 3.1

Find the average atomic weight of chlorine from its natural distribution:

$$\text{chlorine-35} = 75.53\,\%$$

$$\text{chlorine-37} = 24.47\,\%$$

1. Convert the percentage to a decimal value and multiply by the more accurate isotopic mass obtained from a reference table:

$$34.97\ (0.7553) = 26.41$$

$$36.97\ (0.2447) = \ \ 9.04$$

2. The sum of the products is the average atomic weight:

$$\begin{array}{r} 26.41 \\ +\ 9.04 \\ \hline \text{atomic wt.} = 35.45 \end{array}$$

3.8 THE MOLE CONCEPT

It was mentioned previously that an organized table of relative atomic weights can be obtained by the measurement of equally large numbers of atoms of different elements. The validity of larger weights in the same ratio was illustrated in Figure 3–2.

The specific large number is 6.02×10^{23}. This number, known as Avogadro's number, symbolized $\mathcal{N}_A$, when described as being merely "large," does not begin to describe the enormity of 10^{23}. An analogy that might suggest the magnitude of Avogadro's number would be to try to imagine how much of the earth's surface could be covered by 6.02×10^{23} marbles. The answer is that the entire surface of our planet could be covered by Avogadro's number of marbles to a depth of over 50 miles!

Avogadro's number of particles is universally called a *mole*,

Amedeo Avogadro

symbolized **n**, from the Greek word meaning a pile or mound. While Amedeo Avogadro did not experimentally determine the number that bears his name, he has been so honored because his studies of gas behavior indirectly led to the finding of the number. There are several experimental methods that can be used to verify that Avogadro's number is indeed 6.02×10^{23}, but discussions of electrodeposition, x-ray diffraction, and monomolecular layers as examples of these experimental methods are beyond the scope of this text.

Avogadro's number of anything may be specified as a mole; 6.02×10^{23} atoms of lead is a mole of lead *atoms*; 6.02×10^{23} molecules of carbon dioxide is a mole of carbon dioxide *molecules*; and 6.02×10^{23} sneakers (perish the thought) is a mole of sneakers. However, *the internationally recognized standard for a mole is the number of atoms in 12.000 grams of carbon-12.* This number of atoms is, then, Avogadro's number.

The student might wonder why the mole is defined as 6.02×10^{23} particles instead of a simpler expression, such as 1×10^{20} or 1×10^{23}. The reason for 6.02×10^{23} being a naturally useful quantity to chemists is because Avogadro's number of atoms of each element has a mass equal to the relative atomic weight of the element. This relationship is illustrated in Figure 3–5.

The mole concept may be visualized as something akin to a diamond. It is valuable and useful. It has a number of facets, each of which makes its own contribution to the beauty of the gem. Figure 3–3 represents the mole concept and its facets.

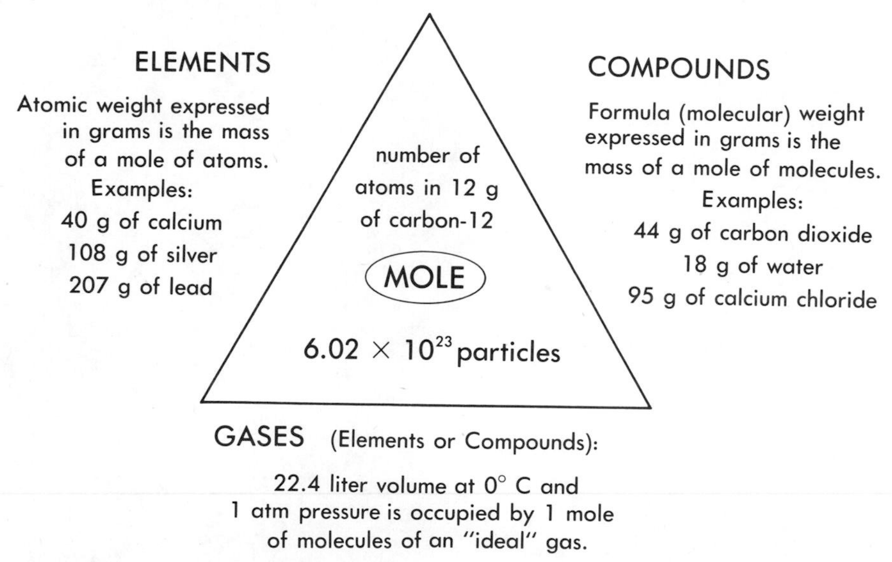

Figure 3–3 Facets of the mole: particle, elements, compounds, and gases.

All the examples in Figure 3–3 are expressions of a mole because each item is composed of 6.02×10^{23} particles, regardless of whether the particles are atoms, molecules, or unit structures in a crystal.

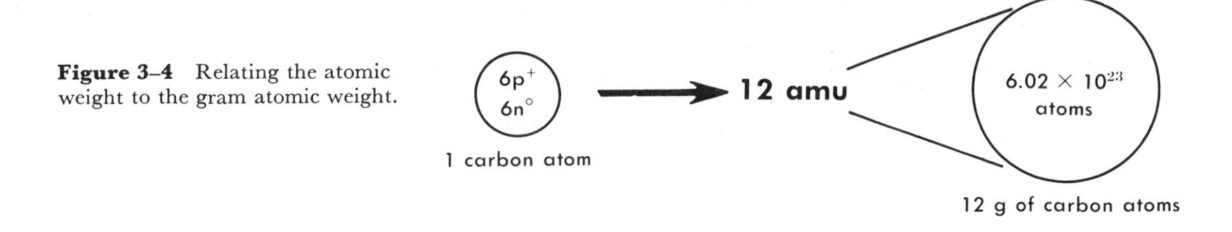

Figure 3–4 Relating the atomic weight to the gram atomic weight.

3.9 MOLES OF ELEMENTS

Figure 3–3 indicates that 40 g of calcium, 108 g of silver, and 207 g of lead are each the mass of onc mole of the respective element. Why? The answer is discovered by another look at the carbon-12 atom, which is the basis for the atomic-weight scale. Observe that the mass number of the carbon-12 atom is the sum of the 6 protons and 6 neutrons which compose its nucleus. The sum of the nuclear units is equal to 12 atomic mass units. This cannot be measured easily, but the mass of 6.02×10^{23} atoms of carbon-12 can be measured with common devices. It is found to be 12 grams. Notice that the sum of protons and neutrons yields a numerical value in arbitrary units that is exactly the same as the numerical mass value of the mole of atoms of C-12 (Fig. 3–4). Comparing other atoms in the same way, it can be demonstrated that the atomic weight in atomic mass units can be multiplied by Avogadro's number to provide a table of relative molar weights in grams, where each value is the mass of a mole of atoms. Figure 3–5 illustrates this comparison.

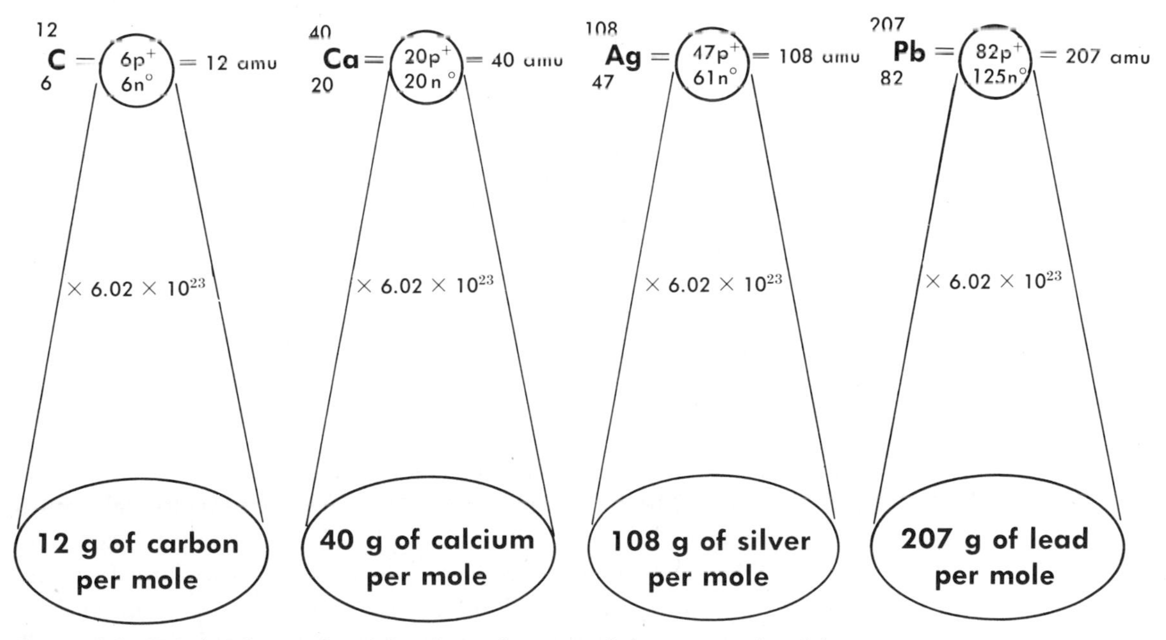

Figure 3–5 Relating the atomic weights of several atoms to their gram atomic weights.

Table 3–6 lists values for a mole of some commonly used elements.

TABLE 3–6

Element	Symbol	Atomic weight (amu)	Grams per mole (rounded off)
aluminum	Al	26.98	27
barium	Ba	137.34	137
*bromine	Br	79.90	80
cadmium	Cd	112.40	112
calcium	Ca	40.08	40
carbon	C	12.01	12
*chlorine	Cl	35.45	35.5
chromium	Cr	52.00	52
cobalt	Co	58.93	59
copper	Cu	63.55	63.5
*fluorine	F	19.00	19
gold	Au	196.97	197
helium	He	4.00	4
*hydrogen	H	1.01	1
*iodine	I	126.90	127
iron	Fe	55.85	56
lead	Pb	207.19	207
magnesium	Mg	24.31	24
*mercury	Hg	200.59	200.5
nickel	Ni	58.71	59
*nitrogen	N	14.01	14
*oxygen	O	16.00	16
*phosphorus	P	30.97	31
potassium	K	39.10	39
silicon	Si	28.09	28
silver	Ag	107.87	108
sodium	Na	22.99	23
strontium	Sr	87.62	88
*sulfur	S	32.06	32
tin	Sn	118.69	119
zinc	Zn	65.37	65

* Thoughtful care must be taken when applying the mole concept to Br, Cl, F, H, Hg, I, N, O, P, and S. The *atomic weight in grams* expresses the mass of a mole of *atoms* of these elements. However, these same elements occur naturally as *polyatomic molecules*: Br_2, Cl_2, F_2, H_2, Hg_2, I_2, N_2, O_2, P_4, and S_8, S_6, S_4, S_2. The *molecular weight in grams* expresses the elemental masses of a mole of *molecules*. For example, 16 g is the mass of a mole of oxygen *atoms*, while 32 g is the mass of a mole of oxygen *molecules*.

3.10 MASS, MOLES, AND THE NUMBER OF ATOMS

The beauty and utility of the mole concept lies in the fact that the masses of elements and compounds, the volumes of gases, and the numbers of particles are all like different facets of the same gem. If there is not an absolutely clear knowledge of which facet is relevant to the solution of a problem involving the mole concept, confusion reigns. The student must firmly decide what he is looking for, what data are available, and what kind of answer makes sense. Dimensional analysis is the surest method to avoid the mistakes that have perennially made the mole concept a horror story for students instead of simplifying the process of problem solving. Consider the following examples.

EXAMPLE 3.2

How many moles of calcium atoms have a mass of 200 grams?

1. Find the weight of one mole of calcium from a chart of atomic weights:

$$^{40}_{20}\text{Ca means } 40 \text{ g mole}^{-1}$$

2. Reasoning that 200 g is the mass of more than one mole, the answer is obtained by dividing the mass of one mole into the total mass given:

$$\text{number of moles} = \frac{\overset{5}{\cancel{200}}\ \cancel{\text{g}}}{\cancel{40}\ \cancel{\text{g}}\ \text{mol}^{-1}}$$

$$\text{number of moles} = 5$$

3. Notice how the gram units cancelled out and $1/\text{mole}^{-1}$ becomes the reciprocal expression, mole, since,

$$\frac{1}{\text{mole}^{-1}} = \frac{\text{mole}^{+1}}{1}$$

Example 3.2 may be generalized to produce a simple but invaluable equation:

$$n = \frac{\text{g}}{\text{g mole}^{-1}}$$

$$n = \text{number of moles}$$

$$\text{g} = \text{grams available}$$

$$\text{g mole}^{-1} = \text{mass of 1 mole, in grams}$$

EXAMPLE 3.3

30.6 mg of sodium is the mass of how many *moles*?

1. Express the available mass of sodium in grams, so that the unit of mass is compatible with the mass of a mole, which is in grams.*

---

* An alternative method is to relate milligram weights to *millimoles*. This method, however, is not recommended until the student gains considerable experience.

2. Find the mass of a mole of sodium from an atomic weight chart:

$$^{23}_{11}\text{Na means 23 grams per mole (g mole}^{-1}\text{)}$$

3. Substitute in the appropriate equation:

$$n = \frac{g}{g\ mole^{-1}} = \frac{30.6 \times 10^{-3}\ \not{g}}{23\ \not{g}\ mol^{-1}} = 1.3 \times 10^{-3}\ mole$$

EXAMPLE 3.4

What is the mass, in grams, of 0.0024 moles of carbon?

1. Find the weight of one mole of carbon:

$$^{12}_{6}\text{C means 12 g mole}^{-1}$$

2. Predict an answer that is much less than 12 g because 0.0024 moles is obviously a very small fraction of one mole. Compare the predicted answer (as a very crude approximation) to the final calculated value.
3. Use scientific notation to express 0.0024 as 2.4×10^{-3}.
4. Substitute in the appropriately modified equation:

$$n = \frac{g}{g\ mole^{-1}} \qquad \text{(multiply the equation by g mole}^{-1}\text{)}$$

$$(g\ mole^{-1})(n) = \frac{g}{\cancel{g\ mole^{-1}}}(\cancel{g\ mole^{-1}})$$

The emerging equation is

$$g = n(g\ mole^{-1})$$

$$g = (2.4 \times 10^{-3}\ \cancel{mol})(12\ g\ \cancel{mol}^{-1})$$

Answer is 28.8×10^{-3} g, preferably expressed as

$$2.9 \times 10^{-2}\ g$$

or

$$0.029\ g$$

EXAMPLE 3.5

How many *atoms* are there in a 20.0 g sample of silver?

1. Find the weight of one mole of silver:

$$^{108}_{47}\text{Ag indicates 108 g mole}^{-1}$$

2. Predict the answer to be a very large number, since even a small sample of matter will be composed of a great many atoms.
3. Since the number of particles (in this example, the particles are atoms) depends on the number of moles, the equation is

$$\text{number of particles} = n \times \mathcal{N}_A$$

$$n = \text{number of moles}$$

$$\mathcal{N}_A = 6.02 \times 10^{23} \text{ particles per mole}$$

Note: Since a particle is not a dimensional unit, it is not included in dimensional analysis. $\mathcal{N}_A$ is expressed as 6.02×10^{23} mole^{-1}, and the nature of the particle is understood, and *not* written:

$$\mathcal{N}_A = 6.02 \times 10^{23} \text{ mole}^{-1}$$

4. Since $n = \dfrac{g}{g \text{ mole}^{-1}}$, the equation may be rewritten:

$$\text{number of particles (atoms)} = \frac{g}{g \text{ mole}^{-1}} (\mathcal{N}_A)$$

5. Substituting in the equation:

$$\text{number of atoms of silver} = \frac{20 \not{g}\,(6.02 \times 10^{23} \not{\text{mol}}^{-1})}{108 \not{g} \not{\text{mol}}^{-1}}$$

Answer is 1.11×10^{23} atoms

EXAMPLE 3.6

What is the mass of 1 million atoms of chromium?

1. Find the mass of one mole of chromium:

$$^{52}_{24}\text{Cr means } 52 \text{ g mole}^{-1}$$

2. Predict the answer to be an incredibly small fraction of a gram since one million atoms is a very small fraction of Avogadro's number.
3. Modify the equation from Example 3.5, to solve for grams:

$$\text{number of particles} = (n)(\mathcal{N}_A)$$

$$\text{number of particles} = \frac{(g)(\mathcal{N}_A)}{g \text{ mole}^{-1}}$$

$$g = \frac{(\text{number of particles})(g \text{ mole}^{-1})}{\mathcal{N}_A}$$

4. Substitute in the equation and solve:

$$g = \frac{(10^6)(52 \text{ g } \cancel{\text{mol}}^{-1})}{6.02 \times 10^{23} \cancel{\text{mol}}^{-1}}$$

Answer is 8.6×10^{-17} g

Note: The care taken with arithmetical detail is critical. An answer of 8.6×10^{17} g looks like a tiny mistake, 10^{17} instead of 10^{-17}. However, it is an error of 34 orders of magnitude, and it says in effect that the mass of a million chromium atoms begins to approach the mass of our planet, instead of being microscopically tiny.

Exercise 3.2

1. Calculate the average atomic weight of silicon from the data below:

Isotope	Isotopic Mass (amu)	Percent of Natural Occurrence
silicon-28	27.98	92.2
silicon-29	28.98	4.7
silicon-30	29.97	3.1

2. What is the mass of 0.06 moles of iron?
3. How many moles of aluminum have a mass of 60.0 mg?
4. What is the mass, in grams, of one lead atom?
5. How many atoms of barium are there in 1.2×10^{-2} moles?

3.11 MOLES OF COMPOUNDS

Since a mole of an element is composed of the Avogadro number of atoms, it naturally follows that a mole of a compound is made of the Avogadro number of molecules (or the smallest number of **ions** in an ionic crystal that expresses the fundamental ratio of the elements in the compound). Ions may be usefully defined as electrically charged particles.

There is a difference between the structure of the molecule and the structure of ionic crystals. Some chemists prefer to emphasize this difference by making a distinction between **molecular weights** and **formula weights.** The molecular weight designation is thus reserved for compounds made of molecules (which are defined as discrete uncharged particles, each composed of a precise number of atoms). A molecule has a describable size, geometry, and mass. It is also described as the smallest unit structure of matter that can exist as an

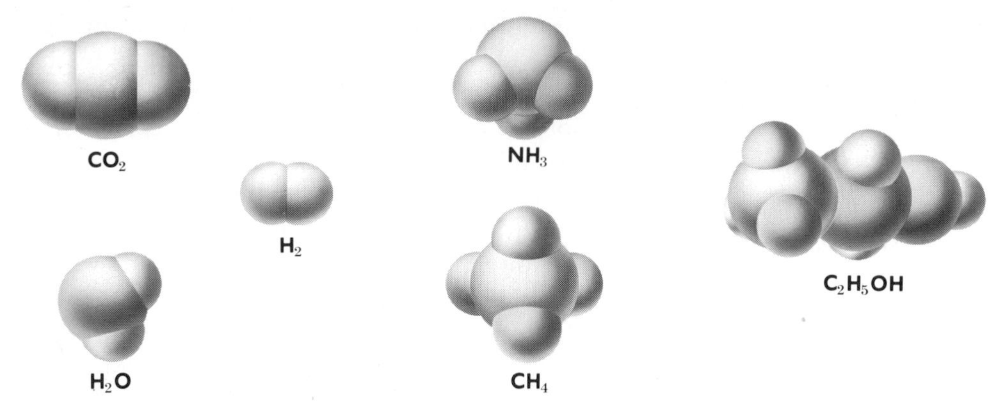

Figure 3–6 Examples of molecules.

individual particle under ordinary conditions. Some examples of molecules are illustrated in Figure 3–6.

The crystal structure formed by the forces of attraction that exist between *ions* is not molecular. The only definite aspects of the ionic crystal are the particular ions involved and their ratios. For example, table salt (sodium chloride) is a structure composed of sodium ions and chloride ions in a one-to-one ratio, held together by electrostatic forces of attraction. This obviously crystalline compound has no precise dimensional limits, and it has no specific number of ions or total mass. Individual sodium chloride units of two ions do not ordinarily exist alone; any visible crystal fragment would contain thousands of ions. The formula, NaCl, is called an **empirical formula,** which means that the formula represents only the simplest ratio of the ions. Figure 3–7 illustrates the arrangement of ions in NaCl. A more systematic discussion of molecular and ionic compounds will be taken up in Chapter 6, Chemical Bonds. However, the application of the mole concept to compounds is uniform regardless of any distinction made between molecular and ionic structures.

Figure 3–7 The crystal structure of sodium and chloride ions in table salt.

Since a molecule of carbon dioxide, for example, is made of one carbon atom and two oxygen atoms (represented by the formula, CO_2), it follows that one mole of CO_2 is composed of one mole of carbon atoms and two moles of oxygen atoms. If a mole of carbon atoms has a mass of 12 grams, and two moles of oxygen atoms have a mass of 2×16 g, the total mass of a mole of CO_2 is 44 g. It may be concluded from this observation that the mass of a mole of a compound is equal to the sum of the atomic weights expressed in grams. This is often called the **gram molecular weight** or the **gram formula weight** of the compound. The term gram formula weight is especially applicable to a nonmolecular compound such as sodium chloride. The one-to-one ratio indicated by the formula NaCl means that the weight of one mole of sodium atoms plus one mole of chlorine atoms is equal to the weight of one mole of NaCl units. The gram atomic weight of Na, 23 g, added to the gram atomic weight of chlorine, 35.5 g, equals the gram formula weight of NaCl, 58.5 g. In other words, 58.5 g of NaCl is the mass of a mole of table salt. Table 3–7 lists the values for a mole of some common compounds.

Some compounds have more complicated formulas than those illustrated in Table 3–7. Calculations of the mass of a mole of such compounds are illustrated in the following examples:

TABLE 3–7

Compound	Formula	Gram Molecular or Gram Formula Weight
Water	H_2O $(2 \times 1) + 16 = 18$	18 g mole^{-1}
Ammonia	NH_3 $14 + (3 \times 1) = 17$	17 g mole^{-1}
Benzene	C_6H_6 $(6 \times 12) + (6 \times 1) = 78$	78 g mole^{-1}
Oxygen	O_2 $(2 \times 16) = 32$	32 g mole^{-1}
Sucrose (table sugar)	$C_{12}H_{22}O_{11}$ $(12 \times 12) + (22 \times 1) + (11 \times 16) = 342$	342 g mole^{-1}
Silver nitrate	$AgNO_3$ $108 + 14 + (3 \times 16) = 170$	170 g mole^{-1}

EXAMPLE 3.7

What is the mass of a mole of $Ca(NO_3)_2$?

1. The subscript 2 outside the parenthesis means that the number of atoms *within* the parenthesis is doubled:

(NO_3)

1 atom of nitrogen ↗ ↖ 3 atoms of oxygen

$(NO_3)_2$

$2 \times 1 = 2$ atoms of nitrogen ↗ ↖ $2 \times 3 = 6$ atoms of oxygen

2. The mass of 1 mole of calcium is added to the total masses of 2 moles of nitrogen and 6 moles of oxygen:

$Ca(NO_3)_2$

$$40 + (2 \times 14) + (6 \times 16) = 164$$

3. The mass of $Ca(NO_3)_2$ is 164 g $mole^{-1}$.

EXAMPLE 3.8

Calculate the mass of a mole of $Fe_4[Fe(CN)_6]_3$.

1. Find the total number of each atom per formula. The subscript outside the bracket multiplies the subscript outside the enclosed parenthesis:

7 atoms of Fe have a mass of $7 \times 56 = 392$ g

18 atoms of C have a mass of $18 \times 12 = 216$ g

18 atoms of N have a mass of $18 \times 14 = 252$ g

Total $= 860$ g

2. Answer is 860 g $mole^{-1}$.

The calculation of the number of moles or the number of molecules of a compound is performed in the same way as was done in the case of elements. The general equation $n = g/g\ mole^{-1}$ is very satisfactorily employed.

EXAMPLE 3.9

How many moles of NaOH have a mass of 5.0 g?

1. Find the mass of one mole of NaOH from the formula:

$$\text{NaOH}$$
$$23 + 16 + 1 = 40 \text{ g mole}^{-1}$$

2. Substitute in the equation:

$$n = \frac{g}{\text{g mole}^{-1}}$$
$$n = \frac{5.0 \text{ g}}{40 \text{ g mol}^{-1}}$$
$$n = 0.125 \text{ or } 1/8 \text{ mole}$$

EXAMPLE 3.10

What is the mass of 0.042 mole of H_2SO_4?

1. Find the mass of 1 mole of H_2SO_4 from the formula:

$$H_2SO_4$$
$$(2 \times 1) + 32 + (4 \times 16) = 98 \text{ g mole}^{-1}$$

2. Develop the appropriate equation and substitute the data:

$$n = \frac{g}{\text{g mole}^{-1}} \qquad \text{(multiply the equation by g mole}^{-1}\text{)}$$
$$(\text{g mole}^{-1})(n) = \frac{g}{\text{g mole}^{-1}}(\text{g mole}^{-1})$$
$$g = n(\text{g mole}^{-1})$$
$$g = (0.042 \text{ mol})\ (98 \text{ g mol}^{-1})$$

3. Answer is 4.12 g.

EXAMPLE 3.11

How many *molecules* are there in 4.0 g of CO_2?

1. Find the mass of a mole of CO_2:

$$CO_2$$
$$12 + (2 \times 16) = 44 \text{ g mole}^{-1}$$

2. Remember from Example 3.5 that the number of particles is equal to the number of moles multiplied by Avogadro's number:

$$\text{number of molecules} = n \times \mathscr{N}_A$$

3. Since $n = g/g\ mole^{-1}$, the equation may be rewritten:

$$\text{number of molecules} = \left(\frac{g}{g\ mole^{-1}}\right)(\mathscr{N}_A)$$

4. Substituting the data:

$$\text{number of molecules} = \frac{4.0\ \cancel{g}\ (6.02 \times 10^{23}\ \cancel{mol^{-1}})}{44\ \cancel{g}\ \cancel{mol^{-1}}}$$

$$\text{number of molecules} = 0.546 \times 10^{23}$$

5. The answer, better expressed, is

$$5.5 \times 10^{22}\ \text{molecules}$$

Exercise 3.3

1. Calculate the mass of a mole of each of the following:
   a. $CaBr_2$
   b. $(NH_4)_2S$
   c. CH_3COOH
   d. $Al_2(SO_4)_3$
   e. $CuSO_4 \cdot 5H_2O$ (Hint: include the mass of 5 moles of water)

2. What is the mass of 2.5×10^{-3} mole of H_2O?
3. How many moles are there in 60.0 mg of NH_3?
4. What is the mass in grams of one molecule of CO_2?
5. How many molecules are there in 6.0 ml of mercury if the density of mercury is 13.6 g ml^{-1}?

3.12 MOLES OF GASES

The introduction to the mole pointed out that one aspect of this multifaceted concept is the fact that a volume of 22.4 liters of any "ideal" gas at 0°C and standard atmospheric pressure contains 6.02×10^{23} molecules. By virtue of this fact, 22.4 liters of gas at STP (standard temperature and pressure)* may be called a **molar volume.** Although a more detailed discussion of the behavior of gases is reserved

* The exact meaning of STP will be discussed in the next chapter.

for the next chapter, the expression of gas volumes in terms of the mole concept serves as a timely introduction.

It was Amedeo Avogadro who hypothesized that equal volumes of different gases, under the same conditions of temperature and pressure, will contain equal numbers of molecules. His hypothesis, now elevated to the status of a scientific law, is entirely consistent with the law of conservation. Figure 3–8 illustrates the relationship between the mass of a mole of gas and the volume it occupies at STP.

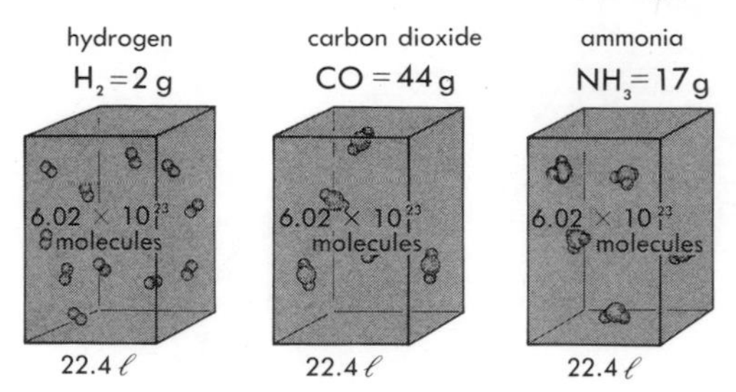

Figure 3–8 The weight of a mole of each of several gases at STP.

Since one mole of any gas may be said to have a volume of 22.4 liters at STP, a variety of problems can be solved. The description of gases under conditions other than standard will be covered in the next chapter. Assume standard conditions for all of the following examples.

EXAMPLE 3.12

How many moles are there in 5.0 liters of oxygen gas?

1. Establish the fact that one mole of the gas occupies 22.4 liters mole^{-1} at STP.
2. The equation that is extracted from the mole concept is

$$n = \frac{\text{liters}}{\text{liters mole}^{-1}}$$

3. Substitute the data and calculate the answer:

$$n = \frac{5.0 \text{ liters}}{22.4 \text{ liters mol}^{-1}} = 0.22 \text{ mole}$$

EXAMPLE 3.13

What volume is occupied by 0.17 mole of a gas at STP?

1. Modify the equation from Example 3.12:

$$n = \frac{\text{liters}}{\text{liters mole}^{-1}}$$

2. liters = n × liters $mole^{-1}$
3. Substitute the data and calculate the answer:

$$\text{liters} = 0.17\ \text{mol}\ (22.4\ \text{liters mol}^{-1})$$

Answer is 3.8 liters.

EXAMPLE 3.14

Approximately how many molecules are there in 400 ml of a gas at STP?

1. The same basic equation previously developed applies here:

$$\text{number of molecules} = n \times \mathcal{N}_A$$

2. Since the number of moles may be expressed

$$n = \frac{\text{liters}}{\text{liters mole}^{-1}}$$

the equation is modified:

$$\text{number of molecules} = \left(\frac{\text{liters}}{\text{liters mole}^{-1}}\right)(\mathcal{N}_A)$$

3. Convert 400 ml to liters:

$$\frac{400\ \text{ml}}{10^3\ \text{ml liter}^{-1}} = 0.4\ \text{liter}$$

4. Substitute in the equation and find the answer:

$$\text{number of molecules} = \frac{0.4\ \text{liter}\ (6.02 \times 10^{23}\ \text{mol}^{-1})}{22.4\ \text{liters mol}^{-1}}$$

Answer is 0.107×10^{23} molecules, better expressed as 1.1×10^{22} molecules.

Exercise 3.4

1. How many moles of NH_3 gas occupy 300 ml at STP?
2. What volume is occupied by 3.2×10^{-3} moles of N_2 gas at STP?
3. Approximately how many molecules of CO_2 gas are there in a 2.0 liter volume at STP?

4. What is the density of CO_2 gas at STP? (Hint: remember that gas densities are expressed as g liter^{-1}.)
5. If 3.0 liters (STP) of an unknown gas have a mass of 5.85 g, what is the molecular weight of this gas?

3.13 THE LAW OF DULONG AND PETIT*

An interesting method for the determination of approximate atomic weights was discovered by Pierre Dulong and Alexis Petit in the early part of the 19th century. The relationship between the specific heat of an element and its atomic weight can easily be observed in the laboratory today.

In effect, Dulong and Petit found that the atomic weight is inversely proportional to the specific heat:

$$\text{atomic weight} \propto \frac{1}{\text{specific heat}}$$

Mathematically expressed:

$$\text{atomic weight} = \frac{\text{k}}{\text{specific heat}}$$

The experimentally determined value for k, the proportionality constant, is about 6.2 cal mole^{-1} °C^{-1}. This means that 6.2 cal is required to raise one mole of an element through 1°C. The value 6.2 is not exact. Measurements made on a variety of elements indicate that 6.2 ± 0.6 would be a more appropriate representation. However, when 6.2 is used as an average, the atomic weight obtained from the calculation is close enough to be convincing.

EXAMPLE 3.15

Find the approximate atomic weight of iron if the specific heat is measured in the laboratory as 0.111 cal g^{-1} °C^{-1}.

1. Set up the equation:

$$\text{atomic weight} = \frac{\text{k}}{\text{specific heat}}$$

2. Substitute and solve:

$$\text{atomic weight} = \frac{6.2\ \cancel{\text{cal}}\ \text{mole}^{-1}\ °\cancel{\text{C}}^{-1}}{0.111\ \cancel{\text{cal}}\ \text{g}^{-1}\ °\cancel{\text{C}}^{-1}}$$

$$\text{atomic weight} = 55.8\ \text{g mole}^{-1}$$

3.14 THE HISTORICAL CONCEPT OF EQUIVALENT WEIGHTS*

While the atomic weights obtained from the method of Dulong and Petit yield only approximate values, they can be modified by other data to increase accuracy. An example of this would be the use of the **equivalent weight** of an element, which can be determined very accurately. The equivalent weight, historically called the **combining weight** of an element, is empirically (by direct experiment and observation) defined as that weight of an element which can combine with 8.0 g of oxygen.* It is equivalent weights that give rise to the concept of **valence,** which refers to the combining values of elements. While there is considerable justification for the use of valence in formula writing, that skill can be developed more understandably after the chapter on atomic structure. Nevertheless, the relationship between equivalent weight and atomic weight can be demonstrated.

EXAMPLE 3.16

Find the equivalent weight of mercury if 4.0000g of an oxide of mercury decomposed by heating leaves a residue of pure mercury that weighs 3.7045 g.

1. Organize the data:

$$\text{mercury oxide} = 4.0000 \text{ g}$$

$$\text{remaining mercury} = 3.7045 \text{ g}$$

$$\text{oxygen lost} = 0.2955 \text{ g}$$

2. Since 3.7045 g of mercury combined with 0.2955 g of oxygen, a simple proportion can be used to find the mass of mercury that would combine with 8.000 g of oxygen. That weight is, by definition, the equivalent weight:

$$x = \frac{3.7045 \text{ g Hg}}{0.2955 \text{ g } \emptyset} \times 8.00 \text{ g eq}^{-1} \emptyset$$

$$x = \text{equivalent weight of Hg} = 100.3 \text{ g eq}^{-1}$$

Note: This equivalent weight of mercury is seen to be 1/2 the atomic weight, so each mole weight of mercury contains two equivalent weights. Just as the equivalent weight of oxygen is 1/2 the atomic weight, the equivalent weight of any element is always a simple fraction (1/2, 1/3, 1/4, 1/5, 1/6, 1/7, 1/8), if not equal to the atomic weight.

---

* It should be pointed out that this is a historical definition gained from experimental data and practically limited to metallic elements. A more expanded discussion of equivalent weights (and probably more significant, too) will be taken up in Chapter 9.

EXAMPLE 3.17

Calculate the atomic weight (to three significant figures) of aluminum from the specific heat, 0.215 cal g^{-1} °C^{-1}, and the equivalent weight, 8.99 g.

1. Use the method of Dulong and Petit to find the approximate atomic weight:

$$\text{atomic weight} = \frac{k}{\text{specific heat}}$$

$$\text{atomic weight} = \frac{6.2\ \cancel{cal}\ mol^{-1}\ {}^{\circ}C^{-1}}{0.215\ \cancel{cal}\ g^{-1}\ {}^{\circ}\cancel{C}^{-1}}$$

$$\text{atomic weight} = 29\ g\ mole^{-1}$$

2. Compare the rounded-off value for the equivalent weight to the approximate value of the atomic weight. This indicates the simple fractional relationship between the two:

$$\frac{9\ \text{equiv. wt.}}{29\ \text{atomic wt.}} = \sim\frac{1}{3}$$

3. Since the equivalent weight is 1/3 of the atomic weight, the accurate atomic weight is 3 × equiv. wt:

$$\text{atomic weight of aluminum} = 3(8.99\ g) = 26.98$$

Two additional topics dealing with weight relationships in chemical formulas that could be discussed in this chapter are percentage composition of compounds and empirical formulas. However, they will be included in another equally appropriate chapter on chemical formulas and nomenclature (Chap. 7).

PROBLEMS AND QUESTIONS

3.1 List five properties of metals that serve to contrast them with nonmetals.

3.2 Write the symbols for the following elements:

a. lithium	f. fluorine
b. barium	g. molybdenum
c. arsenic	h. radium
d. tin	i. iodine
e. bismuth	j. krypton

3.3 Name the following elements:

a. Sc	f. K
b. Hf	g. Sb
c. Cd	h. Mn
d. Ar	i. Sr
e. U	j. V

3.4 Complete the following chart:

Element	Atomic Number	Number of Neutrons	Number of Electrons	Number of Protons	Atomic Weight
$^{91}_{40}Zr$					
Zinc					
$_{78}Pt$			78		
	38				
		74			127

3.5 Define a mole.

3.6 What are isotopes of an element?

3.7 How many moles of magnesium are there in 6.0 grams?

3.8 400.0 mg of copper is the mass of how many moles?

3.9 How many grams are there in 0.0026 moles of silicon?

3.10 What is the mass, in grams, of an atom of gold?

3.11 What is the total mass of 5.2×10^{-2} moles of iron and 1.2×10^{22} molecules of H_2O?

3.12 What volume is occupied by 8.0 g of O_2 at STP?

3.13 Calculate the approximate specific heat of silver.★

3.14 If 495.0 mg of a metallic oxide is decomposed and 78.0 mg of oxygen is collected, find the equivalent weight of the unknown metal.★

3.15 Find the atomic weight of the unknown metal in the previous question if the specific heat is found to be 0.073 cal g^{-1} $°C^{-1}$.★

CHAPTER 4

The society of scientists is simple because it has a directing purpose: to explore the truth. Nevertheless, it has to solve the problem of every society, which is to find a compromise between man and men. It must encourage the single scientist to be independent, and the body of scientists to be tolerant. From these basic conditions, which form the prime values, there follows step by step a range of values: dissent, freedom of thought and speech, justice, honor, human dignity and self-respect.

J. Bronowski, *Science and Human Values**

Behavioral Objectives

At the completion of this chapter, the student should be able to:

1. Describe, in general terms, the effect of changes of temperature and pressure on gases.
2. Define an ideal, or perfect, gas.
3. List the three principal points of the kinetic molecular theory.
4. Find the molecular weight of a gas from its density.
5. List the four parameters, and their symbols, that affect the behavior of gases.
6. State Boyle's and Charles's laws.
7. Define the terms *torr* and *atmosphere*.
8. Calculate the effect of changes in temperature and pressure on a volume of gas by use of the combined gas laws.
9. State Dalton's law of partial pressures.
10. Find the partial pressures exerted by particular gases in a mixture.
11. Calculate gas volumes collected by water displacement.
12. Use the table of water vapor pressures in solving gas law problems.
13. Derive the value (with proper units) of R, the gas constant.
14. Write the equation of state.
15. Solve problems involving the masses, molecular weights, and densities of gases by proper use of the equation of state.
16. Calculate the volume occupied by a given number of moles of gas at various temperature and pressure conditions.

---

* Harper & Row, Publishers, 1965.

THE BEHAVIOR OF GASES

4.1 INTRODUCTION

Reflection on the gaseous state of matter produces a number of fascinating observations. Consider air as an example of a gas. Although air is not a pure compound (a mixture of approximately 80 percent nitrogen and 20 percent oxygen), its behavior is typical. This invisible air, which offers so little resistance to human activity, can nevertheless support airplanes weighing many tons, inflate tires to accommodate cars and trucks, move with the destructive forces of hurricanes and tornados, and thoroughly incinerate objects from outer space by the heat of friction.

The force of the atmosphere can crush sturdy tin cans from which the air has been removed. It can support one end of a slat of wood while the other end is "karate-chopped" off. Air can support the water in an inverted glass (Fig. 4–1), and it can exert sufficient force to hold up a column of dense mercury to a height of 760 mm.

In general, gases are observed to increase in volume when heated. When a balloon is taken from a refrigerator and put into a tub of hot water, the increase in volume is clearly seen (Fig. 4–2). The cooling of a gas volume has the reverse effect, as seen by returning the expanded balloon to the refrigerator. Gases exert a markedly increased force when heated in a container of fixed volume. The printed warnings against incineration of aerosol cans attest to the likelihood of explosions occurring when these containers are heated. Figure 4–3 illustrates the effect of temperature on the pressure exerted by a gas in a container of fixed volume.

Compressibility, as an obvious physical property of gases, has numerous practical applications. For example, cylinders of compressed gases are widely used by laboratories and industrial concerns. The apparatus used in scuba diving offers a more dramatic application of compressed air. Refrigeration and air conditioning systems take advantage of the temperature-volume compressibility relationships of gases. As a gas expands within the metal coils of the refrigerator's interior, it does so by absorbing some of the heat inside, and the compression of the gas volume outside the refrigerator releases the heat into the room. However, regardless of practical applications, an attempt to understand the behavior of gases requires that they be considered within the framework of a unifying concept.

Figure 4–1 Demonstration of atmospheric pressure.

hot water
cool balloon
expanded balloon

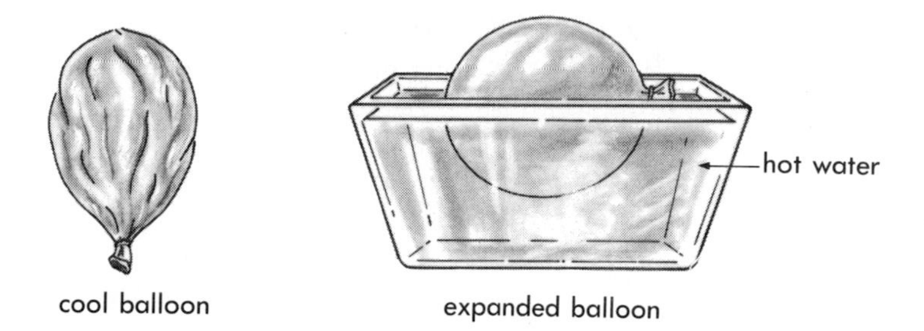

Figure 4–2 The effect of temperature on a gas volume.

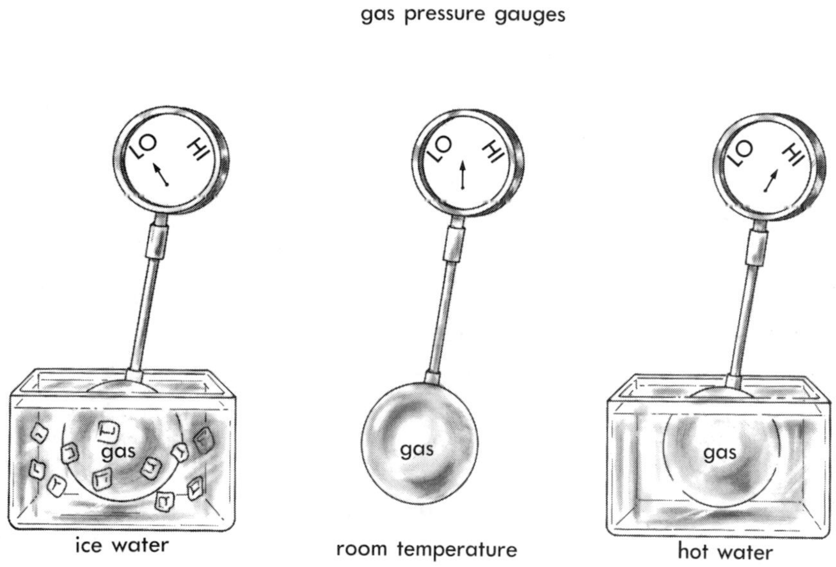

gas pressure gauges
LO
HI
LO
HI
LO
HI
gas
gas
gas
ice water
room temperature
hot water

Figure 4–3 The effect of temperature on the pressure exerted by a gas in a fixed volume.

4.2 THE KINETIC MOLECULAR THEORY OF GASES

About 100 years ago, James Clark Maxwell and Ludwig Boltzmann suggested a model to explain their observations of gas behavior. This explanation is called the *Kinetic Molecular Theory.* The basic statements of the theory are most accurately applied to **ideal,** or **perfect gases.** An ideal gas is one in which the molecules have no attractive forces that might prevent them from having perfectly **elastic collisions.** The attractive forces, essentially electrical in nature, are usually called **van der Waals** interactions. An elastic collision occurs when two objects collide and rebound without causing a net change in their physical structures and total kinetic energies. There is no such thing as an ideal gas in reality, but most gases (especially when they are at low pressures and high temperatures) behave nearly enough like ideal gases to permit the application of the following basic postulates:

1. Gases consist of atoms or molecules so small that their actual volumes are negligible compared to the spaces between them (Fig. 4–4).

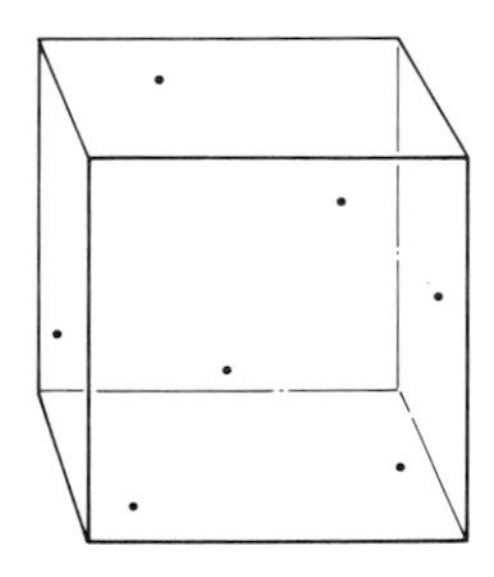

Figure 4–4 Negligible van der Waals forces between tiny particles that are far apart.

2. Gas particles are constantly moving at high speed at ordinary temperatures. Their motion is straight-line and their direction is random. The collisions between the particles are perfectly elastic (Fig. 4–5).
3. The collisions between the gas particles result in no net loss of the average kinetic energies of the particles. The average kinetic energy is reflected by the temperature of the gas (Fig. 4–6).

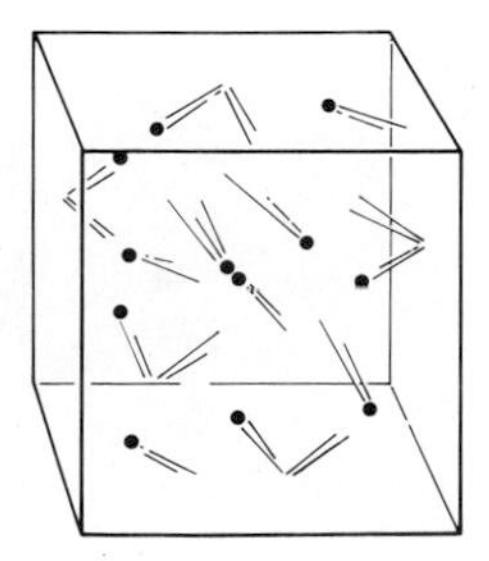

Figure 4–5 The motion, direction, and collisions of gas particles.

Figure 4–6 Average E_k of (A) = average E_k of (B) resulting from the transfer of kinetic energy upon collision.

SLOW
FAST
(A)
before collision

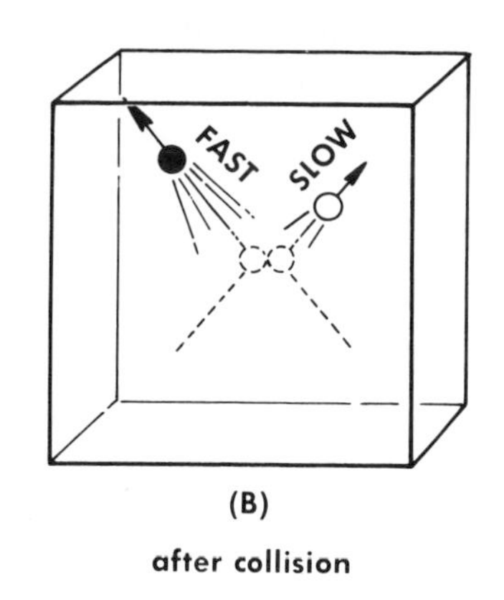

The model of rapidly moving and colliding molecules presents a superb explanation of gas behavior. However, before considering any historically significant topics that are now explained in terms of the kinetic molecular theory, some illustrative data should be set forth to emphasize the statistical approach to the description of gas behavior. Words such as "model" and "average" are pointedly used to indicate that gas measurements should be made on *moles* of gas rather than on individual molecules.

Maxwell and Boltzmann theorized that if a volume of a particular gas is placed in a container that does not permit any energy exchange with the outside, there will be a characteristic *average* velocity among the contained molecules. Some molecules will have velocities, and related kinetic energies, above or below this average, but there is no way of singling out these molecules. The graphical representation of the Maxwell-Boltzmann distribution is illustrated in Figure 4–7.

It is interesting to compare two dissimilar gases in order to see how their behavior relates to the kinetic molecular theory. The comparison will be between the light gas hydrogen (2 g mole^{-1}) and the much heavier gas carbon dioxide (44 g mole^{-1}). The comparisons are made at ordinary room temperature and normal barometric pressure.

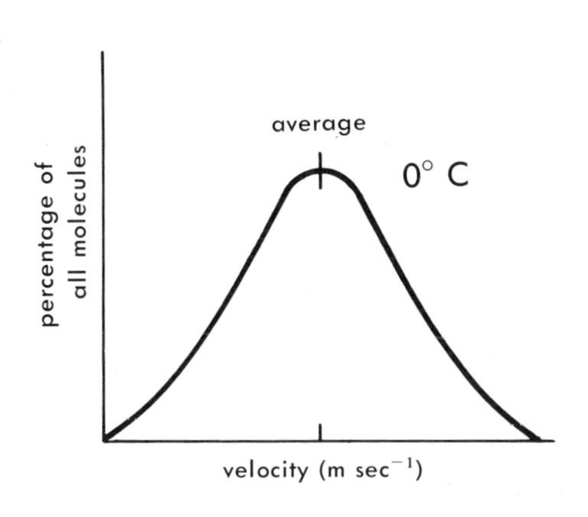

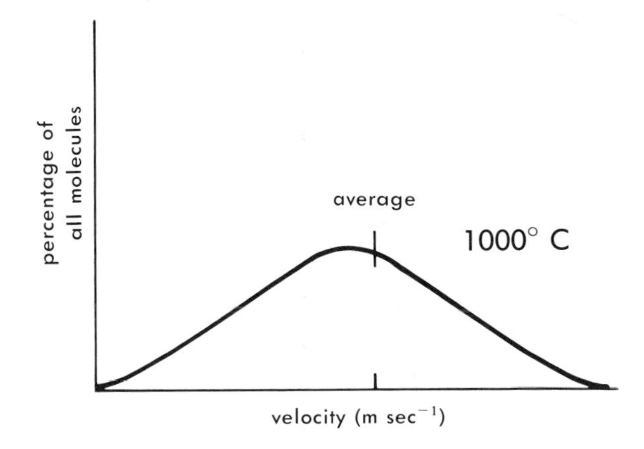

Figure 4–7 Maxwell-Boltzmann molecular velocity distribution curves.

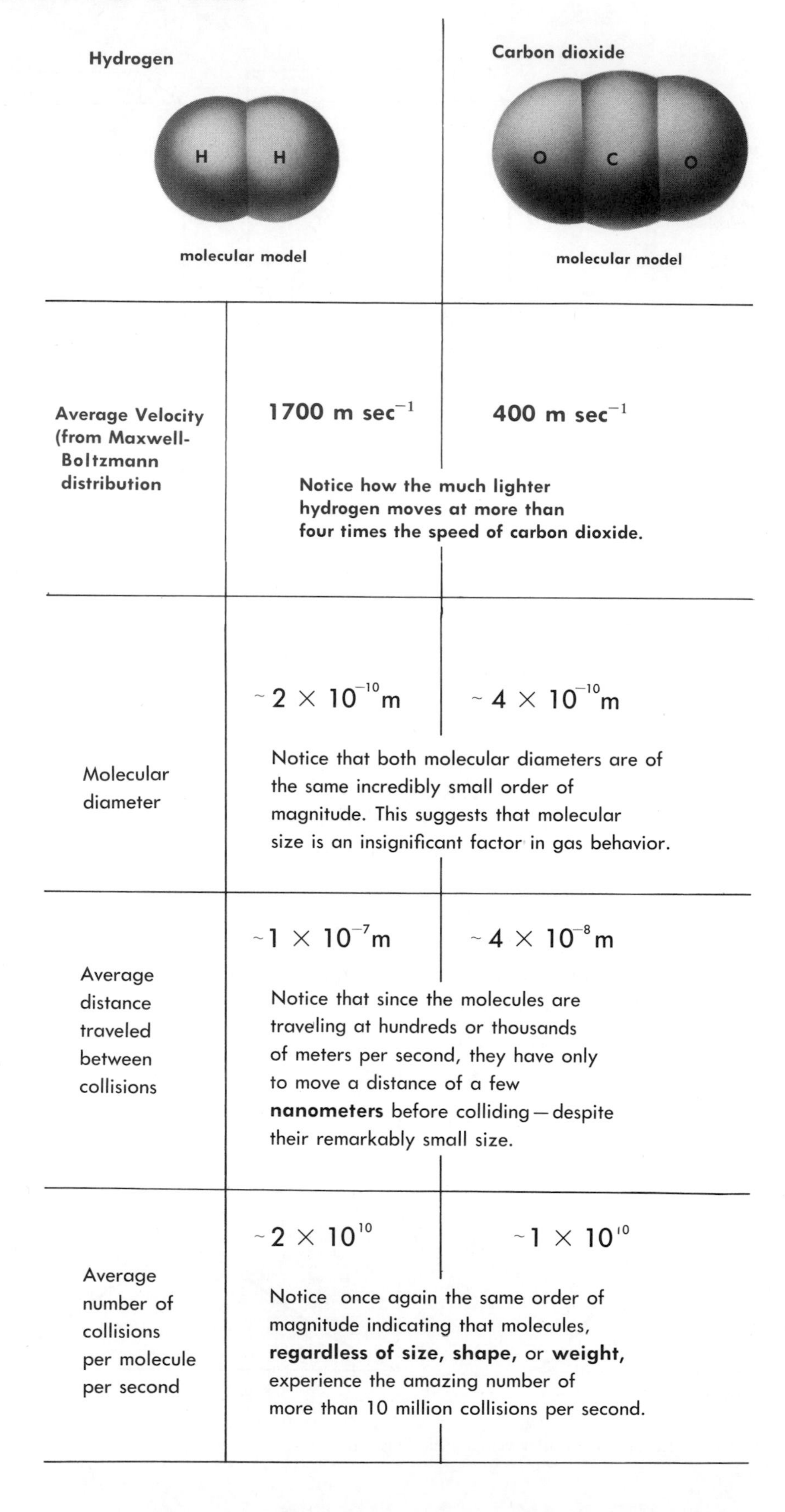

	Hydrogen (molecular model)	Carbon dioxide (molecular model)
Average Velocity (from Maxwell-Boltzmann distribution	1700 m sec^{-1}	400 m sec^{-1}
	Notice how the much lighter hydrogen moves at more than four times the speed of carbon dioxide.	
Molecular diameter	$\sim 2 \times 10^{-10}$m	$\sim 4 \times 10^{-10}$m
	Notice that both molecular diameters are of the same incredibly small order of magnitude. This suggests that molecular size is an insignificant factor in gas behavior.	
Average distance traveled between collisions	$\sim 1 \times 10^{-7}$m	$\sim 4 \times 10^{-8}$m
	Notice that since the molecules are traveling at hundreds or thousands of meters per second, they have only to move a distance of a few **nanometers** before colliding—despite their remarkably small size.	
Average number of collisions per molecule per second	$\sim 2 \times 10^{10}$	$\sim 1 \times 10^{10}$
	Notice once again the same order of magnitude indicating that molecules, **regardless of size, shape,** or **weight,** experience the amazing number of more than 10 million collisions per second.	

4.3 AVOGADRO'S PRINCIPLE

When Avogadro concluded, empirically, that equal volumes of gases under the same conditions of temperature and pressure would contain equal numbers of molecules, he might have welcomed the knowledge that his observations were consistent with the kinetic molecular theory.

Gases which occupy equal volumes at the same temperature and exert the same force on the walls of the container must contain equal numbers of molecules. If this were not true, these gases could not have the same average kinetic energies. Avogadro's principle also says that equal numbers of moles of gases at the same temperature and pressure will occupy the same volume. This observation is one of the cornerstones of the mole concept, which says that a mole of any gas at standard conditions occupies 22.4 liters.

About 50 years later, Stanislao Cannizzaro made use of Avogadro's principle to demonstrate that gas densities could be used to calculate atomic weights. He assigned the atomic weight of 1 to hydrogen and determined a number of other atomic weights using hydrogen as the standard.

An important contribution that just preceded Avogadro's hypothesis was the **law of combining volumes,** presented by J. L. Gay-Lussac in 1808. Gay-Lussac found that volumes of gases in chemical reactions were related in small whole number ratios when measured at fixed temperatures and pressures. In terms of his own principle, Avogadro explained how 1 volume of nitrogen gas could react with 3 volumes of hydrogen gas to produce 2 volumes of ammonia gas and still be consistent with the law of conservation of matter. If the equation

$$N_2 + 3H_2 \longrightarrow 2NH_3$$

is visualized as a simple model, shown in Figure 4–8, the truth of the strange arithmetic $1 + 3 = 2$ can be understood.

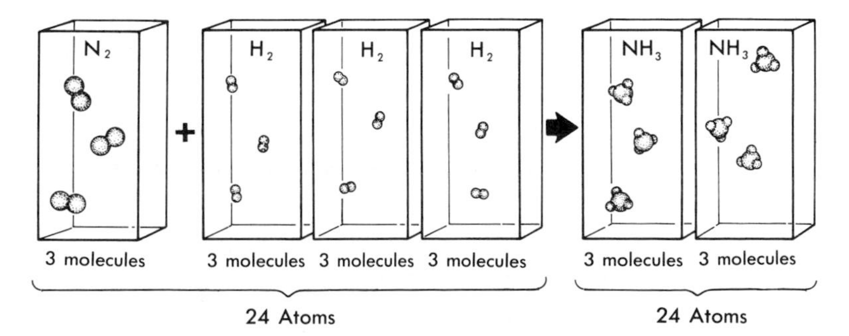

Figure 4–8 Equal numbers of molecules occupy equal volumes at the same temperature and pressure.

Notice in Figure 4–8 that conservation of matter is demonstrated in terms of the number of nitrogen and hydrogen atoms at the beginning and at the end of the reaction. The only difference is in the arrangement of the atoms in molecules. However, the kinetic molecular theory clearly states that the volume occupied by a gas depends on numbers of molecules and *not* their size, shape, or weight. The heavier ammonia molecules simply move more slowly than the lighter nitrogen and hydrogen molecules, but the average kinetic energies are all the same.

The kinetic molecular theory was discussed earlier in this book, with the constant reservation that temperature and pressure were not variables. The specific value for standard temperature and pressure (STP) was introduced still earlier as part of the mole concept. It is now appropriate to consider the interrelationship of the **gas laws,** where temperature, pressure, and number of molecules are realistically treated as variable factors.

4.4 THE GAS LAWS

The kinetic molecular theory plainly suggests that gases must be measured and described in terms of four **parameters** (variables) organized in Table 4–1.

TABLE 4–1 FOUR PARAMETERS USED TO DESCRIBE GASES

Parameter	Symbol	Brief Definition	Unit of Measure
pressure	P	force exerted on container walls by molecular battering	torr atm (atmosphere)
volume	V	volume of container occupied by molecules	ℓ (liter)
number of molecules	n	number of moles of gas in container	mole
temperature	T	absolute temperature of a gas volume	K

4.5 PRESSURE, VOLUME, AND BOYLE'S LAW

Pressure is defined as the force on a unit area. The pressure exerted by gases is due to the collisions between the rapidly moving molecules and the walls of the container. Gas pressure could be accurately described in dynes per cm^2, where a dyne is a unit of force. In physics, force is equal to mass times acceleration:

$$F = ma$$

A dyne is the force needed to change the speed of (accelerate) a mass of 1 gram by 1 cm per sec, each second. More concisely expressed:

$$1 \text{ dyne} = \underbrace{\text{g}}_{\text{mass}} \times \underbrace{\text{cm s}^{-2}}_{\text{cm per sec per sec (acceleration)}}$$

However, the pressure exerted by gases is more conveniently related to the atmosphere, which is pressing in on objects at sea level with a force of 1 million dynes per square centimeter. It so happens that a column of mercury 760 mm high also exerts the same force. If a completely full mercury column (a **barometer**) is inverted into a container of mercury, the relationship between the downward force (weight) of the mercury column and the atmospheric pressure can be observed (Fig. 4–9).

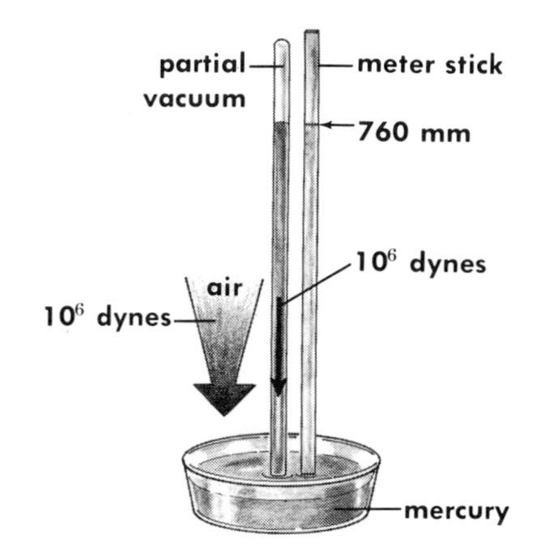

Figure 4–9 Atmospheric pressure equals the force per unit area exerted by a column of mercury 760 mm high.

Of course atmospheric pressure changes, but **standard atmospheric pressure** is officially defined as that which can support a mercury column 760 mm high. In honor of the inventor of the barometer, Evangelista Torricelli, a mm of mercury in such a column (as a unit of atmospheric pressure) is called a **torr.** The alternate unit of pressure is the **atm** (atmosphere), which is equal to 760 torr. Expressed as a conversion factor:

$$760 \text{ torr atm}^{-1}$$

Robert Boyle, in 1662, described the relationship between volume and pressure as being *inverse*:

$$V \propto \frac{1}{P}$$

Mathematically expressed

$$V = \frac{k}{P} \quad \text{or} \quad k = PV$$

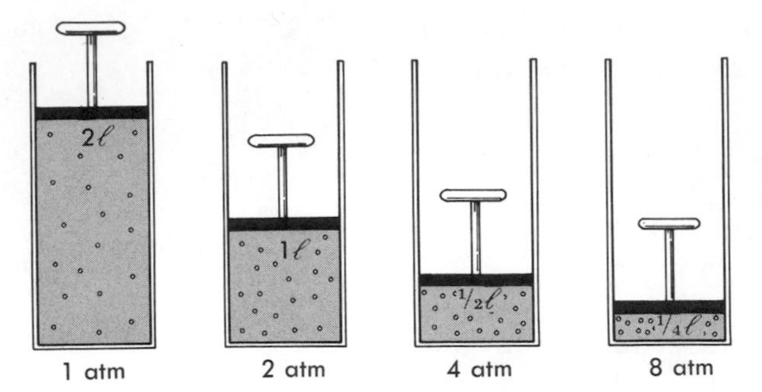

Figure 4–10 A cylinder and piston system showing the relationship between P and V. As the pressure increases, the volume decreases.

When pressure and volume data are graphically illustrated, the proportionality constant may be obtained from the slope of the line. Figure 4–10 illustrates the inverse relationship from which data may be obtained.

When volume is plotted as a function of pressure, the resulting **hyperbola** indicates the inverse relationship. "Straightening out the line" is accomplished by plotting volume as a function of the reciprocal of pressure (Fig. 4–11). Although the proportionality constant in the previous example is 2, the constant in each individual system depends on the initial volume and the temperature.

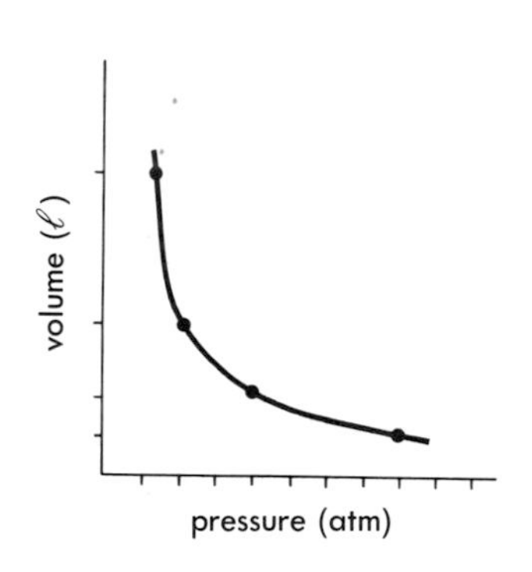

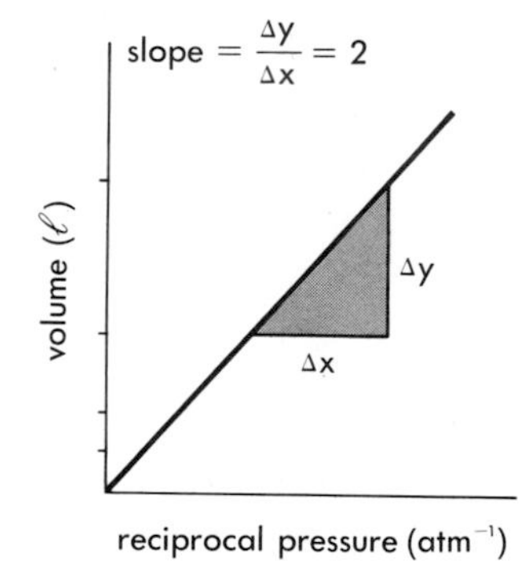

Figure 4–11 Boyle's law: the inverse proportionality of P and V.

EXAMPLE 4.1

If a gas occupies 250 ml at 600 torr, what volume will it occupy at standard pressure if the temperature remains constant?

1. Organize the data:

initial volume $= V_1 = 250$ ml

final volume $= V_2 = ?$

initial pressure $= P_1 = 600$ torr

final pressure $= P_2 = 760$ torr

2. *Reason* that the increase in pressure will cause a corresponding decrease in volume. Simply multiply the original volume by the pressure change in the form of a *common fraction.* Any number multiplied by a common fraction will be reduced:

$$? = 250 \text{ ml} \left(\frac{600 \text{ torr}}{760 \text{ torr}}\right)$$

$$\text{final volume} = 197 \text{ ml}$$

Notice the answer is in *milliliters*

3. The common-sense approach to solving a problem of this type is preferred. However, it can be verified by the equation-substitution method.

Since P and V are the only variables, and

$$PV = k$$

and since k remains the same in this system

$$\underset{\text{initial}}{k_1} = \underset{\text{final}}{k_2}$$

$$P_1V_1 = P_2V_2$$

$$\frac{P_1V_1}{P_2} = \frac{\cancel{P_2}V_2}{\cancel{P_2}} \quad \text{(divide the equation by } P_2\text{)}$$

Therefore

$$V_2 = \frac{V_1P_1}{P_2}$$

Substituting

$$V_2 = \frac{250 \text{ ml } (600 \cancel{\text{torr}})}{760 \cancel{\text{torr}}}$$

$$V_2 = 197 \text{ ml}$$

There is one other problem that must be considered before leaving the discussion of pressure. It is the question of the pressure exerted by a mixture of gases.

4.6 DALTON'S LAW OF PARTIAL PRESSURES

In 1801, John Dalton reported that the total pressure exerted by a mixture of nonreacting gases was the sum of the pressures that each gas in the mixture exerted individually. Figure 4–12 illustrates this simple cause and effect relationship.

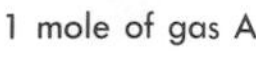

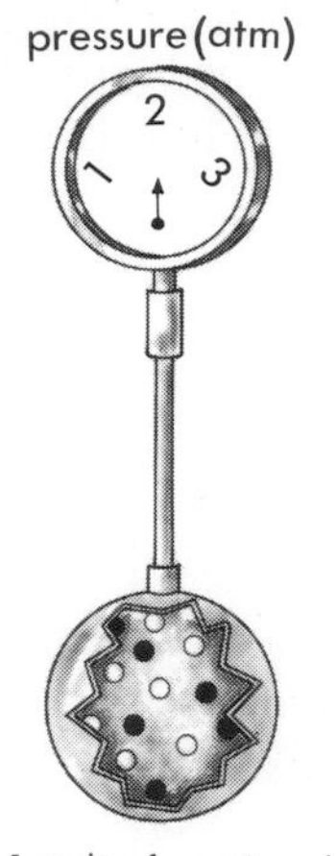

1 mole of gas A and
1 mole of gas B

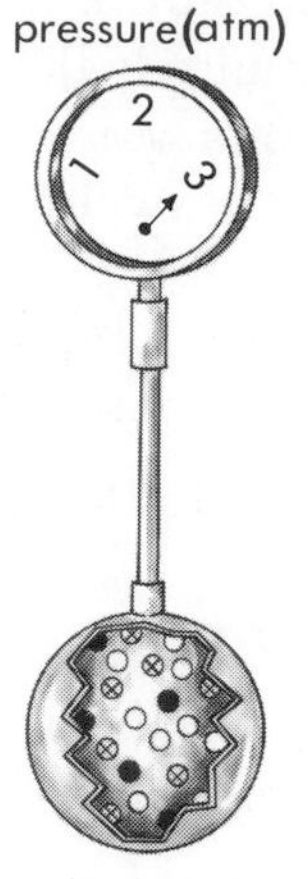

1 mole of gas A and
1 mole of gas B and
1 mole of gas C

Figure 4–12 The total pressure exerted by a mixture of gases is the sum of the individual pressures.

A fairly simple demonstration of Dalton's law can be effected by wetting the inside of a 20 cm test tube with water so that a thin layer of iron filings can adhere. Invert the tube in a beaker of water so that the pressure in the tube nearly equals the room pressure. After the iron filings combine chemically with the oxygen in the air (the rust is obvious), the water level will be observed to have risen about 4 cm. This 20 percent reduction of the gas volume in the tube corresponds nicely to the fact that air is about 20 percent oxygen. Since the iron filings removed 20 percent of the original gas volume, the total pressure was reduced by that same fraction. Figure 4–13 illustrates the 20 percent reduction in pressure.

In summary, Dalton's law of partial pressures may be expressed as

$$P_{total} = P_a + P_b + P_c + \cdots$$

One of the most important applications of Dalton's law occurs when a gas is collected over water. Collecting a gas over water is a laboratory method of collecting a volume of an invisible and insoluble gas by the displacement of water from a bottle. Figure 4–14 illustrates how oxygen may be collected by water displacement.

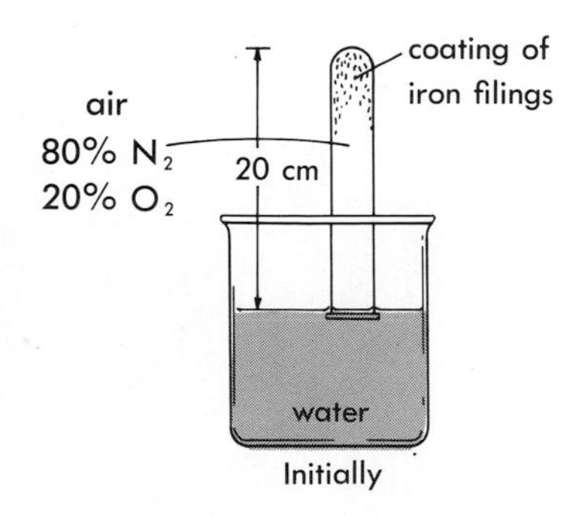

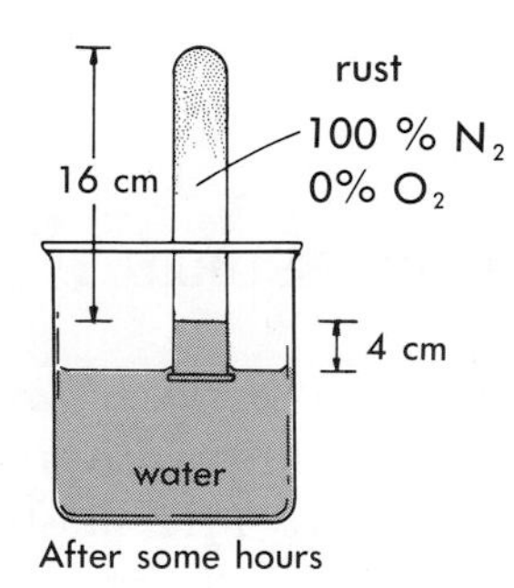

Figure 4–13 Dalton's law of partial pressures. The removal of oxygen from air reduces the pressure by about 20 percent.

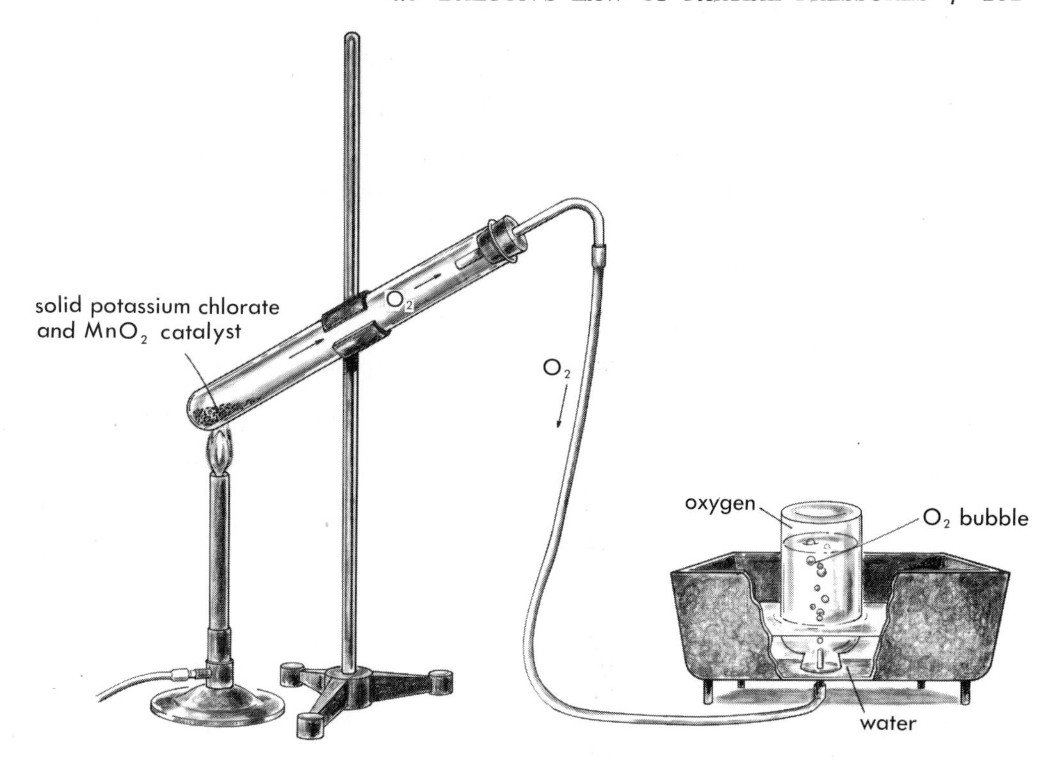

Figure 4–14 Collecting oxygen gas by water displacement.

The bottle of oxygen pictured in Figure 4–14 may be the subject of some useful calculations. One possibly essential measurement is the pressure of the contained gas or gases. This can be quickly determined by equalizing the level of the water in the bottle with the water level in the trough (Fig. 4–15).

If the barometer reads 760 torr in the room, the gas pressure in the bottle is also 760 torr, if the levels are equal. *But, the gas in bottle contains water vapor in addition to oxygen.* The oxygen is exerting a pressure of something less than 760 torr, because the water vapor is also exerting pressure. A correction must be made if the true pressure of dry oxygen gas is to be recorded and used. The equation applicable in this case is

$$P_{total} = P_{oxygen} + P_{water\ vapor}$$

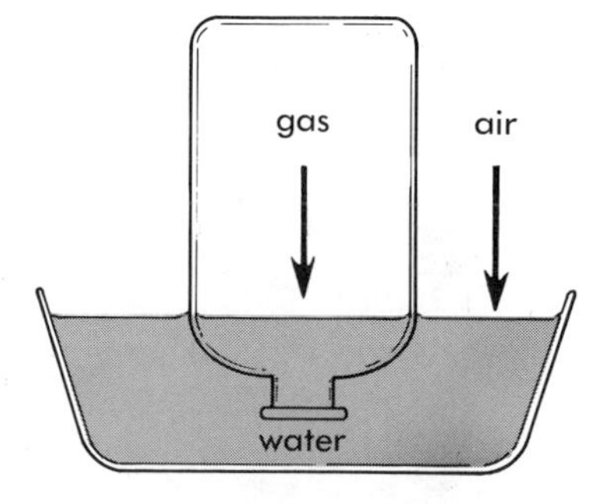

Figure 4–15 Method of equalizing internal and external gas pressures.

To obtain the P_{oxygen}, the equation is rearranged:

$$P_{oxygen} = P_{total} - P_{water\ vapor}$$

The pressure exerted by water vapor depends on the temperature. The hotter the water, the more molecules there are that are able to exist in the vapor state in a particular container, and, consequently, the more pressure will be exerted.

Table 4–2 shows the vapor pressure of water at a variety of temperatures. This table is for pure water. The effect of dissolved particles on the vapor pressure will be discussed in Chapter 9, Solutions.

TABLE 4-2 VAPOR PRESSURE OF WATER

Temperature C	torr	Temperature C	torr
0°	4.6	28°	28.3
5°	6.5	29°	30.0
10°	9.2	30°	31.8
15°	12.8	31°	33.7
16°	13.6	32°	35.7
17°	14.5	33°	37.7
18°	15.5	34°	39.9
19°	16.5	35°	42.2
20°	17.5	40°	55.3
21°	18.6	50°	92.5
22°	19.8	60°	149.3
23°	21.0	70°	233.7
24°	22.4	80°	355.1
25°	23.8	90°	525.8
26°	25.2	100°	760.0
27°	26.7		

Bear in mind that the vapor pressure of water does not depend on the *amount* of water, any more than barometric pressure depends on the diameter of the mercury column. Pressure is always described as force per unit area (Fig. 4–16).

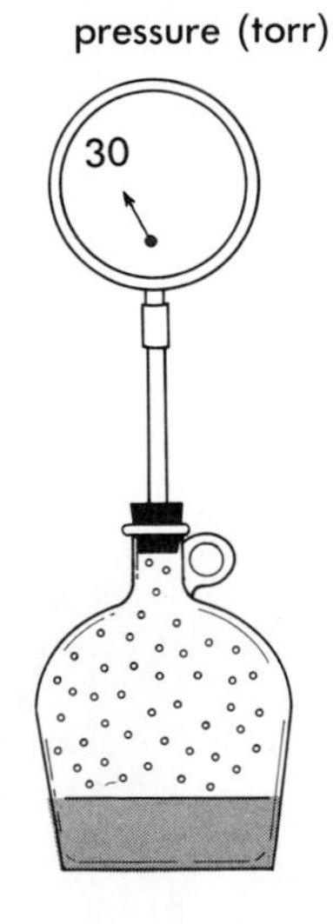

Figure 4–16 Water vapor pressure at 29.0°C.

EXAMPLE 4.2

What is the pressure of a dry gas if the volume collected over water at 26°C is measured at 742.0 torr?

1. Establish the equation:

$$P_{\text{dry gas}} = P_{\text{total}} - P_{\text{water vapor}}$$

2. Find the pressure due to water vapor at 26°C from Table 4–2:

$$\text{vapor pressure @ 26 C} = 25.2 \text{ torr}$$

3. Substitute in the equation and solve for the dry gas pressure:

$$P_{\text{dry gas}} = 742.0 \text{ torr} - 25.2 \text{ torr}$$

$$P_{\text{dry gas}} = 716.8 \text{ torr}$$

Exercise 4.1

1. If a gas occupies 2.3 liters at 710 torr, what volume will it occupy at standard pressure?
2. At 0.82 atm, a gas occupies 172.0 ml. What volume will the gas occupy at 792.0 torr?
3. If the 1.4 liter container of a gas at 10.0 atm is altered so that the volume is increased to 4.2 liters, what is the final pressure in atmospheres?
4. A 350.0 ml sample of gas is collected over water at 40°C and a barometric pressure of 756.7 torr. What is the pressure due to the dry gas at that temperature?

4.7 TEMPERATURE, VOLUME, AND CHARLES'S LAW

Boyle's law was entirely adequate for the description of pressure–volume relationships at constant temperature. But what about temperature changes? Jacques Charles, in 1887, experimentally determined that the volume of a fixed mass is *directly* proportional to the *absolute temperature*. He observed that the volume of a sample of gas at 0°C would expand or decrease by 1/273 of the 0°C volume for each degree of temperature rise or fall.

It is important to note that the volume varies with the *absolute* temperature, or degrees Kelvin, and not the Celsius temperature. One reason for this is that the use of degrees below zero (−5°C, −15°C, or −82°C, for example) would produce the fiction of *negative volumes* in calculations that relate volume and temperature. Another reason involves the definition of absolute zero, which says that it is the temperature at which a gas volume is *theoretically* reduced to zero.* This point can be determined by *extrapolating* from empirical data (Fig. 4–17).

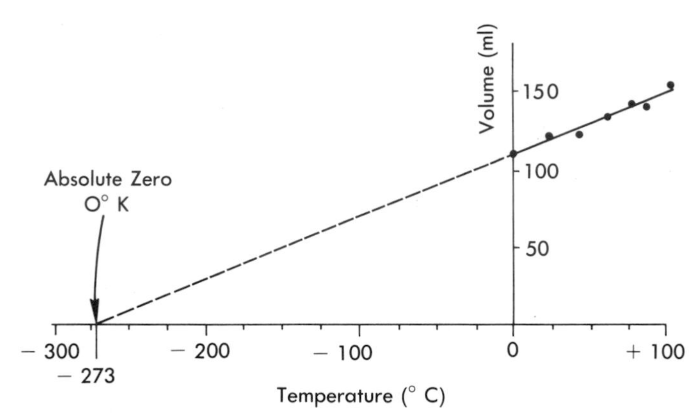

Figure 4–17 The extrapolation of volume-temperature data to absolute zero.

Observe in Figure 4–17 how the volume of the gas decreases with the temperature. This was the essential observation that led Charles to state the direct proportionality of temperature and volume:

$$V \propto T$$

Mathematically stated:

$$V = kT$$

or

$$k = \frac{V}{T}$$

which means that the ratio of a gas volume to the absolute temperature, at a fixed pressure, is constant.

When volume is plotted as a function of the absolute temperature, the proportionality constant may be determined (Fig. 4–18). The value of the proportionality constant depends on the original volume and the atmospheric pressure. A brief summary of the temperature–volume relationship is that a doubling of the absolute temperature results in a doubling of the volume. If the absolute temperature is reduced to one-fourth its original value, the volume is also reduced by that fraction.

---

* Recall from Chapter 1 that there is in fact no substance existing in the gaseous state below 4°K.

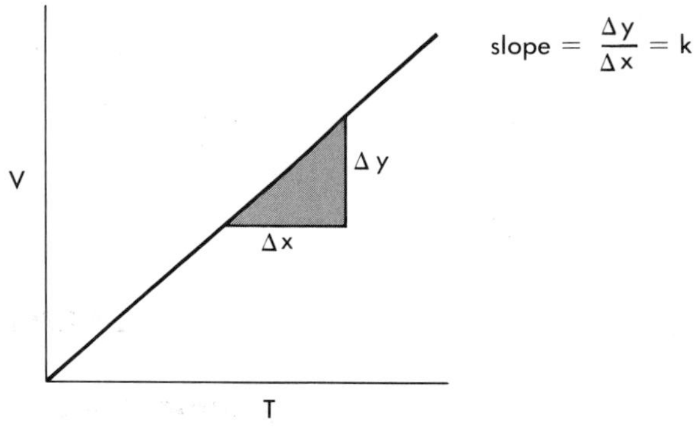

$$\text{slope} = \frac{\Delta y}{\Delta x} = k$$

Figure 4–18 The direct proportionality of V and T.

EXAMPLE 4.3

A gas occupies 200.0 ml at 0°C. What volume will it occupy at 27°C?

1. Organize the data, being sure to convert °C to °K:

$$V_1 = 200.0 \text{ ml}$$
$$V_2 = ?$$
$$K = °C + 273$$
$$T_1 = 0°C = 273 \text{ K}$$
$$T_2 = 27°C = 300 \text{ K}$$

2. Using the common-sense approach, it is clear that a rise in temperature will cause an increase in volume. Therefore, multiply the original volume by the kind of temperature ratio that will produce a larger number. This is an *improper* (larger than unity) fraction:

$$V_2 = 200 \text{ ml} \left(\frac{300 \cancel{K}}{273 \cancel{K}}\right) \leftarrow \text{improper fraction}$$

$$V_2 = 220.0 \text{ ml}$$

3. The formula-substitution method is based on the fact that the value for k does not change in this system:

$$\underset{\text{initial}}{k_1} = \underset{\text{final}}{k_2}$$

Since $k = \frac{V}{T}$, then

$$\frac{V_2}{T_2} = \frac{V_1}{T_1}$$

$$\frac{\cancel{T_2}\, V_2}{\cancel{T_2}} = \frac{V_1\, T_2}{T_1} \qquad \text{(multiply the equation by } T_2\text{)}$$

$$V_2 = \frac{V_1 T_2}{T_1} = \frac{(200.0 \text{ ml})(300 \cancel{K})}{(273 \cancel{K})}$$

$$V_2 = 220.0 \text{ ml}$$

4.8 THE COMBINED GAS LAWS

Solving problems that include changes in both temperature and pressure creates no difficulty whatsoever if it is remembered that pressure and temperature can operate independently. Both P and T may work in concert either to increase or to decrease a volume of gas. However, the system may be manipulated so that P and T may effectively cancel each other out. For example, an increase in pressure caused by pushing a piston down in a cylinder of gas would liberate heat, but the temperature change due to this effect could be made insignificant if the cylinder were placed in either an ice-water bath or a hot-oil bath.

In short, the proper technique to use in the common-sense method is to consider the effects of T and P separately and perform the appropriate calculations.

EXAMPLE 4.4

If 150 ml of gas is collected at 20°C and 700 torr, what volume will it occupy at STP?

1. Organize the data, paying careful attention to the proper units:

$$\begin{aligned} V_1 &= 150 \text{ ml} \\ V_2 &= ? \\ T_1 &= 20°C = 293 \text{ K} \\ T_2 &= 0°C \text{ (standard)} = 273 \text{ K} \\ P_1 &= 700 \text{ torr} \\ P_2 &= 760 \text{ torr (standard)} \end{aligned}$$

2. Consider the temperature change. It is going *down* (293 to 273). Therefore, the temperature ratio must be the kind of fraction that will *reduce* the volume (Charles's law). It is a *common fraction.*
3. Consider the pressure change. It is going *up* (700 to 760).

Therefore, the pressure ratio must be such that it, too, *reduces* the volume (Boyle's law). It is a *common fraction.*

4. Outline the equation (mentally, if not on paper):

$$V_2 = V_1 \times \underset{\substack{\text{temperature}\\ \text{effect}}}{\text{common fraction}} \times \underset{\substack{\text{pressure}\\ \text{effect}}}{\text{common fraction}}$$

5. Substitute in the equation and *estimate* the answer:

$$V_2 = 150 \text{ ml} \left(\frac{273 \text{ K}}{293 \text{ K}}\right)\left(\frac{700 \text{ torr}}{760 \text{ torr}}\right)$$

Estimating: both fractions will cause a reduction of the original volume of 150 ml. The fractions are not extreme, so the answer should be in the neighborhood of 130 ml.

6. Perform the slide rule calculation:

$$V_2 = 129 \text{ ml}$$

This agrees with the estimated answer.

EXAMPLE 4.5

If 67.0 ml of a gas is collected over water at a temperature of 15°C and a pressure of 1.2 atm, what volume will the dry gas occupy at 40°C and 770.0 torr?

1. Organize the data, paying special attention to *compatible pressure units,* to *Kelvin temperature,* and to *correcting wet gas pressure to dry gas pressure.*

   a. Change pressure units to torr:

$$1.2 \cancel{\text{atm}} \underset{\text{conversion factor}}{(760 \text{ torr } \cancel{\text{atm}}^{-1})} = 912 \text{ torr}$$

   b. Correct the total pressure, 912 torr, by subtracting the water vapor pressure at 40°C (from Table 4–2):

   since

$$P_{\text{total}} = P_{\text{gas}} + P_{\text{water vapor}}$$

   then

$$P_{\text{gas}} = P_{\text{total}} - P_{\text{water vapor}}$$

$$P_{\text{gas}} = \underset{\text{(total)}}{912 \text{ torr}} - \underset{\substack{\text{(water vapor}\\ \text{pressure at 40°C)}}}{55.3 \text{ torr}}$$

$$P_{\text{dry gas}} = 857 \text{ torr} \quad \text{(rounded off)}$$

c. Convert the temperatures to °K:

$$K = °C + 273$$
$$15°C = 288\ K$$
$$40°C = 313\ K$$

d. Arrange the data in table form:

$$V_1 = 67.0\ ml$$
$$V_2 = ?$$
$$T_1 = 288\ K$$
$$T_2 = 313\ K$$
$$P_1 = 857\ torr$$
$$P_2 = 770\ torr$$

2. Consider the *individual* effects of temperature and pressure on the original volume of 67.0 ml.

   a. The temperature is *rising*. Therefore, the volume should *expand* proportionately. Multiplying 67.0 ml by an *improper fraction* of the temperatures is indicated.
   b. The pressure on the gas is *decreasing*. This causes an *expansion* of the gas volume. Again, an *improper fraction* is indicated.

3. Arrange the fractions in the equation and *estimate* the approximate answer:

$$V_2 = 67.0\ ml \left(\frac{313\ K}{288\ K}\right)\left(\frac{857\ torr}{770\ torr}\right)$$

   Since both fractions are somewhat larger than unity, the answer might be expected to be in the neighborhood of 80 to 90 ml.

4. Solve the equation by slide rule:

$$V_2 = 81.0\ ml$$

   This answer agrees nicely with the estimate.

EXAMPLE 4.6

If 0.62 g of a gas occupies a 140 ml volume at 100°C and 0.98 atm, what is the density and molecular weight of this gas?

1. In order to use the *mole concept* in the solution of this problem, it is necessary to find the *volume* of the gas at STP.

Caution: A common error made by students is to alter the weight of the gas by the combined gas law equation. This cannot be done. If temperature and pressure changes could alter mass, it would be a gross disregard of the conservation laws. Indeed, it would be a curious panacea for obese people if they could reduce their weight by sitting in a refrigerator. A method for incorporating mass into a gas law equation is made possible by another approach, which is introduced in the next section dealing with the **equation of state.**

2. Organize the data, bearing in mind that the final conditions are STP:

$$V_1 = 140 \text{ ml}$$
$$V_2 = ?$$
$$T_1 = 100°\text{C} = 373 \text{ K}$$
$$T_2 = 0°\text{C} = 273 \text{ K}$$
$$P_1 = 0.98 \text{ atm}$$
$$P_2 = 1.0 \text{ atm}$$

3. Substitute in the equation and predict a marked *decrease* in volume due to the large temperature drop. The pressure change is almost negligible:

$$V_2 = 140 \text{ ml} \left(\frac{273 \text{ K}}{373 \text{ K}}\right)\left(\frac{0.98 \text{ atm}}{1.0 \text{ atm}}\right)$$

sharply decreases volume	slightly decreases volume

Estimate the final answer to be roughly 100 ml.

4. Solve for the final volume by slide rule calculation:

$$V_2 = 100 \text{ ml at STP}$$

5. Calculate the density from the equation. Change 100 ml to liters, since gas densities are reported in grams per liter:

$$d = \frac{M}{V}$$

$$d = \frac{0.62 \text{ g}}{0.1 \text{ liter}}$$

$$d = 6.2 \text{ g liter}^{-1}$$

6. The mole concept states that 1 mole of any gas at STP occupies 22.4 liters. Therefore

$$\text{molecular weight} = (\text{g liter}^{-1})(\text{liter mol}^{-1})$$

$$\text{molecular weight} = (6.2\ \text{g}\ \cancel{\text{liter}}^{-1})(22.4\ \cancel{\text{liters}}\ \text{mol}^{-1})$$

$$\text{molecular weight} = 139\ \text{g mol}^{-1}$$

Exercise 4.2

1. If a gas occupies 1.3 liters at −20°C, what volume will it occupy at 30°C, pressure remaining constant?
2. A gas occupies 220 ml at STP. What volume will this gas occupy at 80°C and 0.04 atm?
3. Correct a dry gas volume for STP if 75 ml is collected over water at 27°C and 705 torr.
4. If 240 ml of a gas collected at 60°C and 1.4 atm weighs 820.0 mg, calculate its density and molecular weight.

4.9 THE IDEAL GAS LAW AND THE EQUATION OF STATE

It has been demonstrated that Charles's law and Boyle's law may be conveniently combined in a common-sense approach to solving problems dealing with the behavior of gases. While this method permits consideration of pressure and temperature as individual effects on volume, they are, nevertheless, related to each other and to the number of molecules in the container. The kinetic molecular theory says that the kinetic energy of particles is related to the absolute temperature, because a higher temperature means greater molecular speed. Although the molecular mass remains constant, the increased velocity causes an increased kinetic energy:

$$E_k = \frac{mv^2}{2}$$

While the product of pressure (force per unit area) and volume is not the same as kinetic energy, the PV product is, nevertheless, proportional to the number of molecules present in a container and their absolute temperature:

$$P = \text{dynes cm}^{-2}$$
$$V = \text{cm}^3$$
$$P \times V = (\text{dynes cm}^{-2})(\text{cm}^3) = \text{dyne-cm} \text{ (energy unit)}$$

Since energy is proportional to the number of molecules (moles) and the absolute temperature (°K)

$$\text{energy} \propto nT$$

The pressure times volume product may be substituted:

$$PV \propto nT$$

Mathematically expressed

$$PV = RnT$$

(R: proportionality constant)

The proportionality constant may be calculated from P, V, n, and T data for 1 mole of any ideal gas at STP:

$$P = 1 \text{ atm}$$
$$V = 22.4 \text{ liters}$$
$$n = 1 \text{ mole}$$
$$T = 0°C = 273 \text{ K}$$

Since $PV = RnT$:

$$R = \frac{PV}{nT}$$

Substituting the STP data:

$$R = \frac{(1 \text{ atm})(22.4 \text{ liters})}{(1 \text{ mol})(273 \text{ K})}$$

$$R^* = 0.082 \text{ liter-atm mole}^{-1} \text{ K}^{-1}$$

The equation, commonly known as the **equation of state** or the ideal gas law, is usually written:

$$PV = nRT$$

---

* Many chemists find it more convenient to express the gas constant as liter-**torr** mole^{-1} K^{-1}. This eliminates the necessity of converting torr to atm for a calculation. Hence, the alternative value for R is 62.4 *liter-torr mole*$^{-1}$ K^{-1}.

The equation of state is a very versatile tool and solves many problems more efficiently than does the common-sense application of the combined laws of Boyle and Charles. It should be emphasized, however, that the equation of state represents an alternative that is definitely related to the fundamental observations of P, V, n, and T relationships. For example

$V \propto T$ (volume is directly proportional to the absolute temperature)

and

$V \propto \frac{1}{P}$ (volume is inversely proportional to the pressure)

and

$V \propto n$ (volume is directly proportional to the number of moles)

Combining these proportionalities

$$V \propto \frac{nT}{P}$$

Mathematically expressed

$$V = \frac{RnT}{P}$$

and rearranging

$$PV = nRT$$

It should be pointed out, for future reference, that since the gas constant has the units of energy (liter-atmospheres), it may have a variety of numerical values if liter-atm are converted to calories, joules, or other energy units. This can be easily accomplished by use of the appropriate energy unit conversion factor. See Example 2.6.

EXAMPLE 4.7

If 80.0 ml of a gas is collected at 720 torr and 5°C, what volume will it occupy at STP?

1. Write the equation of state:

$$PV = nRT$$

Since n and R remain constant in this system, it follows that

$$\underset{\text{initial}}{n_1R_1} = \underset{\text{final}}{n_2R_2}$$

2. Rewrite the equation: since

$$n_1R_1 = \frac{P_1V_1}{T_1} \quad \text{and} \quad n_2R_2 = \frac{P_2V_2}{T_2}$$

then

$$\frac{P_1V_1}{T_1} = \frac{P_2V_2}{T_2}$$

3. Organize the data and substitute in the equation:

$$V_1 = 80 \text{ ml}$$
$$V_2 = ?$$
$$T_1 = 5°C = 278 \text{ K}$$
$$T_2 = 0°C = 273 \text{ K}$$
$$P_1 = 720 \text{ torr}$$
$$P_2 = 760 \text{ torr}$$

$$\frac{(720 \text{ torr})(80 \text{ ml})}{278 \text{ K}} = \frac{(760 \text{ torr})(V_2)}{273 \text{ K}}$$

4. Cross-multiply and solve for V_2 after estimating the answer by scientific notation:

$$V_2 = \frac{(720 \cancel{\text{torr}})(80 \text{ ml})(273 \cancel{\text{K}})}{(278 \cancel{\text{K}})(760 \cancel{\text{torr}})}$$

Estimate:

$$V_2 = \frac{(7 \times 10^2)(8 \times 10^1)(3 \times 10^2)}{(3 \times 10^2)(8 \times 10^2)}$$

$$V_2 = 7 \times 10^1 = {\sim}70 \text{ ml}$$

5. The slide rule answer is

$$V_2 = 74.4 \text{ ml}$$

EXAMPLE 4.8

What volume will 0.2 mole of a gas occupy at 1.3 atm and a temperature of 27°C?

1. Write the equation of state:

$$PV = nRT$$

2. Organize the data:

$$P = 1.3 \text{ atm}$$
$$V = ?$$
$$n = 0.2 \text{ mole}$$
$$R = 0.082 \text{ liter-atm mole}^{-1} \text{ K}^{-1}$$
$$T = 27°C = 300 \text{ K}$$

3. Divide the equation by P, substitute, and solve for V by careful dimensional analysis:

$$\frac{\cancel{P}V}{\cancel{P}} = \frac{nRT}{P}$$

$$V = \frac{nRT}{P}$$

$$V = \frac{(0.2 \cancel{\text{mol}})(0.082 \text{ liter-}\cancel{\text{atm}} \cancel{\text{mol}^{-1}} \cancel{\text{K}^{-1}})(300 \cancel{\text{K}})}{(1.3 \cancel{\text{atm}})}$$

$$V = 378 \text{ liters (tentative)}$$

4. Determine the placement of the decimal point by estimation:

$$\frac{(2 \times 10^{-1})(1 \times 10^{-1})(3 \times 10^{2})}{1} = \sim 6$$

↑

(a one digit number)

Therefore, the answer must be one digit before the decimal point:

$$V = 3.78 \text{ liters}$$

EXAMPLE 4.9

If 0.66 g of a gas occupies 180 ml at 90°C and 740 torr, calculate the molecular weight.

1. Write the equation of state:

$$PV = nRT$$

2. Express n in a more useful form. Remember that

$$n = \frac{g}{g \text{ mole}^{-1}}$$

Since g mole^{-1} is the molecular weight, use the symbol M:

$$n = \frac{g}{M}$$

3. Substitute $\frac{g}{M}$ for n in the equation of state:

$$PV = \frac{gRT}{M}$$

4. Since PV and M constitute a "cross-product" in a proportion

$$\frac{PV}{1} = \frac{gRT}{M}$$

their positions may be exchanged:

$$M = \frac{gRT}{PV}$$

5. Organize the data and substitute in the equation. Perform the dimensional analysis:

$$P = \frac{740\ \text{torr}}{760\ \text{torr atm}^{-1}} = 0.97\ \text{atm}$$

$V = 180\ \text{ml} = 0.18\ \text{liter}$

$g = 0.66\ \text{g}$

$R = 0.082\ \text{liter-atm mole}^{-1}\ \text{K}^{-1}$

$T = 90°\text{C} = 363\ \text{K}$

$M = ?$

$$M = \frac{(0.66\ \text{g})(0.082\ \text{liter-atm mol}^{-1}\ \text{K}^{-1})(363\ \text{K})}{(0.97\ \text{atm})(0.18\ \text{liter})}$$

6. Estimate the answer by scientific notation and solve for M by slide rule:

$$\text{approx. } M = \frac{(7 \times 10^{-1})(8 \times 10^{-2})(4 \times 10^{2})}{(1)(2 \times 10^{-1})}$$

approx. $M = 110\ \text{g mole}^{-1}$

actual $M = 112\ \text{g mole}^{-1}$

EXAMPLE 4.10

What weight of carbon dioxide, CO_2, will be in a 4.0 liter container at 1.8 atm and a temperature of −50°C?

1. Write the equation of state:

$$PV = nRT$$

2. Substitute the form of the equation developed in Example 4.9:

$$PV = \frac{gRT}{M}$$

$$MVP = \frac{gRT\cancel{M}}{\cancel{M}} \qquad \text{(multiply the equation by M)}$$

$$\frac{g\cancel{RT}}{\cancel{RT}} = \frac{PVM}{RT} \qquad \text{(divide the equation by RT)}$$

$$g = \frac{PVM}{RT}$$

3. Organize the data:

$$P = 1.8 \text{ atm}$$
$$V = 4.0 \text{ liters}$$
$$M = CO_2 = 44 \text{ g mole}^{-1} \qquad \text{(from the formula weight)}$$
$$R = 0.082 \text{ liter-atm mole}^{-1} \text{ K}^{-1}$$
$$T = -50°C = 223 \text{ K}$$
$$g = ?$$

4. Substitute the data in the equation, estimate the decimal point placement by scientific notation, and find the actual answer on the slide rule. Perform the dimensional analysis:

$$g = \frac{(1.8 \cancel{\text{atm}})(4.0 \cancel{\text{liters}})(44 \text{ g } \cancel{\text{mol}}^{-1})}{(0.082 \cancel{\text{liter}}\text{-}\cancel{\text{atm}} \cancel{\text{mol}}^{-1} \cancel{\text{K}}^{-1})(223 \cancel{\text{K}})}$$

Estimate:

$$\frac{(2)(4)(4 \times 10^1)}{(1 \times 10^{-1})(2 \times 10^2)} = \sim 16$$

(a two digit number)

The slide rule answer is

17.3 g

2 digits before the decimal point

EXAMPLE 4.11

What is the density of nitrogen gas, N_2, at 30°C and 2.1 atm pressure?

1. Write the equation of state:

$$PV = nRT$$

2. Substitute the form of the equation that uses $\frac{g}{M}$ in place of n:

$$PV = \frac{gRT}{M}$$

3. Since density is mass divided by volume

$$d = \frac{M}{V}$$

rearrange the equation so that

$$\frac{g \text{ (mass)}}{V \text{ (volume)}}$$

may be labeled density, d:

$$PV = \frac{gRT}{M}$$

$$MPV = \frac{gRT\cancel{M}}{\cancel{M}} \quad \text{(multiply the equation by M)}$$

$$gRT = PVM$$

$$\frac{g\cancel{RT}}{\cancel{RT}V} = \frac{P\cancel{V}M}{RT\cancel{V}} \quad \text{(divide the equation by RTV)}$$

Therefore

$$\frac{g}{V} = d = \frac{PM}{RT}$$

4. Organize the data and solve for d:

$$P = 2.1 \text{ atm}$$

$$M = N_2 = 28 \text{ g mole}^{-1} \quad \text{(from the formula weight)}$$

$$R = 0.082 \text{ liter-atm mole}^{-1} \text{ K}^{-1}$$

$$T = 30°C = 303 \text{ K}$$

$$d = ?$$

$$d = \frac{(2.1 \cancel{\text{atm}})(28 \text{ g } \cancel{\text{mol}}^{-1})}{(0.082 \text{ liter-}\cancel{\text{atm}} \cancel{\text{mol}}^{-1} \cancel{\text{K}}^{-1})(303 \cancel{\text{K}})}$$

Estimate:

$$\frac{(2)(30)}{(.1)(300)} = \sim 2 \text{ g liter}^{-1}$$

(one digit number)

Solve by slide rule:

$$d = 2.36 \text{ g liter}^{-1}$$

Exercise 4.3

1. How many moles of gas will occupy a 750 ml volume at 25°C and 0.42 atm pressure?
2. If the argon gas in a light bulb exerts a pressure of 0.54 atm at 20°C, what pressure will it exert at 160°C?
3. What is the molecular weight of a gas if 0.21 g occupies a 0.13 liter volume at 80°C and 0.92 atm pressure?
4. How many grams of ammonia gas, NH_3, are there in a 600 ml volume at 1260 torr and −13°C?
5. What is the density of methane gas, CH_4, at 120°C and a pressure of 0.93 atm?

QUESTIONS AND PROBLEMS

4.1 What are the four parameters used to describe the behavior of gases?

4.2 Briefly summarize the main points of the kinetic molecular theory.

4.3 What is an ideal, or perfect, gas? Under what conditions will an ordinary gas behave most nearly like an ideal gas?

4.4 What does the Maxwell-Boltzmann distribution say about the relationship between the temperature and the average kinetic energies of gas molecules?

4.5 What is Avogadro's principle?

4.6 Illustrate the law of combining volumes diagrammatically by showing how 2 liters of hydrogen can combine with 1 liter of oxygen to produce 2 liters of water vapor.

4.7 State Boyle's law in words and algebraically.

4.8 If 0.2 mole of nitrogen gas, 0.4 mole of oxygen gas, and 0.7 mole of carbon dioxide exert a total pressure of 750 torr, what is the partial pressure of the oxygen?

4.9 State Charles's law in words and algebraically.

4.10 What are two reasons for relating gas volumes to degrees Kelvin rather than to degrees Celsius?

4.11 If 350 ml of a gas is collected over water at 50°C and 740 torr, what volume will the dry gas occupy at STP?

4.12 What is the molecular weight of a gas if it has a density of 1.2 g $liter^{-1}$ at 20°C and 0.88 atm pressure?

4.13 What volume will be occupied by 0.025 mole of a gas at 27°C and 0.84 atm pressure?

4.14 Calculate the molecular weight of a gas if 1.3 g occupies a volume of 460 ml at 10°C and 770 torr pressure.

4.15 What is the density of chlorine gas, Cl_2, at −73°C and 4.0 atm pressure?

CHAPTER 5

The Physical Sciences are, in fact, an integral part of our civilization, not only because our ever increasing mastery of the forces of nature has so completely changed the material conditions of life, but also because the study of these sciences has contributed so much to clarify the background of our own existence.

Niels Bohr *

Behavioral Objectives

At the completion of this chapter, the student should be able to:

1. Explain why the Rutherford atomic model is not satisfactory.
2. Describe the photoelectric effect, and state which model of light it supports.
3. Calculate light energies per photon or per mole, given the frequency or wavelength.
4. Describe the Bohr model of the atom.
5. Write and define each term in the Bohr frequency rule equation.
6. Define ionization energy.
7. Calculate frequencies and wavelengths of emitted radiation that is due to shifts in the electron energy level.
8. Summarize the contributions of de Broglie and Schrödinger that led to a wave model of the atom.★
9. State the uncertainty principle and explain its significance.★
10. Calculate the maximum number of electrons that can occupy each principal energy level.
11. List the symbols that represent energy sublevels in atoms.
12. Write the maximum number of electrons that can occupy each sublevel.
13. Sketch the probability regions about an atomic nucleus that illustrate the first two sublevels.
14. State the exclusion principle.
15. Write the electronic configuration of an atom by using orbital notation.
16. Illustrate the electronic configuration of atoms by pictorial representation.
17. Define and properly apply Hund's rule.
18. Explain what is meant by the term *periodicity* as applied to the elements.
19. Sketch the arrangement of groups and periods on the periodic table.
20. Illustrate the positions of related groups of elements on the periodic table.
21. Sketch the generalized characteristics of elements on a periodic table.

---

* Nature, 143: 268, 1939.

ATOMIC STRUCTURE AND THE PERIODIC TABLE

5.1 THE RUTHERFORD ATOM

After experimentally proving the reality of the nuclear model of the atom in 1911, Ernest Rutherford suggested that an atom was composed of a tiny, dense, positively charged nucleus about which negatively charged electrons whirled in circular orbits. He postulated that the coulombic force of attraction between the oppositely charged particles was just enough to hold the electron in its orbital track despite the centrifugal force that constantly tended to pull it away (Fig. 5–1).

This was a beautiful model. Its logical simplicity was impressive, since it seemed to be perfectly analogous to the relationship between our sun and the planets of our solar system. However, the planets and the sun are not electrically charged particles. Gravitational interaction could not be described by the same laws of classical physics as were orbiting electrical charges.

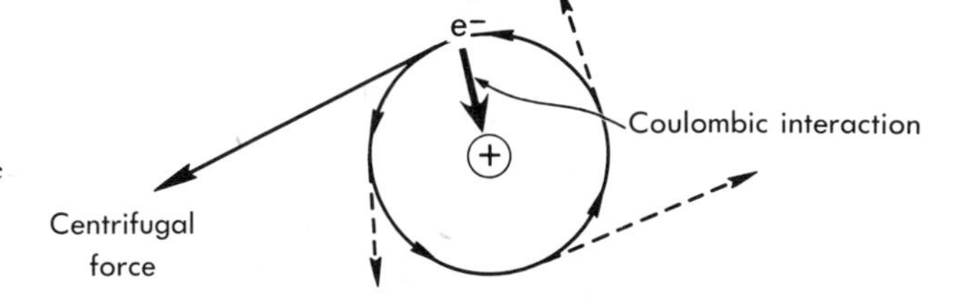

Figure 5–1 The Rutherford planetary model of the atom.

As early as 1879, James Clerk Maxwell demonstrated that light, magnetism, and electricity were related quantitatively, and his work indicated that oscillating electrical charges emit electromagnetic radiation as they lose energy. The amount of energy that is radiated, as low as alternating current or as high as gamma radiation, depends on the frequency of the oscillations. Heinrich Hertz proved Maxwell's conclusions by using oscillating electrical charges to produce radio waves.

The fact that oscillating electrical charges must lose energy in the form of electromagnetic radiation destroyed Rutherford's theoretical basis for his planetary model. As the electron revolved about the nucleus it would have to emit energy. The only way in which the electron could compensate for this energy loss would be to make a smaller orbit—a decrease in orbital radius. But the more energy lost, the tighter the circle would be until the electron crashed into the nucleus, resulting in the destruction of the atom. (Fig. 5–2).

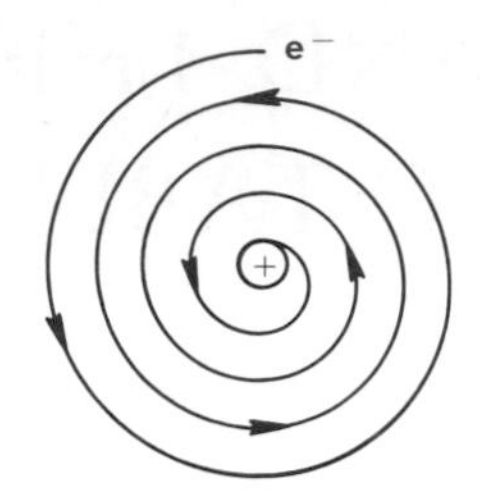

Figure 5–2 The spiral of atomic self-destruction.

The atom, however, in a manner characteristic of all matter, does not tend toward self-destruction. The dilemma Rutherford's model created was *how* could this most logical system be explained, since the electron did indeed behave like a particle oscillating about the nucleus.

5.2 ORIGIN OF THE QUANTUM THEORY

It was Max Planck, in 1901, who daringly proposed that the accepted physical laws of Maxwell, Newton, and others did not necessarily apply to heat radiation. Planck assumed that radiant energy was not a continuous ill-defined blur like a rainbow of color, but instead was a stream of distinct, or discrete, units of energy called **quanta**. The word **quantum** suggests that each discrete packet of energy has a very specific energy content that is proportional to its frequency (rate of oscillation).

Max Planck

Planck's **quantum theory** of electromagnetic radiation effectively resolved a dilemma that arose when scientists attempted to explain light in terms of the Maxwell-Boltzmann distribution for the energies of gas molecules at various temperatures: The average kinetic energies of gas molecules are a statistical average at a given temperature. The distribution of the energy available to all of the molecules present, accepting the fact that many molecules will have more or less than the average, is called the **equipartition principle**. While the principle of equipartition of heat energy among gas molecules works nicely, it must be remembered that there are a finite number of gas molecules in any container. The distribution of energy can go just so far. Light, however, is another matter. The possible number of frequencies over which radiant energy can spread, if the equipartition principle is applied, is infinite. The equipartition principle simply cannot be applied to electromagnetic radiation. If it could, then our friendly reading lamp would bathe us in a small portion of deadly gamma radiation. Planck's quantization of light resolved this problem. A single quantum of light is restricted to a definite amount of energy that depends only upon its frequency.

The discovery of the **photoelectric effect** by Heinrich Hertz and the explanation of it by Albert Einstein supported the particle model of light when they discovered that a quantum of light had to have a very specific frequency to displace an electron from a metal surface (Fig. 5–3).

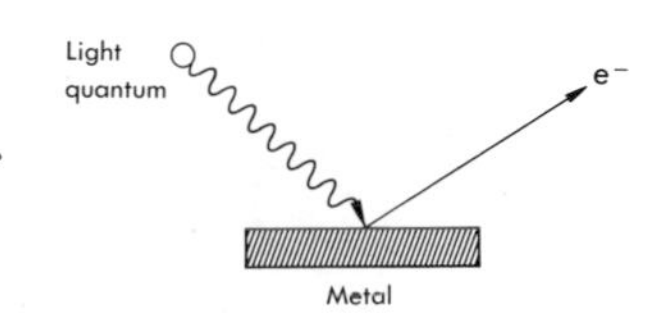

Figure 5–3 The photoelectric effect.

Einstein proposed that the photoelectric effect clearly demonstrated a fundamental relationship between the energy of a quantum of light, also called a **photon,** and its frequency:

$$E \propto \nu$$

energy is proportional to frequency

The proportionality constant that relates the two is known as **Planck's constant** in tribute to the work of Max Planck, and its symbol is h:

$$E = h\nu$$

5.3 THE BOHR ATOM

In his search for a theoretical justification of the Rutherford planetary model, the Danish physicist Niels Bohr brilliantly wedded

Planck's quantum theory of electromagnetic radiation to the studies of the hydrogen spectrum made by Johannes Balmer in 1885.

When an electric discharge is passed through a tube of hydrogen gas at low pressure, a peculiar light is emitted. When viewed through a prism or diffraction grating, the visible light is observed to be made of several distinct lines of color. This **line spectrum** produced by the gas discharge tube must be the result of hydrogen atoms interacting with the high energy electrons moving through the tube (Fig. 5–4). As the high energy electrons strike the H_2 molecules, the molecules are split into atoms. The electrons of the hydrogen atoms seem to absorb some of the available energy, since they exhibit new patterns of oscillation that result in the emission of electromagnetic radiation. The frequencies of these emitted photons must be very specific in order to produce the fine lines of color that characterize the hydrogen spectrum. The wavelength and, therefore, the frequency ($\nu = c/\lambda$) can be precisely determined with the aid of a **spectrometer.**

Niels Bohr explained the Balmer lines of the hydrogen spectrum by postulating that the single electron of the hydrogen atom was able to have only very distinct energy values. In effect, he *quantized* the mechanical energy of the electron "particle" as Planck had previously quantized the energy of the photon "particle."

In other words, Bohr said that the electron of a hydrogen atom was

Niels Bohr

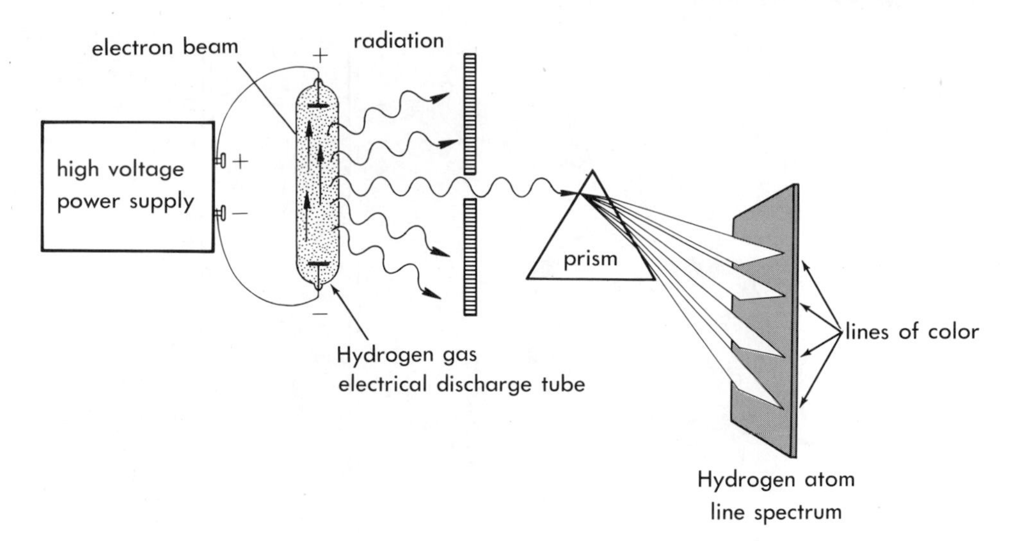

Figure 5–4 Production of the hydrogen spectrum.

restricted to very specific **energy levels**. Electrons having these energy levels were described as following circular orbits, with each orbital radius a specific distance from the nucleus. Each energy level was assigned a definite value for the total energy of the electron. These values could be used to calculate the precise amounts of energy that an electron would have to absorb in order to move to higher energy levels; i.e., to make **quantum jumps.** This process is called *excitation* of the electron.

If the electron moved from a higher energy level to a lower one, some or all of the excitation energy would be emitted as light. The energy of the resulting photon of light would be exactly equal to the difference between the energy values assigned to the energy levels.

Bohr's energy level concept is somewhat analogous to a man on a ladder. Each rung may represent an energy level. The lowest rung, where the man has the least potential energy, is like the orbit of the normal or unexcited electron. It is the orbit nearest the nucleus. The lowest energy level to which an electron can "fall" is called the **ground state** of that electron. As the man moves up and down the ladder, it must be from rung to rung, since standing between the rungs is physically impossible, or "forbidden." Bohr's energy levels also have their "forbidden" regions, which is to say that the electron can only move from one energy level to another—all other orbital radii being forbidden. Figure 5–5 represents the energy levels of the Bohr atom and the corresponding ladder analogy.

The energy of the emitted photon produced by an electron as it makes a nearly instantaneous all-or-nothing shift from a higher to a lower energy level is expressed by an equation known as the **Bohr frequency rule:**

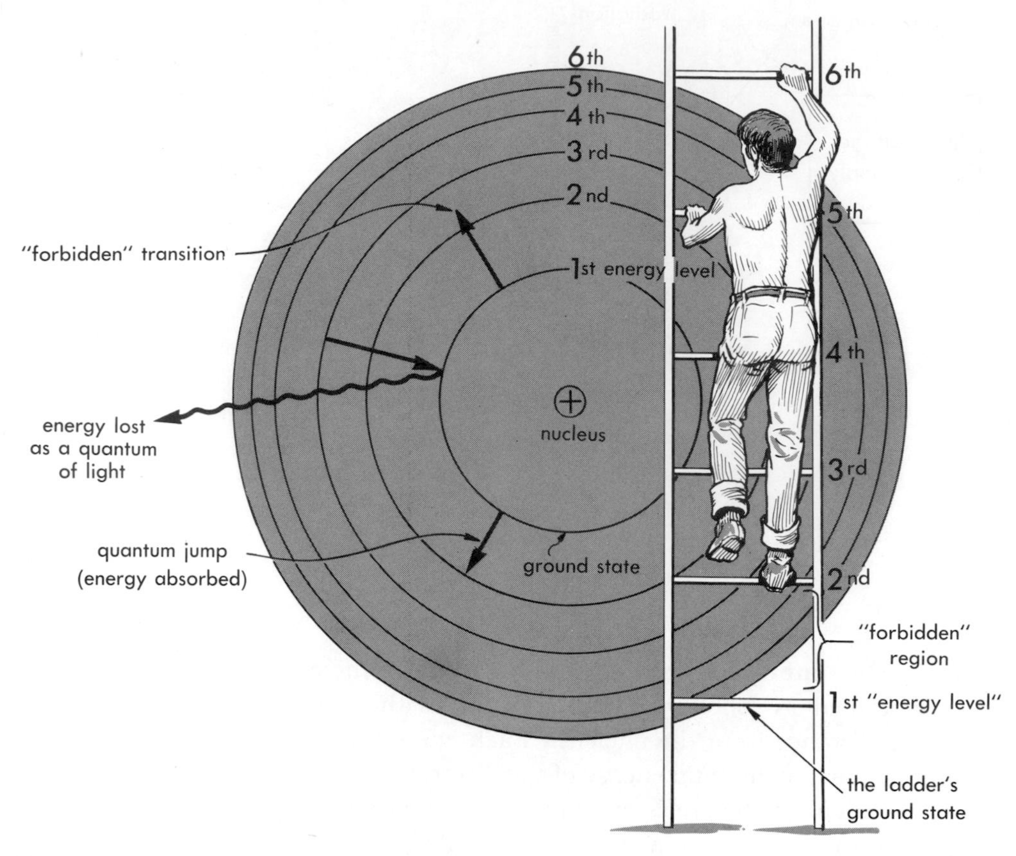

Figure 5–5 Hydrogen atom energy levels and a ladder analogy.

$E_2 - E_1 = h\nu$

E_2 = the energy of an electron on the higher energy level

E_1 = the energy of an electron on the lower energy level

$h\nu$ = the energy of the emitted photon

Bohr was able to calculate the energy of the electron for each energy level by equating the mechanical energy of the orbiting electron to the balancing force of attraction of the positive nucleus. He found the energy of each level was equal to the total energy of the electron divided by the square of the small whole number assigned to each level. This number, called the **principal quantum number,** is symbolized by the letter n. For the ground state, n = 1, and then n successively equals 2, 3, 4, 5, ... The equation is

$$E_n = -\frac{\text{total energy}}{n^2}$$

The negative sign indicates an increasing loss in potential energy as the electron moves to orbits closer to the nucleus. The total energy of the hydrogen electron calculated by Bohr, and experimentally verified by measuring the energy needed to remove the electron from the

TABLE 5–1 THE ENERGY LEVELS OF HYDROGEN ATOMS

n (The Energy Level)		Energy $E = -313.6/n^2$ (kcal $mole^{-1}$)
1 (ground state)		$-\frac{313.6}{1^2} = -313.6$
2	("excited states")	$-\frac{313.6}{2^2} = -78.4$
3		$-\frac{313.6}{3^2} = -34.9$
4		$-\frac{313.6}{4^2} = -19.6$
5		$-\frac{313.6}{5^2} = -12.5$
6		$-\frac{313.6}{6^2} = -8.7$

atom (**ionization energy**), is −313.6 kcal $mole^{-1}$, or −13.6 eV per atom, or -21.8×10^{-12} erg per atom. The energy levels, called **stationary states*** by Bohr, are illustrated in Table 5–1.

The data from Table 5–1 is pictorially represented in Figure 5–6 to show the relationship between the Bohr frequency rule and the spectral lines produced by the hydrogen electron oscillating between energy levels.

In order to illustrate mathematically the relationship between the Bohr frequency rule and the Einstein equation, examine the data associated with the blue line produced when electrons move from the 4th energy level to the 2nd in the hydrogen atom (Fig. 5–7).

EXAMPLE 5.1

Calculate the energy of the n = 4 to n = 2 ($4 \xrightarrow{E} 2$) electron transition illustrated in Figure 5–7. Use the Bohr frequency rule equation.

1. Write the equation:

$$E_4 - E_2 = h\nu = E \text{ (emitted energy)}$$

2. Substitute the data and solve:

$$-19.6 - (-78.4) = E$$

$$E = 58.8 \text{ kcal mole}^{-1}$$

---

* The name "stationary state" refers to the absence of radiation while the electron remains in a particular orbit.

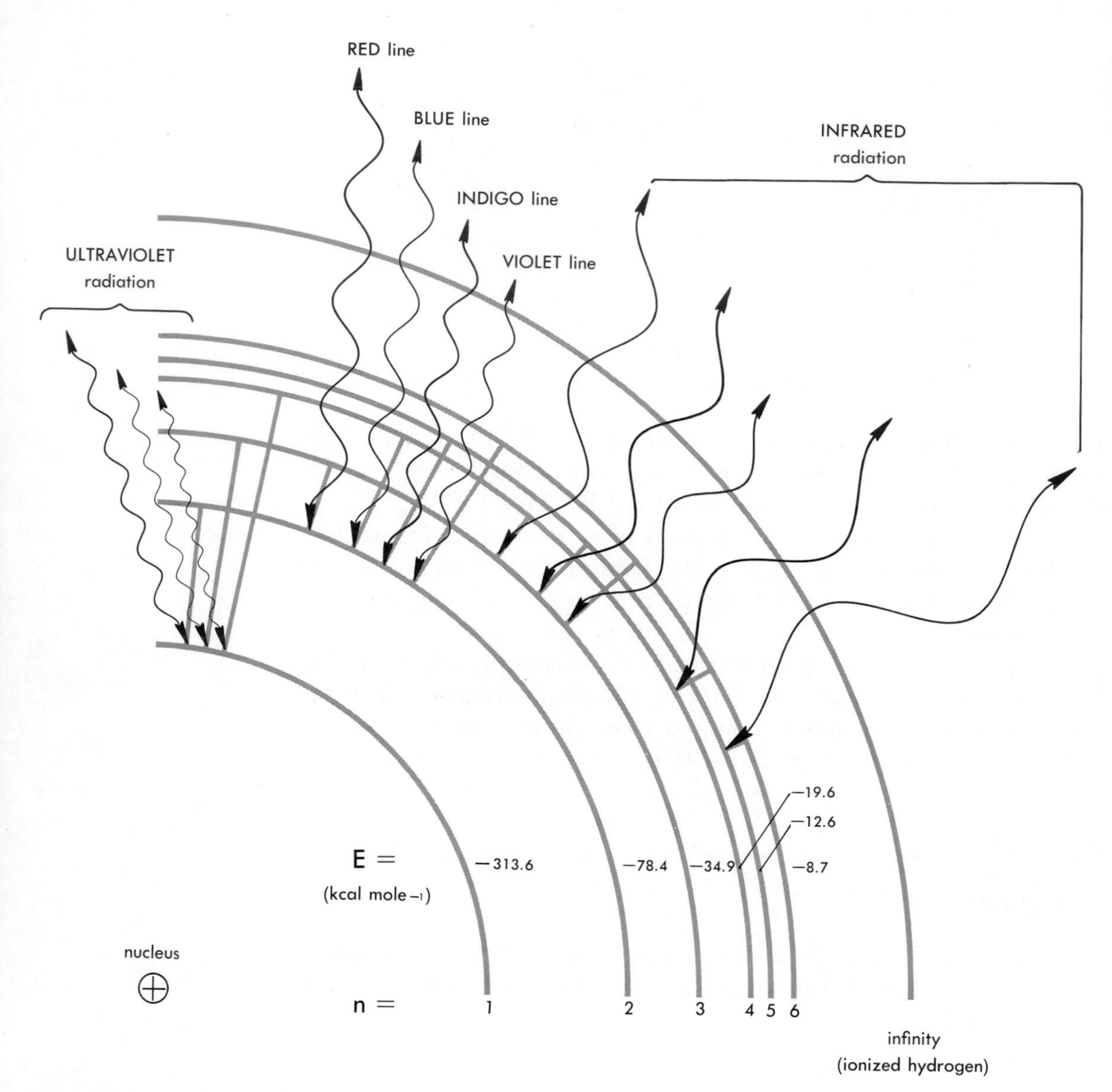

Figure 5–6 Energy level diagram of the hydrogen atom.

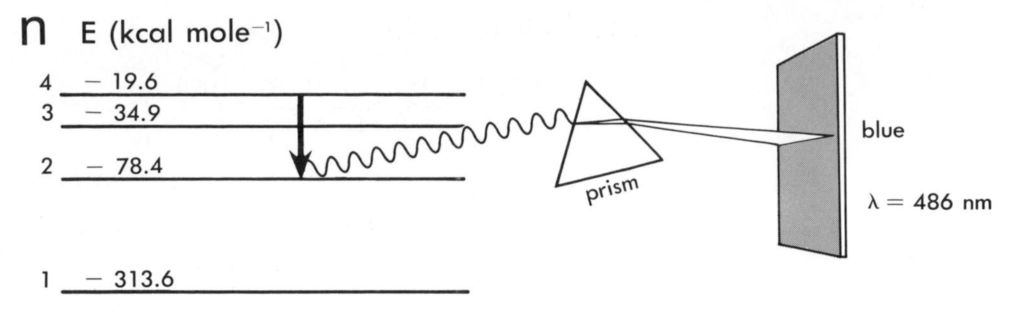

Figure 5–7 Electron transition of an electron of hydrogen.

EXAMPLE 5.2

Check the energy value of the emitted blue line and see how accurately the Einstein equation proves that the energy of the emitted photons equals the energy lost by the electrons in the $4 \xrightarrow{E} 2$ shift.

1. Write the equation:

$$E = h\nu \mathcal{N}_A$$

(energy per mole)

2. Organize the data, substituting

$$\frac{c}{\lambda} \text{ for } \nu \quad \text{since, } \nu = \frac{c}{\lambda}$$

$$E = h\frac{c}{\lambda}\mathcal{N}_A$$

$$E = ?$$

$$h = 1.6 \times 10^{-37} \text{ kcal-s}$$

$$c = 3 \times 10^{10} \text{ cm s}^{-1}$$

$$\mathcal{N}_A = 6.02 \times 10^{23} \text{ mole}^{-1}$$

$$\lambda = 486 \text{ nm} = 4.86 \times 10^{-5} \text{ cm}$$

(blue line)

Note: The wavelength must be expressed in cm in order to be compatible with the cm units in the speed of light, c:

$$\frac{486 \cancel{\text{nm}}}{10^7 \cancel{\text{nm}} \text{ cm}^{-1}} = 486 \times 10^{-7} \text{ cm} = 4.86 \times 10^{-5} \text{ cm}$$

3. Substitute in the equation and solve for the energy of the blue line. Perform the dimensional analysis:

$$E = \frac{(1.6 \times 10^{-37} \text{ kcal-}\cancel{\text{s}})(3 \times 10^{15} \cancel{\text{cm}} \cancel{\text{s}^{-1}})(6.02 \times 10^{23} \text{ mol}^{-1})}{4.86 \times 10^{-5} \cancel{\text{cm}}}$$

4. Estimate the answer:

$$\frac{(2 \times 10^{-37})(3 \times 10^{10})(6 \times 10^{23})}{5 \times 10^{-5}} = \sim 7 \times 10^{1}$$

$$= \sim 70 \text{ kcal mole}^{-1}$$

5. The slide rule calculation provides the two digit answer:

$$E = 59.4 \text{ kcal mole}^{-1}$$

Observe the results for Examples 5.1 and 5.2:

In Example 5.1, E = 58.8 kcal mole^{-1}

In Example 5.2, E = 59.4 kcal mole^{-1}

Allowing for the rounding off of numbers and the normal slide rule error, these values are in good agreement. Taken as a whole, Bohr's attempt to explain the structure of the hydrogen atom in terms of Planck's theory of the quantization of energy, the supporting evidence of spectrum analysis, and the emerging model were very exciting in 1913. Although Bohr's model, with the added refinement of elliptical orbits, was found to be inadequate when applied to many-electron atoms, his principal quantum numbers remain a permanent fixture in the modern atomic model. His concept of the orbit also persists as the word **orbital,** but the meaning is different.

Exercise 5.1

1. What is the frequency of the emitted photon if an electron oscillates between the 6th and 2nd energy levels in a hydrogen atom?
2. What precise amount of energy in kcal mole^{-1} must be absorbed by a hydrogen electron in order to make a quantum jump from the 1st to the 3rd energy level?
3. If a hydrogen atom electron shifts from the 2nd energy level to the ground state which has an energy value of -21.7×10^{-12} erg, what will be the wavelength and frequency of the emitted photon? (Hint: $E_n = -21.7 \times 10^{-12}$ erg/n^2.)
4. Calculate the energy in joules and eV (per atom) released by an oscillating electron if it produces a line of violet color having a measured wavelength of 410 nm.

5.4 MATTER WAVES*

In the early 1920's, improved spectroscopic instruments and the effects of external magnets on gas discharge tubes caused the atomic scientists of the day to begin moving in a new direction. It was a movement toward a consideration of the electron as having a dual nature, that of both particle and wave.

In 1924 Louis de Broglie made a very startling proposal. He reasoned that since the quantum theory so beautifully permitted the wave phenomenon of light to be endowed with the properties of particles, variously called "discrete packets," "corpuscles," "quanta," and "photons," then why not consider electrons (generally accepted as particles of matter) in terms of a wave model? In other words, de Broglie suggested that the electron might actually have a wavelength that would be inversely proportional to its **momentum**. Momentum, symbolized by the Greek letter rho (ρ), is a property of matter. It is the product of a particle's mass and velocity:

$$\rho = mv$$

By manipulating the energy, mass, and wavelength relationships demonstrated by Einstein, de Broglie showed that the proportionality constant relating the electron's wavelength and momentum was Planck's quantum of action, h:

$$\lambda_{e^-} = \frac{h}{mv}$$

λ_{e^-} = the wavelength of the electron
h = Planck's constant

$$\left.\begin{array}{l} m = \text{mass} \\ v = \text{velocity} \end{array}\right\} m \times v = \rho$$

While there was abundant evidence supporting the dual (wave-particle) model for light, de Broglie's interesting hypothesis did not have empirical origins. The photoelectric effect, in addition to the experiment of Arthur Compton, clearly showed that photons of light behaved like particles when they transferred kinetic energy upon colliding with electrons. Compton's observations were made when high energy x-ray photons transferred kinetic energy to electrons, as shown in Figure 5–8. However, just three short years after de Broglie's proposal, experimental evidence was provided in support of wave properties of electrons.

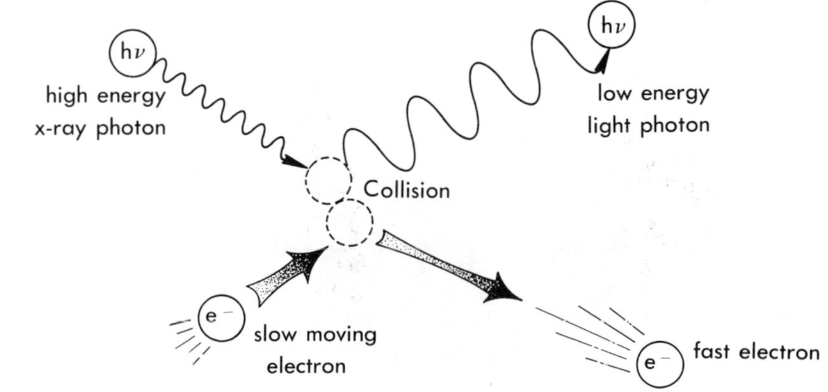

Figure 5–8 The Compton effect.

Of course, de Broglie's equation could not be $E = hc/\lambda$, because matter is not electromagnetic radiation. As particles of matter oscillate at various frequencies, radiation can be emitted. While "matter waves" possess the typical properties of waves (ν, λ, amplitude) they cannot move at the speed of light and should not be confused with light.* The de Broglie relation between wavelength and momentum is illustrated in Figure 5–9.

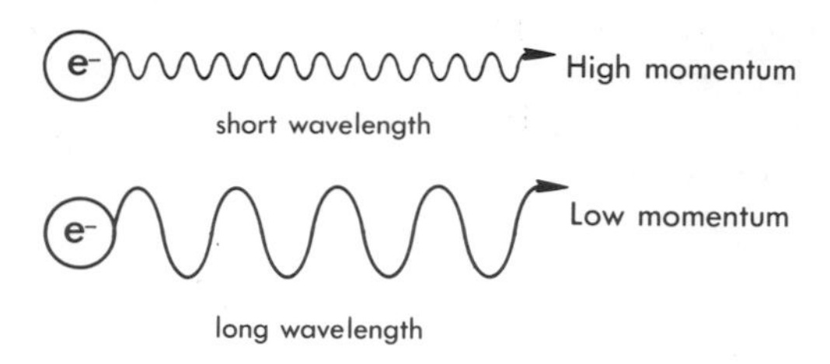

Figure 5–9 An aspect of the de Broglie hypothesis.

5.5 THE WAVE MODEL OF THE ATOM*

As a direct consequence of the experimentally supported wave character of electrons, Erwin Schrödinger, in 1926, described the distribution of electrons in atoms by application of the mathematics used to describe waves. While de Broglie fitted his wave concept of electron motion to Bohr's orbits,

Erwin Schrödinger

saying in effect that the number of waves would be equal to the principal quantum number (i.e., the number of the energy level), as illustrated in Figure 5–10, Schrödinger departed from the notion of circular and elliptical electron orbits and suggested three-dimensional wave forms.

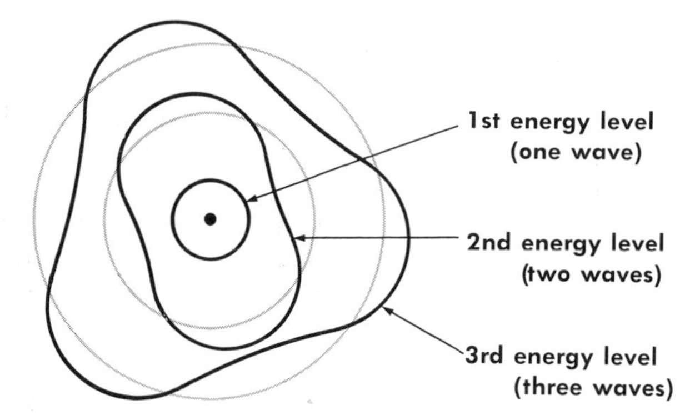

Figure 5–10 The de Broglie electron waves.

Schrödinger's **wave mechanics** (mathematical description of wave motion) quantitatively described the electron waves so that they agreed with the energy characteristics of Bohr's frequency rule for the hydrogen atom. Schrödinger's equations did indeed produce the correct values for the frequencies and intensities of the observed spectral lines.

The vibrational **nodes** (i.e., the points at which a vibrating object does not move) fitted well with Bohr's "forbidden" zones between the energy levels (Fig. 5–11). However, to some extent Schrödinger's results worked for many-electron atoms while the Bohr model did not.

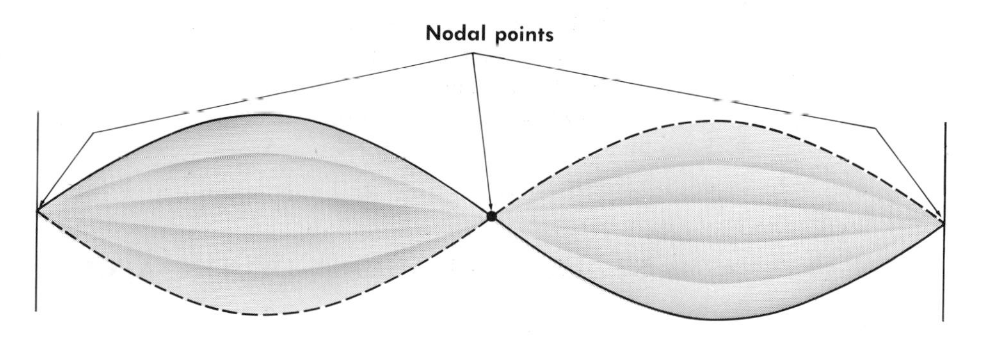

Figure 5–11 Fixed, or nodal, points in a vibrating string.

A significant contribution that Bohr did make at this time was his **principle of complementarity,** which helped resolve the argument that was raging over the nature of the electron. Bohr suggested that the model of the electron that probably described its nature most accurately arose from a duality concept. In other words, rather than having the particle model stand in conflict with the wave model, he suggested that both be accepted as mutually complementary concepts. Therefore, the electron, as in the case of light, was generally considered to have *both* the properties of matter and the properties of waves.

5.6 THE UNCERTAINTY PRINCIPLE

Shortly after Schrödinger's wave-mechanical explanation of electron behavior in atoms, Werner Heisenberg* submitted a mathematical answer to the problem of the still-imperfect description of the atom. In his **uncertainty principle,** he said that it is impossible to measure simultaneously the path (orbital trajectory) and momentum of an orbiting electron. This uncertainty is due to the fact that the physical act of observing and measuring an object will alter, to some degree, those properties that are being measured. When a meter is used to measure the voltage of a dry cell, for example, the very fact that the meter is made a part of the system being measured will change the true voltage. Admittedly, the meter introduces a negligible error by virtue of its design. However, the uncertainties that accompany measurements on atoms are enormous in relation to the smallness of what is being measured. Heisenberg calculated that the product of the uncertainties of path and momentum was about equal to Planck's constant:

$$(\Delta x)(\Delta \rho) \approx h$$

(uncertainty of path) (uncertainty of momentum)

When data, such as the electron's mass and velocity, are substituted in Heisenberg's equation, it is found that the uncertainty of the path of an electron of definite energy is larger than the whole atom, or the uncertainty of the velocity of an electron on a given path exceeds the speed of light! Obviously, the position and the velocity of the electron of a hydrogen atom could not be specified separately, nor indicated in a drawing.

The effect of accepting the uncertainty principle was to abandon any attempt to develop a perfect planetary model of the atom. The original model of Bohr in which electrons described clearly defined orbits about the nucleus had served its purpose. A new direction was needed.

5.7 THE NEW MODEL OF THE ATOM

In 1928 Max Born suggested that Schrödinger's wave equations be interpreted as describing the regions around a nucleus where specific electrons could be found in terms of *probability*. The uncertainty principle had clearly demonstrated the futility of gaining a precise knowledge of electron behavior.

Schrödinger's calculations could be applied to statistical probability distributions as well as represent a particular electron-wave

* An interesting historical note is that while the new quantum theory is commonly linked to Schrödinger's name, Heisenberg simultaneously announced the same results. However, his mathematical format was entirely different from Schrödinger's.

amplitude as was originally intended. The wave amplitude is called the **psi function** and is symbolized by the Greek letter psi, ψ.

It must be mentioned, however, that improved spectroscopic apparatus (as well as studies of magnetic effect on electron behavior) indicated that the original energy levels were, in fact, subdivided. What Bohr saw as a single line of color in a line spectrum was usually found to be made of several lines of color very close together. An external magnet often served to fractionate these lines even further (Fig. 5.12). As a result of these fine-line observations, it was generally concluded that the principal energy levels would have to be subdivided into secondary or **sublevels** (often called **subshells**) occupied by electrons of differing total energy. The energy sublevels are symbolized by letters taken from the spectrum terminology outlined in Table 5–2.

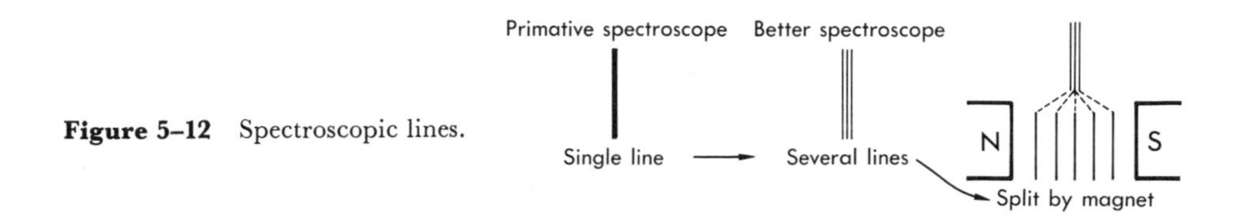

Figure 5–12 Spectroscopic lines.

It was the discovery of sublevels that 10 years earlier led Bohr and Arnold Sommerfeld to suggest elliptical orbits.

Schrödinger, however, was able to incorporate the sublevel distinctions and magnetic field interpretations into a refined method of describing probability distributions or wave amplitudes in which his ψ function was modified to a ψ^2 function (squaring a value eliminates negative signs). The ψ^2 function can be related to the probability of finding an electron at a particular distance (r) from the nucleus. The shapes of the probability regions are described by simulating electron density clouds around three dimensional axes called **Cartesian coordinates** (Fig. 5–13).

The *f* region is much more difficult to represent diagrammatically. Seven probability regions are required to accommodate a maximum of 14 electrons.

The Bohr energy level diagram can be put to good use in illustrating the progressive subdivisions of the principal energy levels (Fig. 5–14).

TABLE 5–2

Spectrum Term	Symbol	Maximum e^- Population
sharp	*s*	2 (1 pair)
principal	*p*	6 (3 pair)
diffuse	*d*	10 (5 pair)
fundamental	*f*	14 (7 pair)

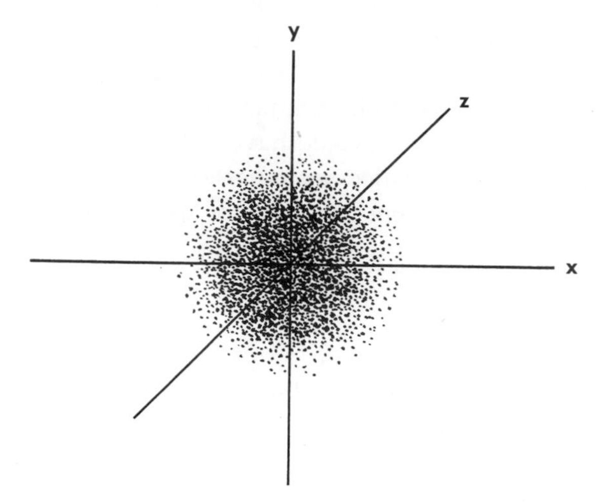

Figure 5–13A An *s* probability region—Always a spherical region of probability around the nucleus.

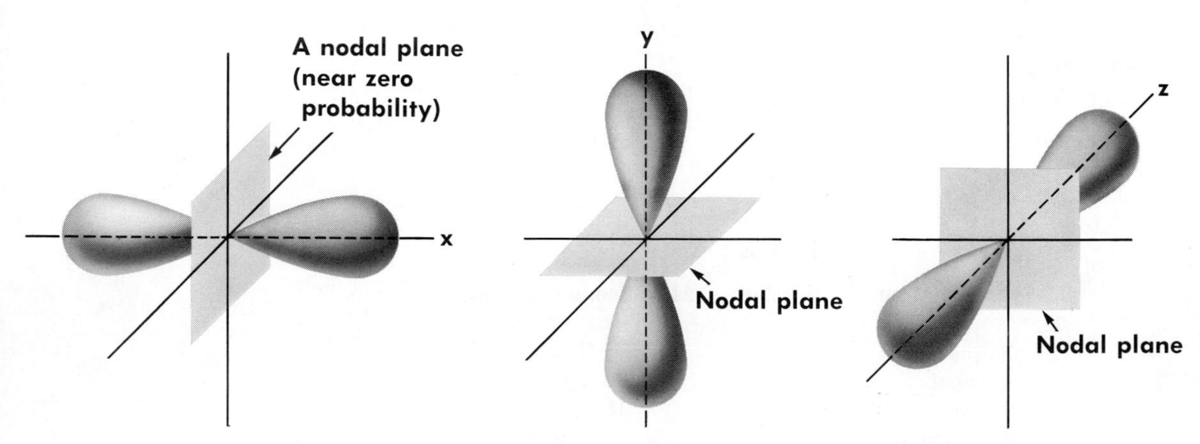

Figure 5–13B The three *p* probability or probability regions optionally called *px*, *py*, and *pz* to indicate difference in directional orientation as well as slight difference in a magnetic field.

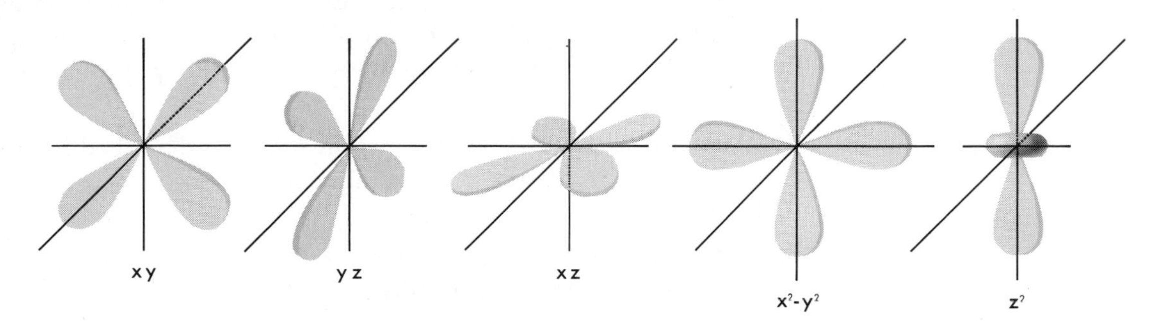

Figure 5–13C The five *d* probability regions that may optionally be distinguished by the subscripts indicated in the diagram.

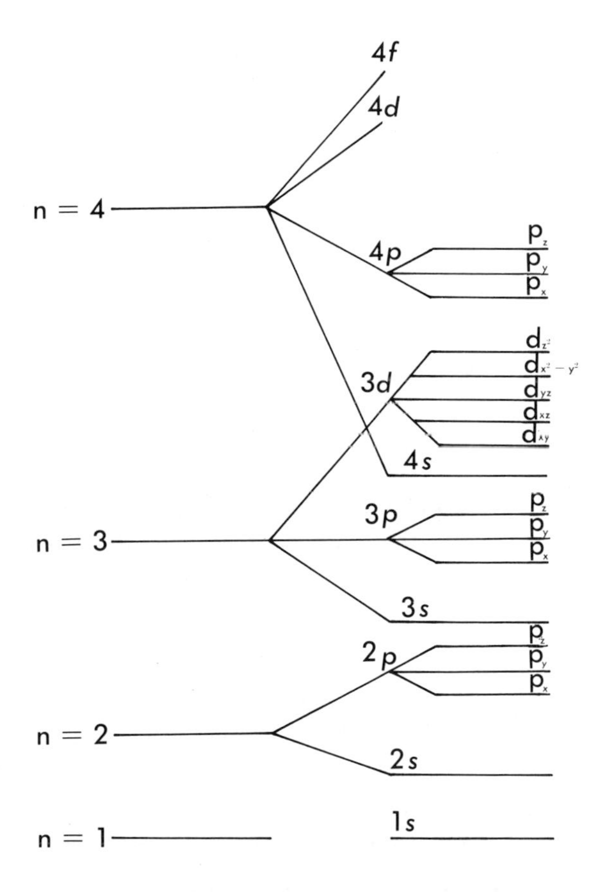

Figure 5–14 The subdivisions of energy levels.

The following points should be observed in Figure 5–14:

1. The number of sublevels is equal to the number assigned to the energy level (the principal quantum number). The first energy level has one subdivision, the *s* probability region. The second energy level has two subdivisions, the *s* and *p* probability regions.
2. There are usually not more than four sublevels described in any case. The *s*, *p*, *d*, and *f* probability regions are the practical limits.
3. The writing of a principal energy level number plus the *s*, *p*, *d*, or *f* symbol describes an **orbital.**

For example

A 1*s* orbital means a region of spherical probability close to the nucleus.
A 3*s* orbital means a region of spherical probability further from the nucleus. (See Fig. 5–15.)

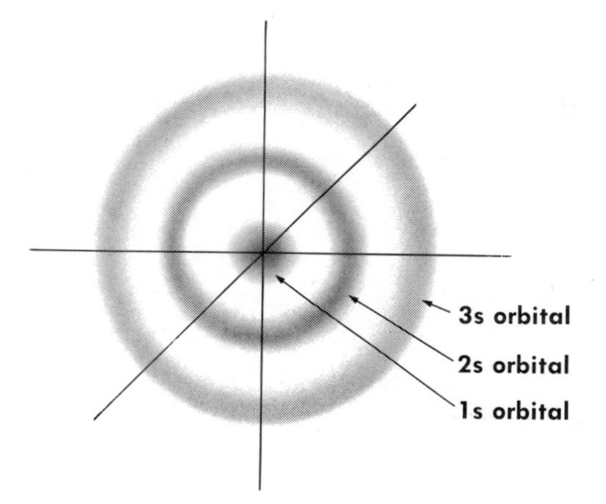

Figure 5–15 Orbital notation.

A 2*p* orbital means an electron probability distribution along the x, y, or z axes of the Cartesian coordinates. A specific x axis orientation could be written $2p_x$.

4. When the maximum electron population of any orbital is related to the number of probability regions, it is observed that there is *one* probability region for every *two* electrons. For example, a *p* orbital has three probability regions and a maximum of six electrons that can be accommodated. Other probability regions are illustrated in Table 5–3.

5.8 THE EXCLUSION PRINCIPLE

The exclusion principle was explained by Wolfgang Pauli in 1929, and it helped to solve some remaining problems in the new atomic

TABLE 5-3

Orbital	Number of probability regions	Maximum number of electrons
s	1 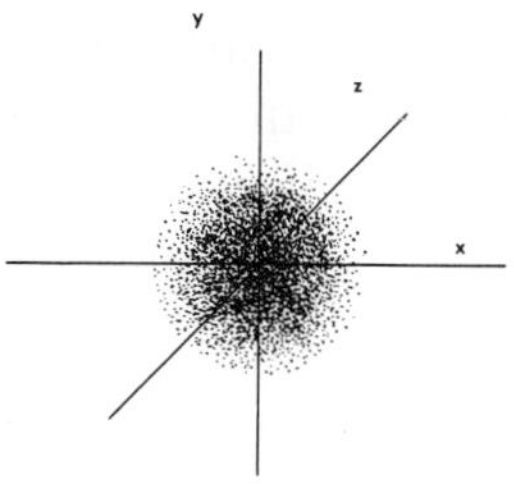 y z x	2
p	3 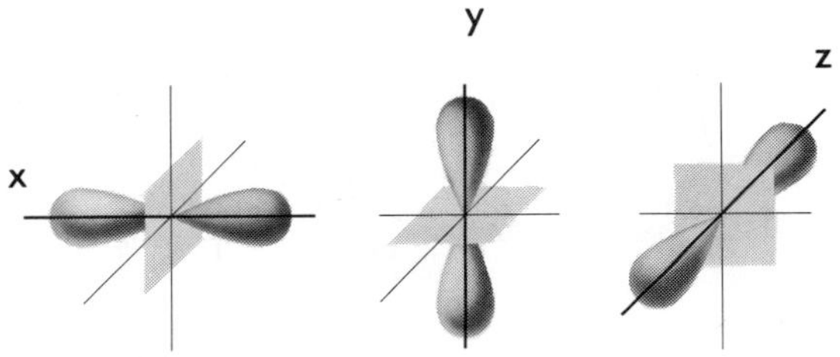 x y z	6
d	5 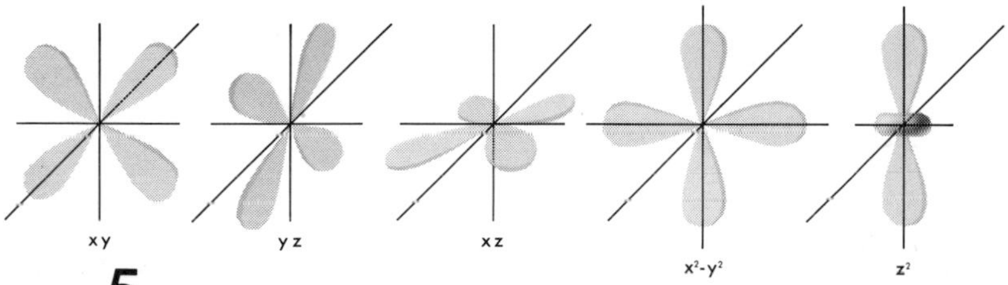 xy yz xz x²-y² z²	10
f	7 probability regions	14

model. In his **exclusion principle,** Pauli stated that no more than two electrons may occupy any single probability region—and then only if they have opposite "spin." A model of **electron spin** suggests that the electron be visualized as a spherical particle rotating about an axis. The direction of the spin may be described as clockwise or counter clockwise. Since a rotating, electrically charged particle behaves like a magnet that has a positive and a negative pole, the type of spin is alternately described as being *parallel* or *antiparallel* to the direction of the field of an external magnet (Fig. 5–16). The total energies of the electrons differ slightly, depending on their parallel or antiparallel orientation to the external magnet. It is the opposite polarization of the electrons that makes the model so attractive. By having their north and south poles reversed, the electrons can approach each other more closely as they occupy a single region (Fig. 5–17).

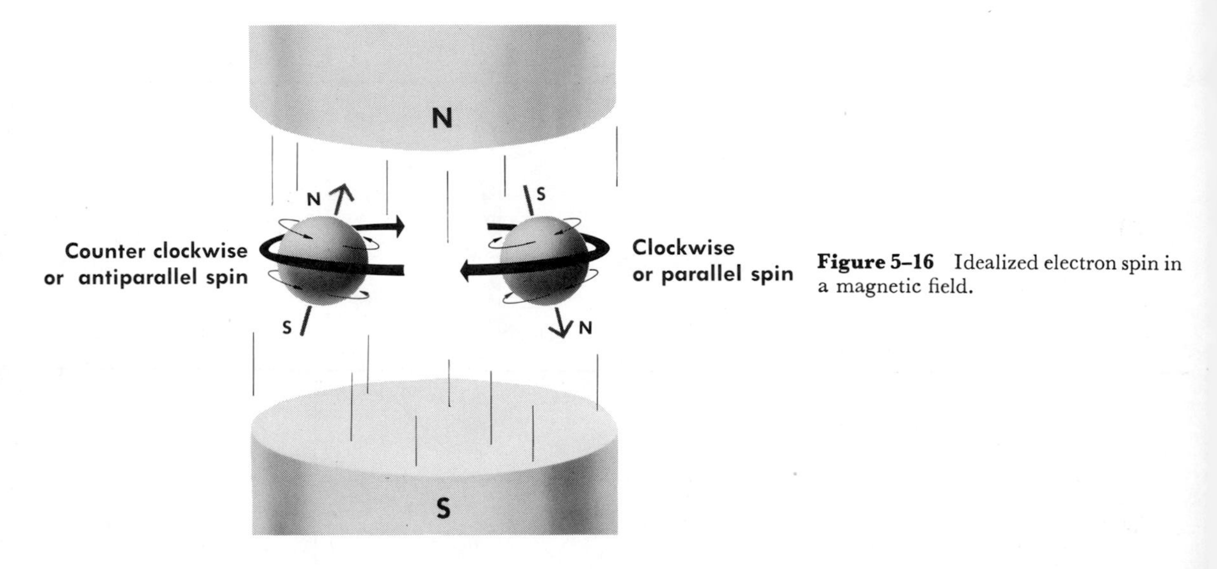

Figure 5–16 Idealized electron spin in a magnetic field.

It should be emphasized, however, that the concept of electron spin is only a model used to explain the slight energy differences between two electrons in the same probability regions. The electron is not a ball and it does not rotate about an axis. However, the rotating ball model does enable the chemist to cope with the mysterious electron in a meaningful way, as long as the model is not confused with reality.

When Schrödinger multiplied the psi squared function, ψ^2, by the formula for spherical volumes, he was able to plot the probability distributions of the electrons in various orbitals (Fig. 5–18).

5.9 THE DISTRIBUTION OF ELECTRONS IN ATOMS: ELECTRONIC CONFIGURATION

The distribution of an atom's complement of electrons in the energy levels is a critical matter. It will be seen that the chemical and

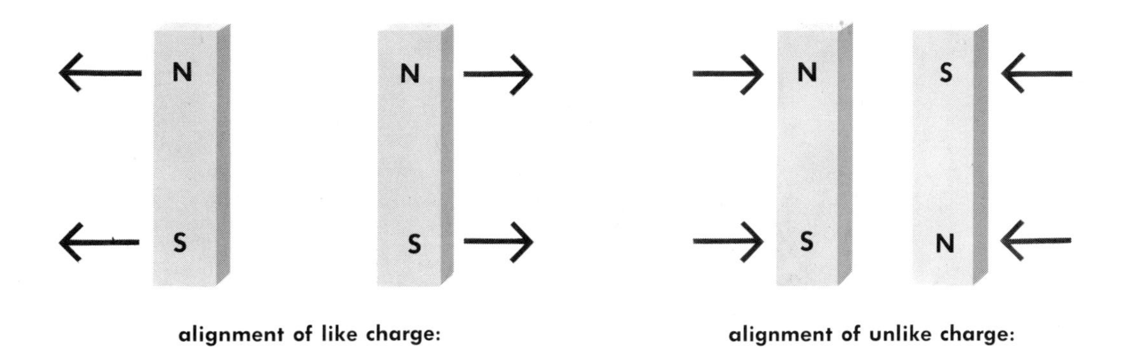

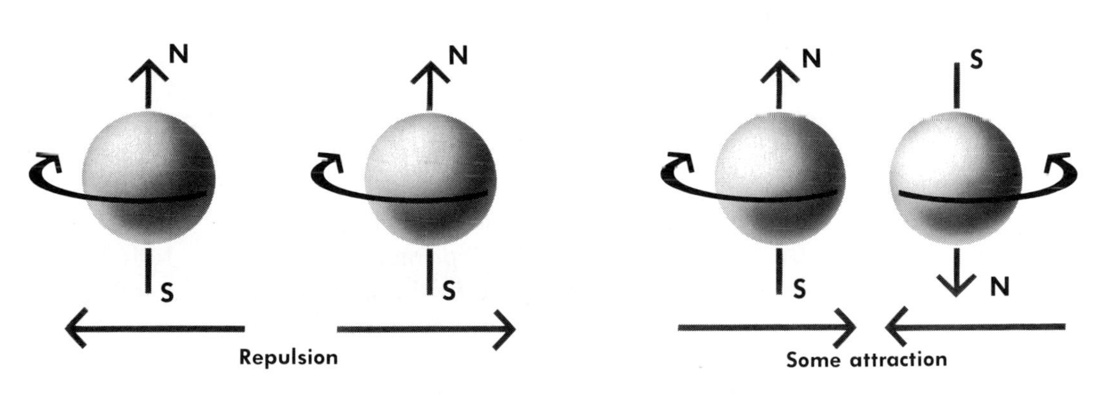

Figure 5–17 Electron spin and magnet analogy.

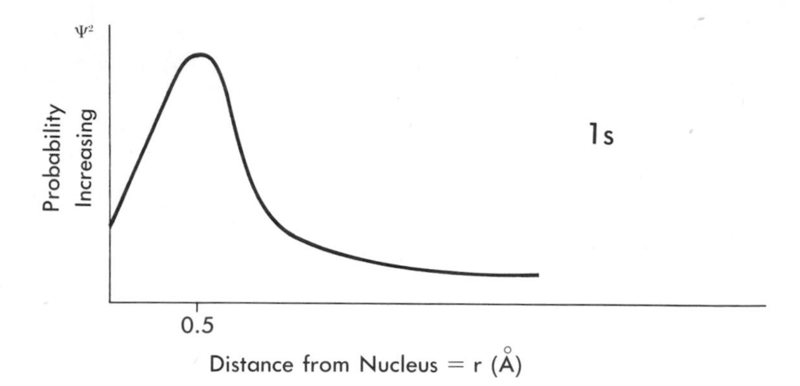

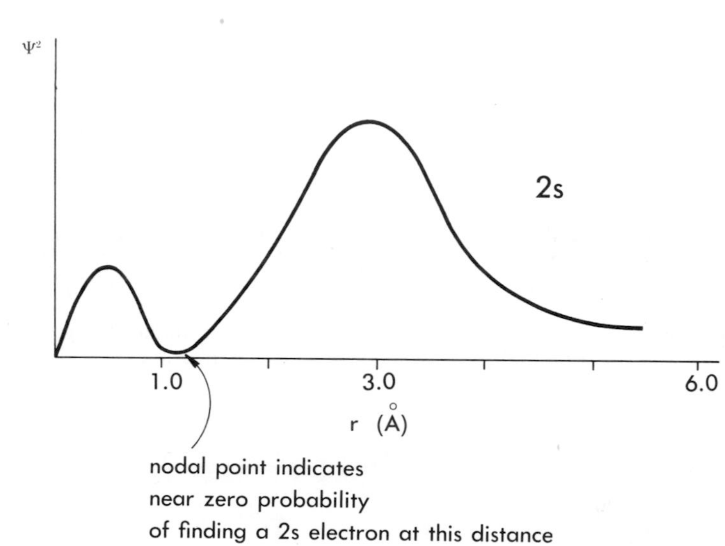

Figure 5-18 Probability distribution plots of various orbitals.

(Figure 5-18 continues on the opposite pag

physical properties of the elements depend on how many electrons are in the various energy levels (classically referred to as **shells**). The determination of the number of electrons in the *outermost* shell, or the highest energy level, is of primary importance. It is the outermost electrons that are involved in the sharing and transferring activities of atoms when chemical bonds are formed among atoms.

The order of mentally "building" the structure of atoms by the distribution of the electrons follows what is called the **aufbau principle.** The aufbau principle says that electrons are arranged by assigning the first ones to the lowest energy levels and then moving toward higher energy levels in known order until all the electrons are distributed among the orbitals.

It is important to note that the order of "filling" the orbitals does not proceed by simply completing one energy level and then moving to the next higher energy level. The Bohr formula for maximum energy level population, $2n^2$, could be misleading, because there are other considerations. Table 5-4 illustrates the maximum electron populations according to the formula.

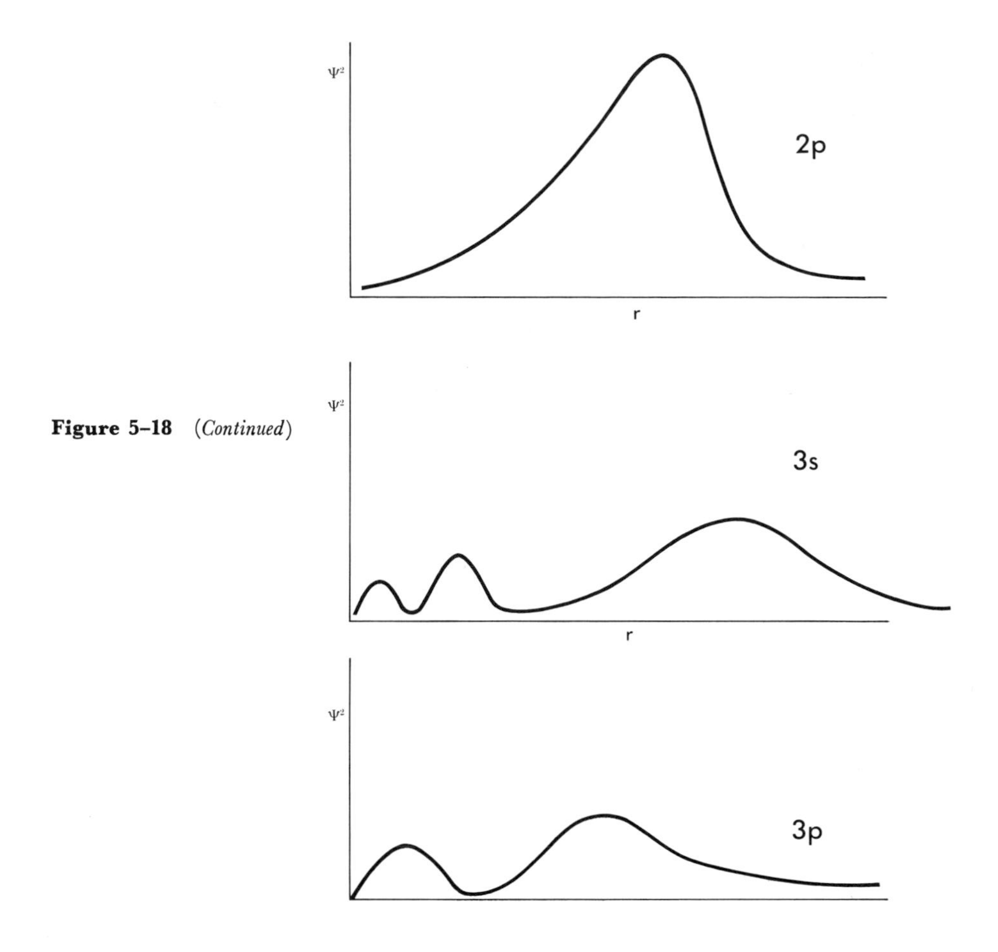

Figure 5–18 (*Continued*)

In point of fact, the total energy required for an electron to occupy a *d* or *f* orbital of a lower principal energy level may be *higher* than the energy of an electron in the *s* or *p* orbital of the principal energy level above it. For example, referring to Figure 5–14, it can be seen that the total energy of a 3*d* electron is *more* than that of a 4*s* electron. Figure 5–19 illustrates a kind of "road map" for distributing electrons according to the aufbau principle. The arrows indicate the proper order of electron filling.

TABLE 5–4

Energy Level, n	$2n^2$	Maximum Electron Population
1	$2(1)^2$	2
2	$2(2)^2$	8
3	$2(3)^2$	18
4	$2(4)^2$	32
5	$2(5)^2$	50

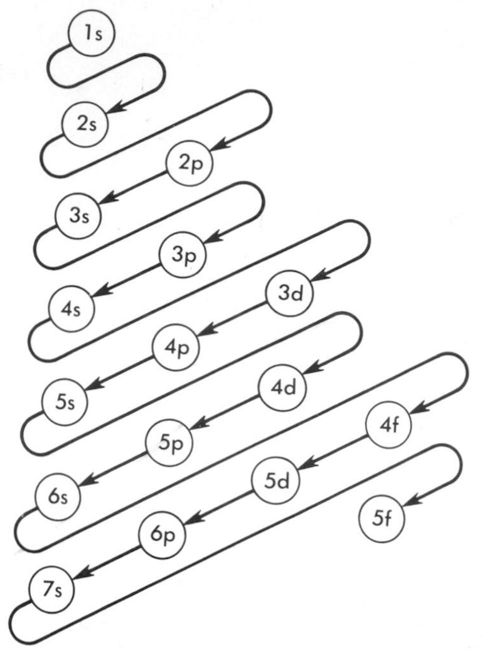

Figure 5–19 The sequence of filling atomic orbitals in atom "building."

The method used to describe the electronic configuration of atoms is to write the principal energy level number (n), then the orbital letter designation (*s*, *p*, *d*, or *f*), and finally a superscript—not to be confused with an exponent—to indicate the number of electrons in that orbital. For example,

$$3d^7$$

means third energy level, *d* orbital, occupied by 7 electrons. The first ten elements of the periodic table are illustrated in Table 5–5.

A very helpful device for visualizing the distribution of electrons

TABLE 5–5 THE ELECTRONIC CONFIGURATION OF THE FIRST TEN ELEMENTS

Symbol and Atomic Number	Element	Electronic Configuration
$_1H$	hydrogen	$1s^1$
$_2He$	helium	$1s^2$
$_3Li$	lithium	$1s^2\ 2s^1$
$_4Be$	beryllium	$1s^2\ 2s^2$
$_5B$	boron	$1s^2\ 2s^2\ 2p^1$
$_6C$	carbon	$1s^2\ 2s^2\ 2p^2$
$_7N$	nitrogen	$1s^2\ 2s^2\ 2p^3$
$_8O$	oxygen	$1s^2\ 2s^2\ 2p^4$
$_9F$	fluorine	$1s^2\ 2s^2\ 2p^5$
$_{10}Ne$	neon	$1s^2\ 2s^2\ 2p^6$

in more detail is to make a **pictorial representation.** This method uses a *platform* ___, *box* □, or *oval* ○ for each orbital, and the electrons are illustrated by *arrows* pointing up and down, ⇅, to represent the "spin" direction.

An important rule governing the pictorial representation is known as **Hund's rule,** or the **principle of maximum multiplicity,** which says that *in the filling of the* p, d, *and* f *orbitals there must be at least one electron in each probability region before pairing begins* for the atoms in the ground state. This means that in regions of equivalent probability, electrons require less energy to occupy orbitals singly than they do to pair off with those of opposite "spin."

Examine nitrogen as an example of pictorial representation and Hund's rule.

$_7N$ $1s^2$ $2s^2$ $2p^3$ electronic configuration

⇅	⇅	↑	↑	↑
		x axis	y axis	z axis

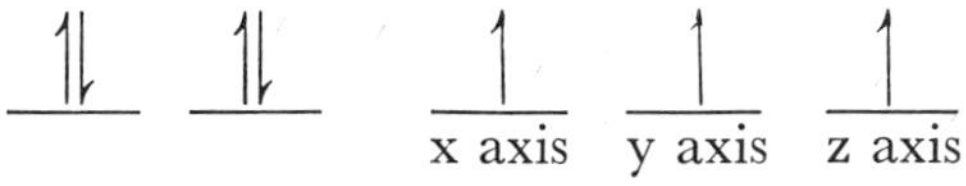

correct pictorial representation

It would be wrong, according to Hund's rule, to write:

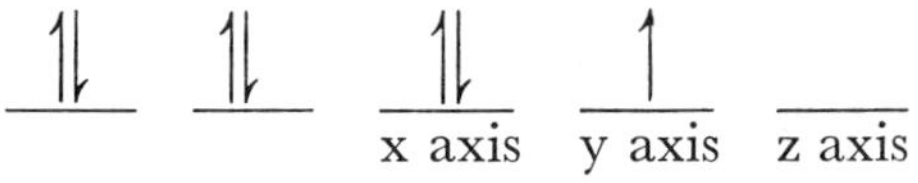

⇅	⇅	⇅	↑	
		x axis	y axis	z axis

(The pairing of electrons in the p_x orbital should not occur until there is at least one electron in each of the three equivalent p orbitals.)

It might also be helpful to visualize the distribution of nitrogen's electrons in a simulated three-dimensional model as shown in Figure 5–20.

Consider the electronic configuration and pictorial representation of a larger atom, iron. The atomic number, 26, indicates the distribution of 26 electrons according to the aufbau principle. It may be useful to employ the orbital "road map" shown in Figure 5–19.

$_{26}Fe$	$1s^2$	$2s^2$	$2p^6$	$3s^2$	$3p^6$	$4s^2$	$3d^6$
	⇅	⇅	⇅	⇅	⇅	⇅	⇅
			⇅		⇅		↑
			⇅		⇅		↑
							↑
							↑

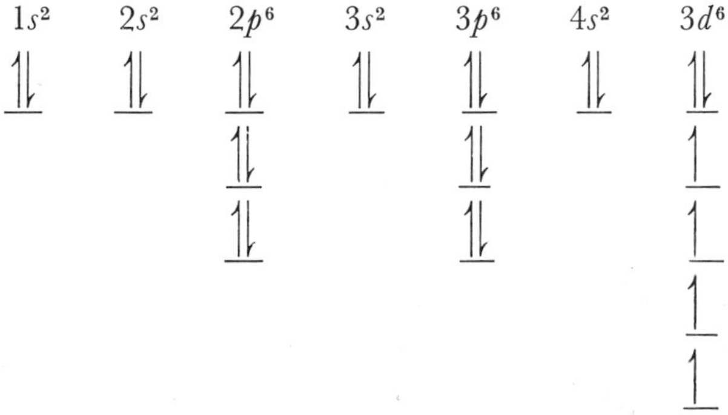

Notice the application of Hund's rule in the $3d$ orbital.

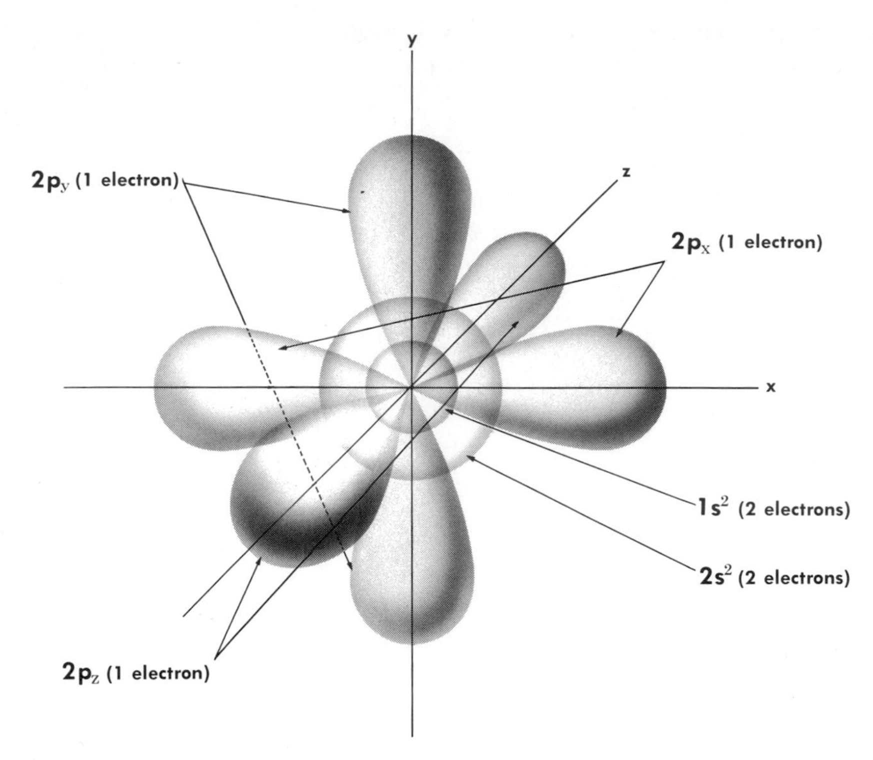

Figure 5–20 Three-dimensional model of electron distribution in nitrogen.

In order to avoid the necessity of drawing the many arrows required by the larger atoms, a commonly used short cut involves the **noble gases** that are found at the end of each **period** (horizontal row) on the periodic table of the elements. Iron, for example, is in the 4th period. The noble gas that marks the end of the period *above* iron is argon. Therefore, the first 18 electrons of iron, which is the total number of electrons for argon, is replaced by the notation "argon core." The pictorial representation for iron may then be written as follows:

$_{26}Fe$ $1s^2$ $2s^2$ $2p^6$ $3s^2$ $3p^6$ | $4s^2$ | $3d^6$
Ar core | ⇅ | ⇅ ↑ ↑ ↑ ↑

EXAMPLE 5.3

Write the electronic configuration and pictorial representation of sulfur.

$_{16}S$ $1s^2$ $2s^2$ $2p^6$ | $3s^2$ | $3p^4$
Ne core | ⇅ | ⇅ ↑ ↑

EXAMPLE 5.4

Write the electronic configuration and pictorial representation of cadmium.

$_{48}Cd\ 1s^2\ 2s^2\ 2p^6\ 3s^2\ 3p^6\ 4s^2\ 3d^{10}\ 4p^6$ | $5s^2$ | $4d^{10}$

Kr core | ↑↓ | ↑↓ ↑↓ ↑↓ ↑↓ ↑↓

EXAMPLE 5.5

Write the electronic configuration and pictorial representation of a zinc (II) ion.

1. The zinc (II) ion is a zinc atom that has lost two electrons, so the net electrical charge on a zinc ion is 2+.

2. The number of electrons to be distributed is 2 less than the atomic number. The number of electrons to be distributed is therefore 28 electrons:

$_{30}Zn^{2+}\ 1s^2\ 2s^2\ 2p^6\ 3s^2\ 3p^6$ | $4s^2$ | $3d^8$*

Ar core | ↑↓ | ↑↓ ↑↓ ↑↓ ↑ ↑

Exercise 5.2

Write the electronic configurations and pictorial representations of the following species:

a. oxygen atom
b. scandium atom
c. barium atom
d. bismuth atom
e. iodine atom
f. lead atom
g. Ca^{2+} ion
h. F^- ion
i. Mn^{2+} ion
j. arsenic atom

5.10 GENERALIZATIONS ON THE PERIODIC TABLE

One of the most significant observations that can be made from the electronic configurations of the atoms is that the *s* and *p* orbitals of a lower energy level are filled to the maximum before electrons begin occupying a higher energy level. The total number of electrons in the *s* and *p* orbitals is *eight.* While this "octet" portends no special magic, it is nevertheless characteristic of the outer energy level of the remarkably stable noble gases. Since the changes in matter that lead toward

* In reality, the two electrons lost by the zinc atom will be from the 4*s* orbital. The explanation, however, will be more appropriately presented in Chapter 6, Chemical Bonds.

increased stability (and lower energy) appear to be a fundamental driving force behind chemical change in general, the "octet' configuration of the noble gases will be a focal point in discussing the nature of the chemical bond in the next chapter.

Just as the "octet" arrangement of the outer energy level characterizes the noble gases, the electron distributions in the outer shells of other related groups of atoms bear directly on their physical and chemical properties. One of the most useful tools available to chemists to help them see the orderly relationships among the atoms, and in a larger sense among the elements they compose, is the periodic chart. The periodic chart, or the periodic table, which can be related to the structure of the atoms, presents a vast amount of fundamental information in a highly organized format.

The table itself is an expression of the **periodic law,** originally presented in its familiar form by Dmitri Mendeleev and later modified as a result of the work of Henry Moseley. The law says, in effect, that *the properties of the elements, when arranged according to increasing atomic number, differ until a noble gas appears, and then the same properties (with some modification) appear again.* This is what chemists call **periodicity**—hence the title periodic table. An overview of the entire chart, shown in Figure 5–21, points up questions to be discussed in more detail.

Dmitri Mendeleev

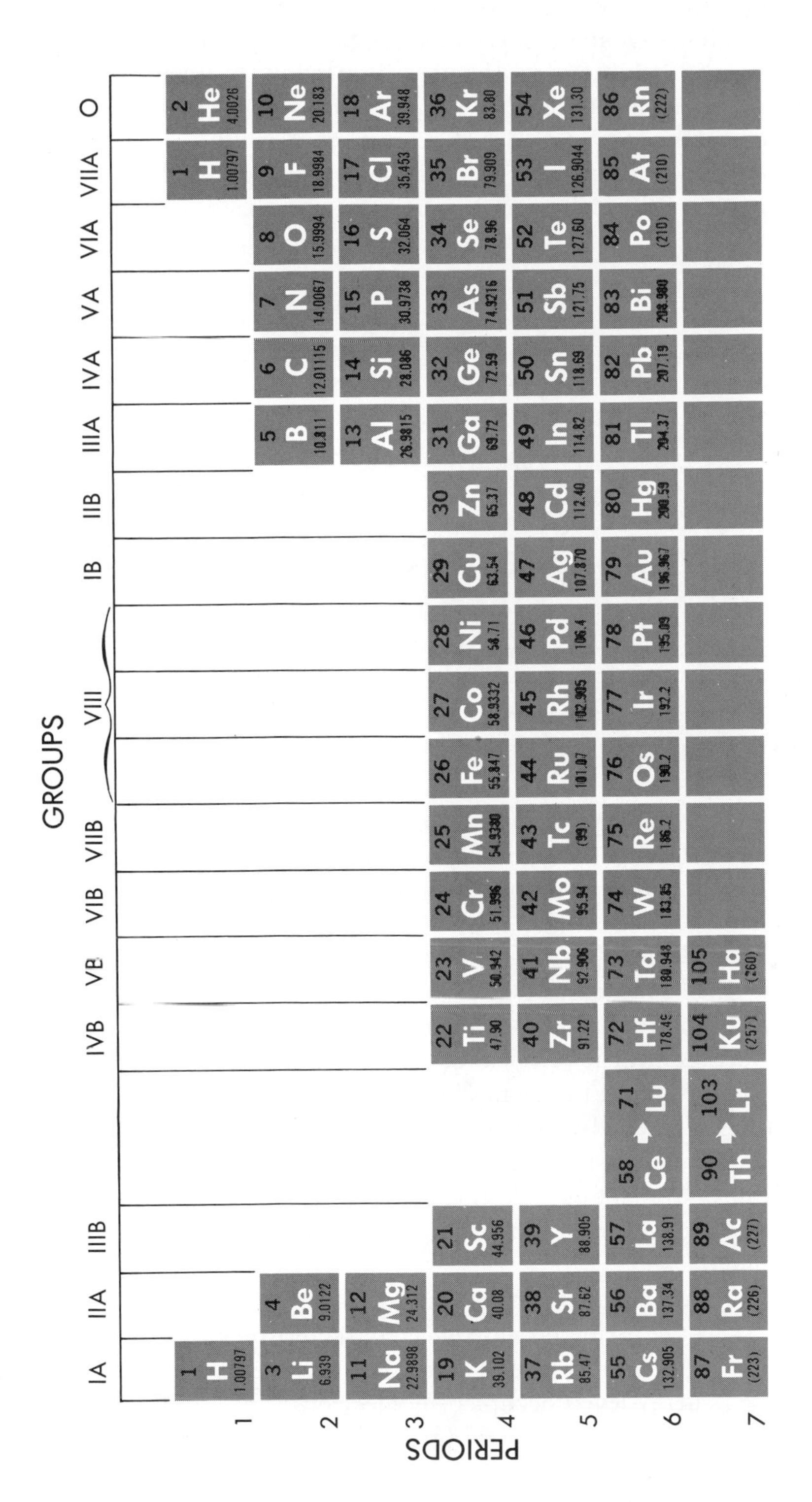

Figure 5-21 The periodic table. (From *Theoretical Inorganic Chemistry* by Day and Selbin © 1969 by Litton Educational Publishing, Inc. Reprinted by permission of Van Nostrand Reinhold Company.)

The periodic chart, in much the same way as a map, can be divided into convenient regions based on similarities and contrasts. The fullest advantage of the chart is gained by making broad generalizations in addition to detailed inspection of particular elements. Broad generalizations provide the perspective that shows regional relationships, and the detailed inspection yields instant information about the distribution of electrons in the energy levels. The following diagrams of the chart are designed to emphasize the broad generalizations. Figure 5–22 illustrates the principal regions of the periodic table.

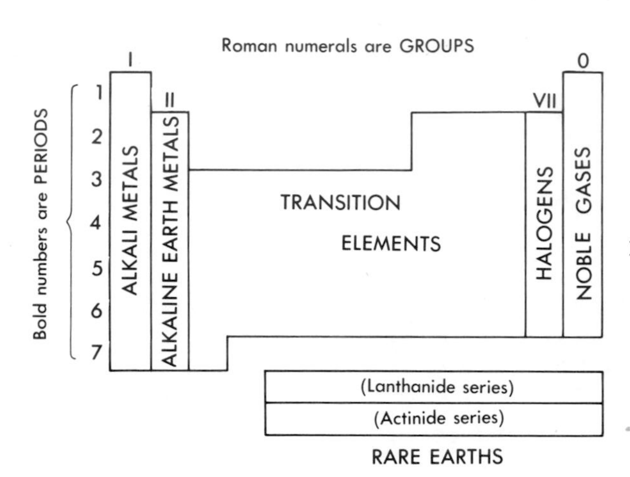

Figure 5–22 The "geography" of the periodic table.

From this point on, the special section labeled **rare earths** will be omitted. This is a unique group of very similar elements that have limited practical value in an introductory chemistry laboratory.

The Roman numerals at the top of the chart indicate **groups,** while the subheadings of A or B indicate the related **families** that subdivide the group. The elements within a family often have marked similarities in their physical and chemical properties.

For example, the metals in group IA, the **alkali metals,** tend to be similar in physical appearance and chemical behavior. The key to this similarity lies in the number of electrons in the highest energy level (Table 5–6). Not only do the alkali metals all have one electron in the highest energy level, but several other significant relationships should be noted:

1. Each alkali metal occurs immediately after a noble gas. This is the periodic recurrence characteristic of the chart.
2. The number of the *period* is the same as the total number of energy levels occupied by electrons.
3. The Roman numeral (group I) agrees with the usual charge on the alkali metal ions.

Group VII, the halogens, provide an interesting contrast. Although the physical properties of the decidedly nonmetallic halogens are

TABLE 5–6 THE PRINCIPAL ENERGY LEVEL DISTRIBUTION OF ELECTRONS IN THE ALKALI METALS

Period	Element		Energy Level Electron Distribution						
2	lithium	$_{3}Li$	2)	)①					
3	sodium	$_{11}Na$	2)	8)	)①				
4	potassium	$_{19}K$	2)	8)	8)	)①			
5	rubidium	$_{37}Rb$	2)	8)	18)	8)	)①		
6	cesium	$_{55}Cs$	2)	8)	18)	18)	8)	)①	
7	francium	$_{87}Fr$	2)	8)	18)	32)	18)	8)	)①

(francium is too rare to have been studied in detail)

not as similar as those of the alkali metals, their chemical behavior is much alike. Astatine is eliminated because it is rare and not yet well described. Both astatine and francium (group I) isotopes produced in the laboratory undergo rapid radioactive decay. Table 5–7 compares the principal energy levels of the halogens.

In addition to the observation that each halogen has seven electrons in the highest energy level, the following points are noted again.

1. Each halogen occurs on the chart just *before* a noble gas.
2. The number of the period is still the same as the number of principal energy levels occupied.
3. The Roman numeral (group VII) points to the number of electrons in the highest energy level and indirectly indicates the number of electrons needed for the completion of the "octet"; when the halogens gain the one lacking electron, this determines the common charge on the ion, 1−.

TABLE 5–7 THE PRINCIPAL ENERGY LEVEL OF ELECTRONS IN THE HALOGENS

Period	Element		Energy Level Electron Distribution				
2	fluorine	$_{9}F$	2)	)⑦			
3	chlorine	$_{17}Cl$	2)	8)	)⑦		
4	bromine	$_{35}Br$	2)	8)	18)	)⑦	
5	iodine	$_{53}I$	2)	8)	18)	18)	)⑦
6	astatine	$_{85}At$	(not readily available)				

The boldface numbers to the left on the periodic chart represent the periods, as mentioned before. The main characteristic within a period is *change*; this starts with metals appearing among the families on the left and gradually proceeds toward nonmetal families on the right. As one moves downward on the chart, the atoms are observed to be larger. Figure 5–23 compares the increasing size of the alkali metal atoms as the period increases from 2 to 6. The outermost electrons on the larger atoms are more easily "lost" (usually via the attractive force exerted by nonmetallic elements), since they are furthest from their own nuclei. This tendency to "lose" electrons (oxidation) is a characteristic of metals.

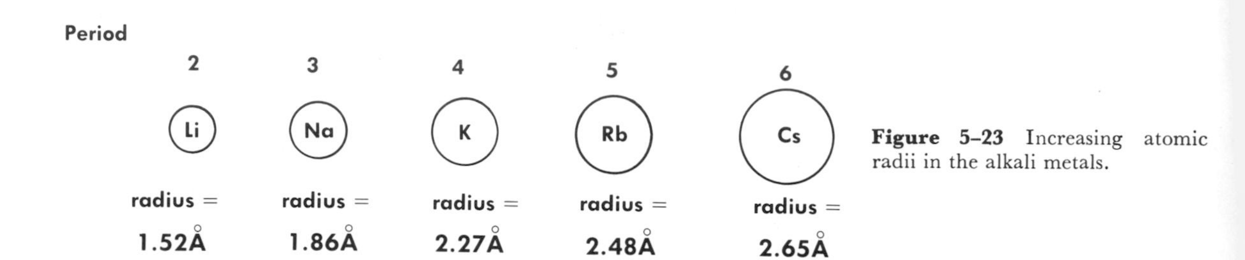

Figure 5–23 Increasing atomic radii in the alkali metals.

5.11 IONIZATION ENERGY

The amount of energy required to separate an electron from an isolated atom is called the **ionization energy.** A table of ionization energies of selected elements is found in Table 5–8. Notice the great difference between the ionization energies of group I and group VII elements. A classic example is found in the comparison between sodium (group I metal) and chlorine (group VII nonmetal).

Both sodium and chlorine have electrons occupying three energy levels. However, sodium has only an 11-proton positive charge holding the single electron in the 3rd shell against a repulsive force of the 10 electrons that occupy the two inner shells. Compare this to chlorine, which has a 17-proton attractive force for the 3rd shell electrons against the same 10-inner-electron force of repulsion. This recitation of the structural differences is summarized pictorially in Figure 5–24.

Hydrogen is a unique element. Although it is placed in group IA, it is definitely not an alkali metal. Its reason for being there is the fact that it has one electron in its normal energy level. However, hydrogen needs only two electrons for the completion of its outer shell, so there is no concern for an "octet" arrangement. The single electron is also held very strongly by its single-proton nucleus. This means that hydrogen's ionization energy is considerably higher than that of the alkali metals.

The transition elements are another special case. It was mentioned before that the progression across a period from group I toward

TABLE 5-8 IONIZATION ENERGIES OF GASEOUS ATOMS (KCAL MOLE^{-1} AND eV)

Atomic Number	Element	First Electron eV	First Electron kcal	Second Electron eV	Second Electron kcal	Third Electron eV	Third Electron kcal
1	H	13.6	314				
2	He	24.6	567	54.4	1254		
3	Li	5.4	124	75.6	1744	122	2823
4	Be	9.3	215	18.2	420	154	3548
5	B	8.3	191	25.2	580	37.9	875
6	C	11.3	260	24.4	562	47.9	1104
7	N	14.5	335	29.6	683	47.4	1094
8	O	13.6	314	35.2	811	54.9	1267
9	F	17.4	402	35.0	807	62.7	1445
10	Ne	21.6	497	41.1	947	64.0	1500
11	Na	5.1	118	47.3	1091	71.7	1652
12	Mg	7.6	176	15.0	347	80.1	1848
13	Al	6.0	138	18.8	434	28.4	656
14	Si	8.2	188	16.3	377	33.5	772
15	P	11.0	254	19.7	453	30.2	696
16	S	10.4	239	23.4	540	35.0	807
17	Cl	13.0	300	23.8	549	39.9	920
18	Ar	15.8	363	27.6	637	40.9	943
19	K	4.3	99	32.2	734	47.8	1100
20	Ca	6.1	141	11.8	274	50.9	1181

Figure 5–25 summarizes the generalized characteristics of the periodic table.

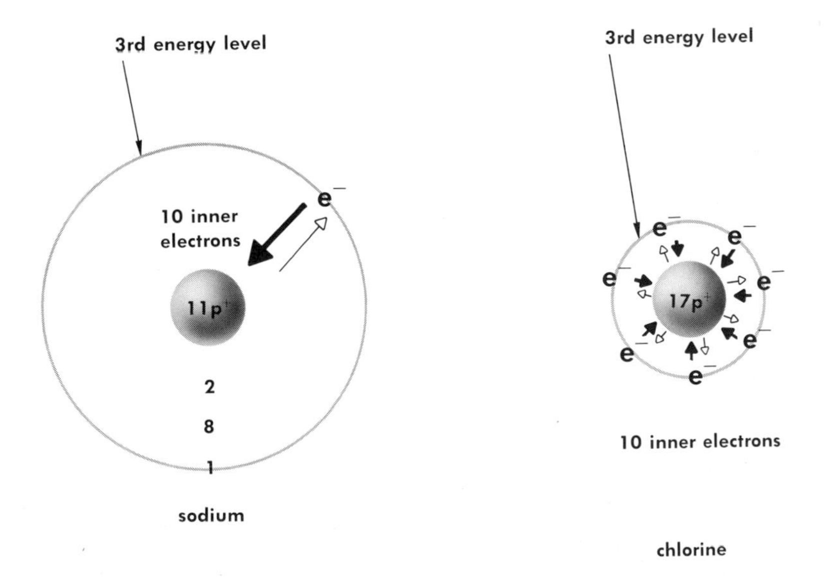

Figure 5–24 Forces of attraction and repulsion on valence electrons.

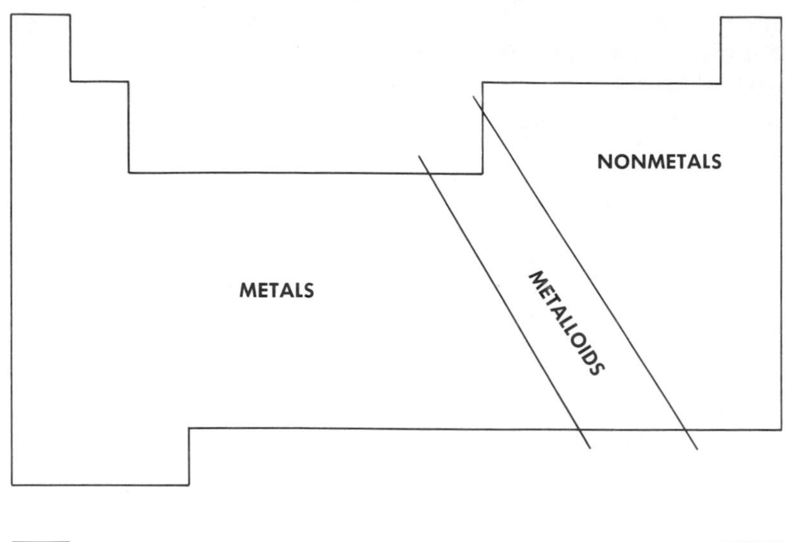

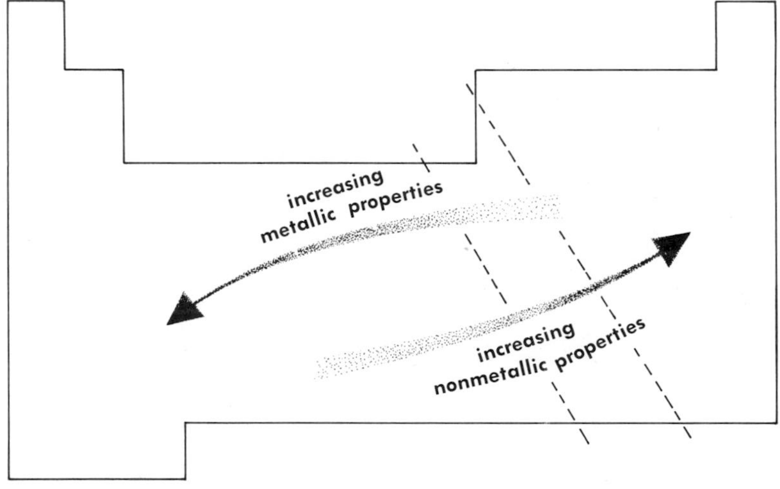

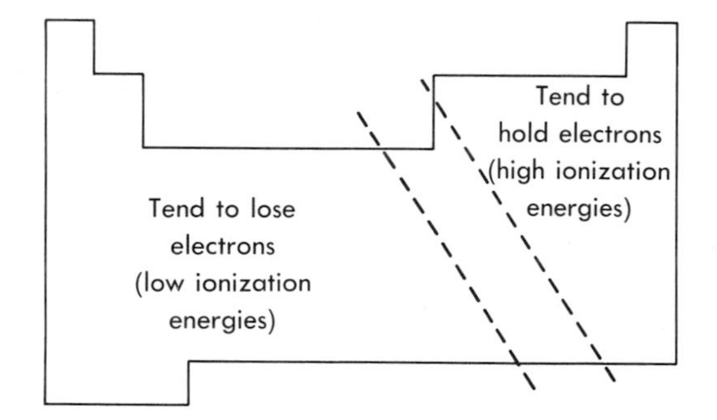

Figure 5–25 Some generalized characteristics of the periodic table.

the noble gases is characterized by change. However, the transition elements have noteworthy similarities within their periods. This apparent contradiction may possibly be explained by the observation that while the transition elements are increasing in the number of electrons as the atomic number increases, the electrons in the valence shells are "shielded" from the nucleus to such an extent that their ionization energies are very similar. Although the properties of an element are mainly dependent on the number of electrons in the highest energy level, the relationship is not quite so clear-cut in the case of the transition elements. Because of this slight difference in energy among the orbitals above the $3p$ level, the so-called valence shell often includes more than one principal energy level.

Another notable characteristic of the transition elements is the variability of the charges on the ions. Because of the slight energy difference between the electrons of two outer energy levels, a transition element such as iron might "lose" either two or three electrons. Experimental evidence indicates that a number of elements form ions of variable charge.

An understanding of the fundamentals of modern atomic structure, and the systematic arrangement of the elements according to the properties that stem from the varieties of electronic configuration will form the basis for probing into the nature of the chemical bond.

QUESTIONS AND PROBLEMS

5.1 What was wrong with the Rutherford planetary model of the atom?

5.2 What is the equipartition principle?

5.3 What model of light evolved from the photoelectric effect?

5.4 What equation did Einstein derive that relates energy and frequency of electromagnetic radiation?

5.5 Calculate the wavelength of a photon having an energy of 2.0×10^{-21} joules.

5.6 State the Bohr frequency rule in words.

5.7 What is a "stationary state" in an atom?

5.8 Calculate the energy, in kcal $mole^{-1}$, of an electron transition in hydrogen atoms where the electrons shift from the 8th to the 2nd energy level.

5.9 What is the energy, in eV, of an emitted photon if its wavelength is 20.0 nm?

5.10 Rearrange the de Broglie equation so that it is ideally set up to solve for the mass of an electron.*

5.11 What was the Compton effect?*

5.12 How did the Heisenberg uncertainty principle serve to modify Schrödinger's contribution?*

5.13 What conflict was resolved to some extent by the principle of complementarity?

5.14 What is the meaning of Schrödinger's psi squared function?

5.15 What is the Pauli exclusion principle?

5.16 Write the electronic configuration and pictorial representation of cobalt.

5.17 What is Hund's rule?

5.18 What is the meaning of periodicity as it relates to the periodic table?

5.19 Why does the metal cesium (group I) have such a low ionization energy?

5.20 Draw a rough outline of a periodic table. Label the groups and periods. Shade the area containing the elements that have the lowest ionization energies.

CHAPTER 6

Science is the century-old endeavor to bring together by means of systematic thought the perceptible phenomena of this world into as thorough-going an association as possible.

A. Einstein, *Out of My Later Years**

Behavioral Objectives

At the completion of this chapter, the student should be able to:

1. Describe chemical bonding by the atomic orbital method.
2. Write electron-dot formulas for atoms and chemically combined atoms.
3. Describe ionic bonds.
4. Define and distinguish between the terms electronegativity and electron affinity.
5. Define and give illustrative examples of oxidation number, oxidation, and reduction.
6. State the rules governing the formation of ionic bonds.
7. Describe covalent bonds.
8. State the octet principle and cite exceptions to it.
9. Explain the use of the concept of resonance.
10. Distinguish between polar molecules and polar bonds.
11. Demonstrate the relationship between electronegativity and percent of ionic character.
12. Predict the degree of polarity of molecules on the basis of the symmetry or asymmetry of covalent bonds.
13. Describe the special characteristics of metallic, network, and hydrogen bonds.
14. Predict the geometry of selected molecules.
15. Identify the type of orbital hybridization associated with the structure of specific molecules.★
16. Sketch a periodic chart so that contrasting regions of electronegativities and ionization energies are illustrated.

---

* New York, Philosophical Library, Inc., 1950.

6.1 THE NATURE OF THE CHEMICAL BOND

A chemical bond may be generally defined as an interaction between atoms or ions resulting in a structure that has greater stability and lower energy than the isolated atoms. A bond is also described as a pair of electrons jointly shared by two atoms. The most probable distance between two bonded atoms is such that the attractive forces between the positively charged nuclei and the negatively charged electron pair are greater than the forces of repulsion between the electrons and between the nuclei. Figure 6–1 illustrates the balanced conflict in forces between two hydrogen atoms in a molecule of hydrogen gas.

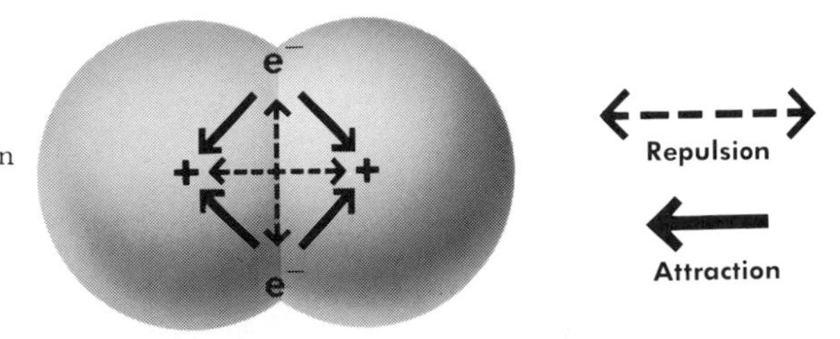

Figure 6–1 The forces of attraction and repulsion in tension between the two atoms of the hydrogen molecule.

The ideal structure of the hydrogen molecule evolves from the loss in energy as two isolated hydrogen atoms approach each other in the drive toward greater stability. Figure 6–2 illustrates the loss in energy as the two atoms move closer and closer together until they stop at an equilibrium distance where the attraction is balanced by a sharply increasing force of repulsion.

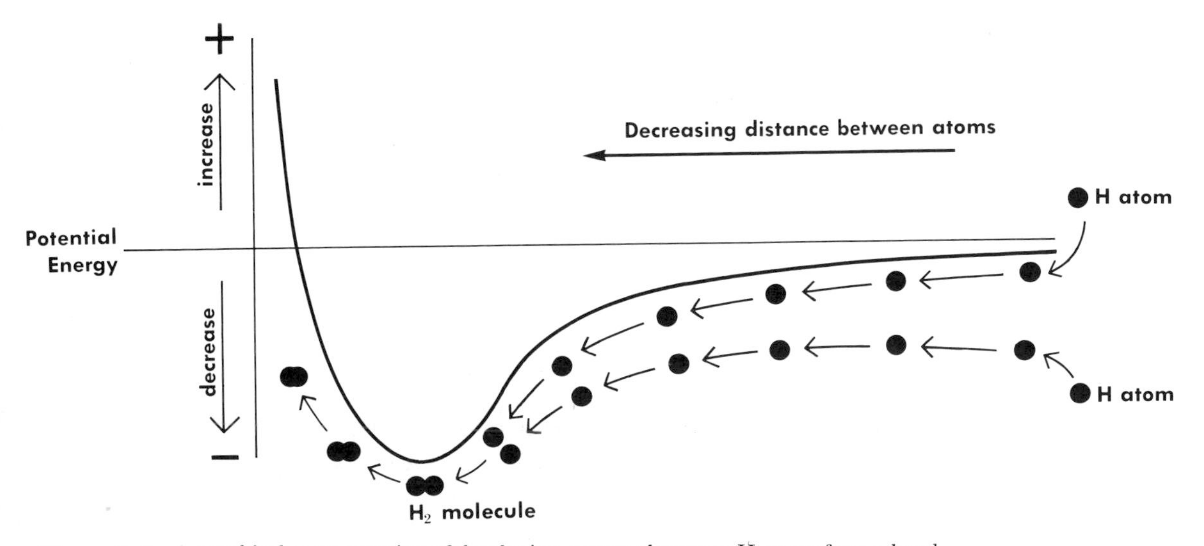

Figure 6–2 A graphical representation of developing energy change as H atoms form a bond.

While the hydrogen molecule illustrates the fundamental nature of the chemical bond, it does not present the various means by which atoms become bonded.

6.2 THE ATOMIC-ORBITAL METHOD OF DESCRIPTION

The **atomic orbital method** is the simplest way of describing chemical bonding. This method focuses on the valence shell of the atom and the tendency of the atom to achieve the "*octet*" configuration of electrons. The special significance of the octet principle is seen in the remarkable stability of the noble gases, all of which, with the exception of helium, have eight electrons in their highest energy level ("valence" shell):

		Number of Electrons in the Valence Shell
$_{2}He$	$1s^2$	
$_{10}Ne$	$1s^2\ \underline{2s^2\ 2p^6}$	8
$_{18}Ar$	$\cdots\ \underline{3s^2\ 3p^6}$	8
$_{36}Kr$	$\cdots\ \underline{4s^2\ 4p^6}$	8
$_{54}Xe$	$\cdots\ \underline{5s^2\ 5p^6}$	8
$_{86}Rn$	$\cdots\ \underline{6s^2\ 6p^6}$	8

The apparent stability of the noble gases lies in the fact that the *s* and *p* orbitals composing their outer energy levels are filled.

The atomic orbital theory of bonding is based on the observation that atoms generally achieve a degree of stability somewhat akin to the stability of the noble gases when they become **isoelectronic** with the noble gases. This means that if an atom can achieve the octet (thus having the same number of electrons in the valence shell as does a noble gas—i.e., isoelectronic), it will gain a marked increase in stability.

The drive toward the octet configuration may involve losing, gaining, or sharing electrons. The specific mechanism used to describe a particular chemical bond depends on the types of atoms involved. The periodic table of the elements is an invaluable aid in the chemist's effort to categorize the elements with regard to the types of bonds they tend to form.

6.3 THE ELECTRON-DOT FORMULA

Before a systematic investigation of types of bonds is presented, it would be advantageous to adopt a symbolism developed by G. N. Lewis. This notation relates the valence shell of atoms to the periodic table groups by representing valence electrons with *dots*. This method is a convenient alternative to elaborate diagrams. The Lewis electron-dot method may be expanded so that the valence electrons of one atom can be distinguished from those of another atom by the use of *dots*, *x's*, and *circles*:

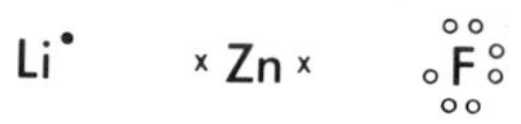

Li· ×Zn× $\overset{\circ\circ}{\underset{\circ\circ}{\circ F \circ\circ}}$

Figure 6–3 shows the electron-dot formulas of a number of atoms in the order in which the elements appear in the periodic table. Note the correlation between the Roman numeral of the group and the number of dots on the symbol.

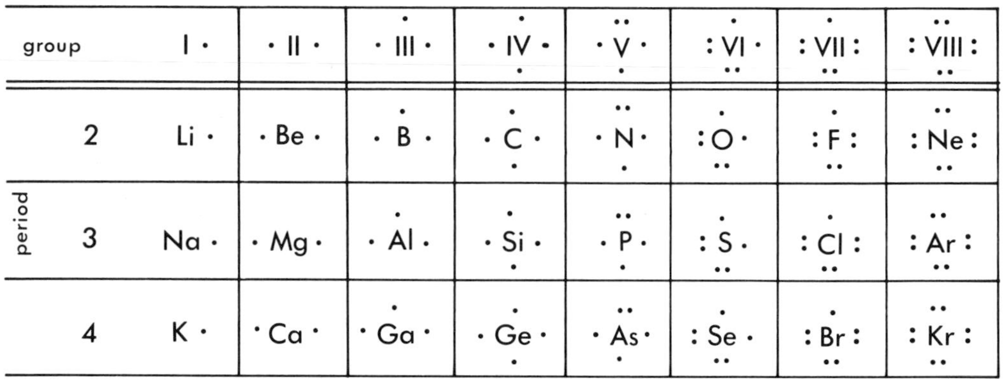

group		I ·	· II ·	· III ·	· IV ·	· V ·	: VI ·	: VII :	: VIII :
period	2	Li ·	· Be ·	· B ·	· C ·	· N ·	: O ·	: F :	: Ne :
	3	Na ·	· Mg ·	· Al ·	· Si ·	· P ·	: S ·	: Cl :	: Ar :
	4	K ·	· Ca ·	· Ga ·	· Ge ·	· As ·	: Se ·	: Br :	: Kr :

Figure 6–3 Electron-dot formulas of selected elements in groups I through VIII.

Notice how the number of dots can be determined from the electronic configuration:

		Number of Electrons in Valence Shell	Electron-Dot Formula
$_{11}$Na	Ne core $3s^1$	1	Na·
$_{20}$Ca	Ar core $4s^2$	2	·Ca·
$_{27}$Al	Ne core $3s^2\ 3p^1$	3	·Al·
$_{6}$C	He core $2s^2\ 2p^2$	4	·C·
$_{15}$P	Ne core $3s^2\ 3p^3$	5	·P·
$_{16}$S	Ne core $3p^2\ 3p^4$	6	·S:
$_{17}$Cl	Ne core $3s^2\ 3p^5$	7	·Cl:

The hydrogen molecule can be represented by an electron-dot formula:

$$\underset{\text{atom}}{H^{\bullet}} + \underset{\text{atom}}{H^{\times}} \qquad \underset{\text{molecule}}{H \overset{\bullet}{\times} H}$$

The placement of the dots near the symbol for the element is purely a matter of convenience and has nothing whatsoever to do with the actual positions of the electrons in the atom. With electron-dot formulas as a useful tool, the various types of chemical bonds can now be investigated.

6.4 IONIC BONDS

Ionic bonds result from the force of attraction between oppositely charged ions. Ease of ionization is a characteristic of metal atoms. Ionic bond formation comes about through the interaction between metal atoms and nonmetallic atoms or groups of nonmetallic atoms. The metal atoms having low ionization energies tend to *transfer* one or two electrons to nonmetals that have a high degree of **electron affinity.** Electron affinity, which implies an *attraction* for electrons, is a property of nonmetals. A nonmetal characteristically has a small radius of negative charge compared to a large positive nuclear charge. Nonmetals may be assigned numerical values that represent the amount of energy liberated when an electron is "captured." This is the source of electron affinity data. Fluorine, which has the largest electron affinity of all elements, is illustrated in the model shown in Figure 6–4.

Just as ionization energies were summarized on the periodic chart in Chapter 5, electron affinities of the elements may be treated similarly (Fig. 6–5).

When a metallic atom having a low ionization energy transfers an electron or two (depending on the number in the valence shell) to a nonmetallic atom of high electron affinity, strongly attracting ions result. These ions are likely to be isoelectronic with noble gases. A typical example of ionic bond formation may be investigated in the production of sodium chloride, NaCl, which is common table salt:

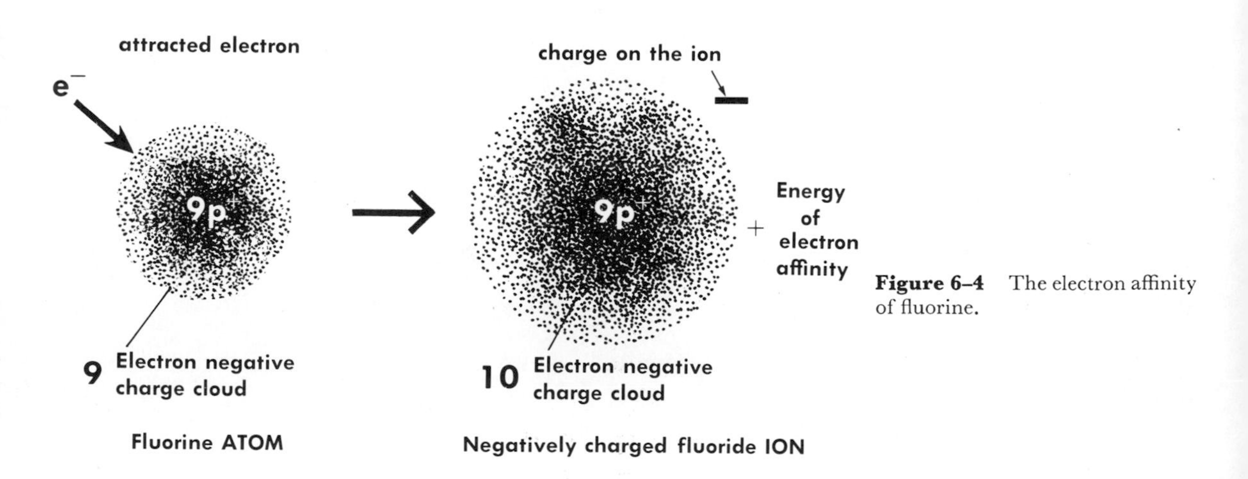

Figure 6–4 The electron affinity of fluorine.

HIGH
electron
affinity

increasing

LOW
electron
affinity

Figure 6–5 The periodicity of electron affinity.

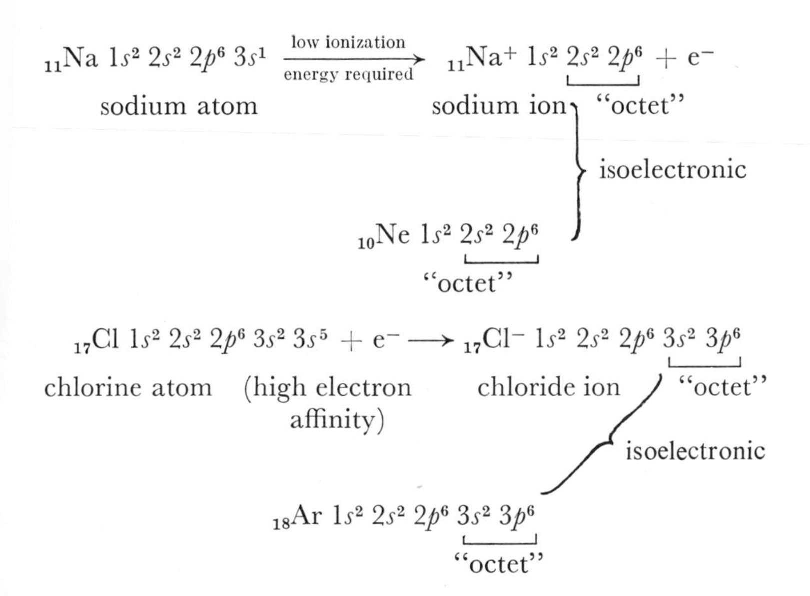

Figure 6–6 illustrates the sodium-chlorine bond formation more diagrammatically.

The sodium ion has 11 protons (units of positive charge) and 10 electrons (units of negative charge). The net difference is 1 positive charge, which is both the charge on the ion and the **common oxidation number,** or **valence.** The chloride ion, by contrast, has a charge (and oxidation number) reflecting the presence of the extra electron.

Oxidation may be defined as a *loss* of electrons. The common oxidation number states the usual number of electrons that an atom *loses* in the formation of a chemical bond. The fact that chlorine, as a typical nonmetal, gains an electron (**reduction**—the opposite of oxidation) is indicated by a *negative* oxidation number. Familiarity with common oxidation numbers is essential for the efficient writing and naming of compound formulas, which will be discussed in the next chapter. The term "valence" is usually taken to be a synonym of the term "oxidation number." While this usage is in common practice, it is not accurate. Valence is a historical method of noting the combining values of atoms. Valences, which are obtained from empirical data, do not have positive and negative signs. For example, the common oxidation number of calcium is 2+ (plus two), but the valence of

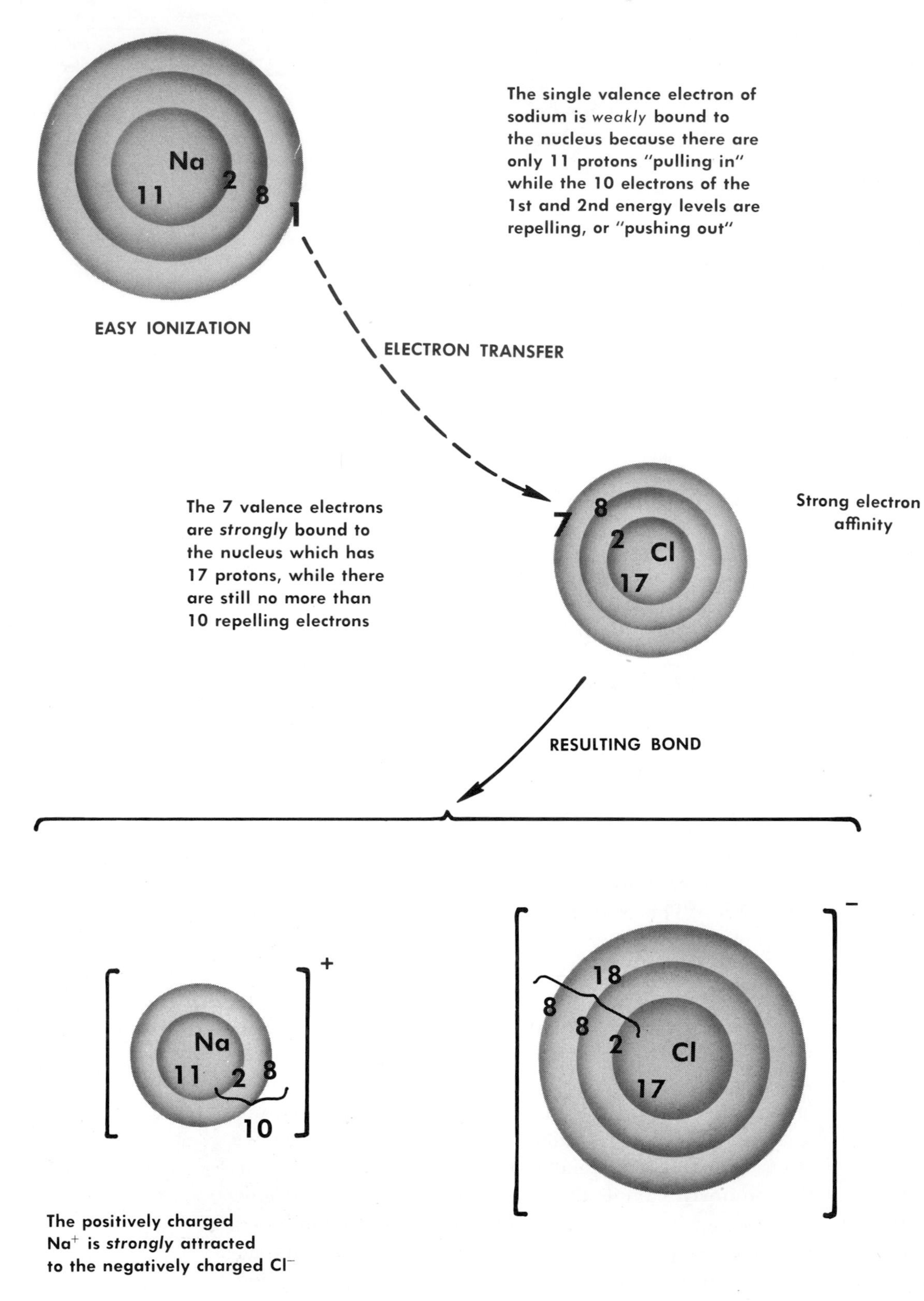

Figure 6–6 Diagrammatic representation of $Na^{+}Cl^{-}$ ionic bond formation.

calcium is 2. Once this distinction is appreciated, the terms "valence" and "common oxidation number" may be used interchangeably. A table of common oxidation numbers for atoms and *polyatomic ions** appears in the next chapter (p. 197). The concept of polyatomic ions will be discussed shortly.

The interaction between sodium and chlorine may be illustrated by the electron-dot method:

transfer

$$\mathrm{Na}\cdot \;+\; \overset{\circ\circ}{\underset{\circ\circ}{\circ\mathrm{Cl}\,{}^{\circ}_{\circ}}} \longrightarrow \mathrm{Na}^{+},\quad \left[\overset{\circ\circ}{\underset{\circ\circ}{\dot{\circ}\mathrm{Cl}\,{}^{\circ}_{\circ}}}\right]^{-}$$

The resulting pair of ions (Na^+, Cl^-) is not meant to suggest the formation of a molecule, because *ionic compounds are not molecular* in structure. A simple (empirical) formula for an ionic compound represents nothing more than the unit structure of a crystal that is a gigantic arrangement of many millions of ions. The **crystal lattice** structure results from the stable alignment of the ions in which positive sodium ions and negative chloride ions surround each other as illustrated in Figure 6–7. The three-dimensional aspect can also be represented by another diagram (Fig. 6–8), where the lines are meant only to show direction and do not indicate specific bonds between atoms.

* A polyatomic ion may be operationally defined as a group of atoms that bears a net electrical charge and functions as a single unit.

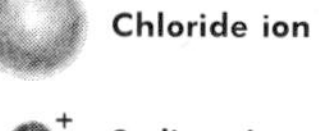

Figure 6–7 The crystal structure model of sodium chloride.

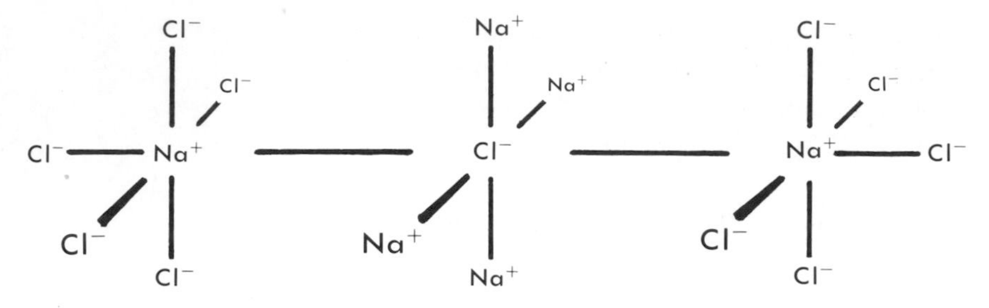

Figure 6–8 The NaCl crystal lattice structure.

The crystal lattice structure for sodium chloride serves as a model for many other ionic compounds, even though the ratios of the ions involved and the basic geometry may vary. It is a matter of convenience to organize compounds into various categories. The rules involved should be considered as "rules of thumb" and exceptions will be expected. In fact, chemists often describe some compounds as **covalent** but with a great deal of ionic character. For purposes of clarity, the examples selected will be very typical members of their type. **Rule:** *ionic bonds are principally formed between metals exhibiting common oxidation numbers of 1+ or 2+, and nonmetals.*

Some selected examples of ionic bonding are illustrated in Table 6–1:

TABLE 6–1

Low Ionization Energy Metal		High Electron Affinity Nonmetal		Electron-Dot Formula
$K^{\times}$	+	$\cdot\ddot{F}:$	→	$K^+, [\overset{\times}{.}\ddot{F}:]^-$
potassium	+	fluorine	→	potassium fluoride
$Li^{\times}$	+	$\ddot{S}:$		$2Li^+, [\overset{\times}{\times}\ddot{S}:]^{2-}$
$Li^{\times}$				
lithium	+	sulfur		lithium sulfide
$Ca\overset{\times}{\times}$	+	$\cdot\ddot{I}:$		$Ca^{2+}, 2[\overset{\times}{.}\ddot{I}:]^-$
		$\cdot\ddot{I}:$		
calcium	+	iodine	→	calcium iodide
$Fe\overset{\times}{\times}$	+	$\cdot\ddot{Br}:$		$Fe^{2+}, 2[\overset{\times}{.}\ddot{Br}:]^-$
		$\cdot\ddot{Br}:$		
iron	+	bromine	→	iron (II) bromide
$Sr\overset{\times}{\times}$	+	$\ddot{S}:$		$Sr^{2+}, [\overset{\times}{\times}\ddot{S}:]^{2-}$
strontium	+	sulfur	→	strontium sulfide

TABLE 6–2

K^+, F^-	positive one and negative one = zero
$2Li^+$, S^{2-}	two ions at positive one and one ion at negative two = zero
Ca^{2+}, $2I^-$	one ion at positive two and two ions at negative one = zero
Fe^{2+}, $2Br^-$	one ion at positive two and two ions at negative one = zero
Sr^{2+}, S^{2-}	positive two and negative two = zero

Notice the net effect of electrical neutrality in every example. **Rule:** *the sum of the positive and negative charges must equal zero in a compound.* Also, the sum of all the oxidation numbers in a compound must equal zero. This rule applies to *all* compounds whether they are ionic or not (Table 6–2).

6.5 COVALENT BONDS

Covalent bonds are formed as pairs of electrons are *shared* between atoms. The extent of the sharing may be unequal, since one atom can have a much stronger attraction for the electron pair, but this is still not the same as the complete transfer of electrons characteristic of ionic bonds.

Most compounds that exhibit covalent bonding involve the atoms and groups of atoms that have oxidation numbers higher than 2+. When the transition elements combine in the 2+ oxidation state, they are usually classified as ionic, but in the 3+ or higher state, they are more likely to be described as covalent (with some degree of ionic character). Another rule of thumb is that *covalent compounds are formed between nonmetallic atoms.* The other types of bonds will be described briefly after the topic of polyatomic ions. A distinguishing characteristic of covalent compounds is that they do exist in the form of distinct molecules.

Some typical examples of covalent compounds are described in Table 6–3. Note that the sharing process produces the "octet" configuration for all the atoms given, with the exception of hydrogen.

It is sometimes difficult to write the electron-dot formulas without additional information. Some molecules have bent shapes, while many others deviate from the octet principle. Some examples are as follows.

Water—H_2O could be illustrated incorrectly without the information that it is not a **linear** molecule:

```
  ..            ..
H°O:H          :O°H
  ..            °.
                H
incorrect    a better formula
```

TABLE 6–3

hydrogen bromide

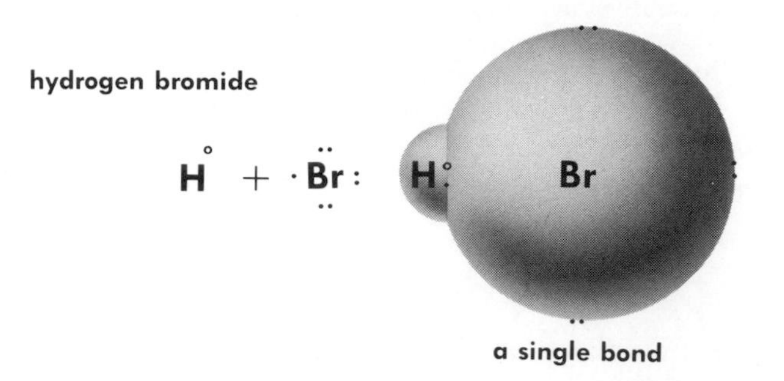

carbon dioxide

$\cdot C \cdot + O : O$

O C O

two double bonds

ethylene

$2 \cdot \dot{C} \cdot + 4H$

H H C C H H

a double bond

acetylene

$2 \cdot \dot{C} \cdot + 2H$

H C C H

a triple bond

Borane—BH_3 apparently does not fit the octet principle. The formula appears to be:

$$\cdot\dot{B}\cdot + H^{\times} \longrightarrow \underset{\overset{\times\cdot}{H}}{H \overset{\times}{\cdot} B \underset{\times}{\cdot} H}$$

But appearances can be deceiving and a more advanced model does satisfy the rule. Borane is not unique. There are many molecules that seem to be exceptions to the octet principle. However, until the atomic orbital method fails to "work" for a vast number of bonds, it is unlikely that it will be discarded for an alternative.

Sulfur Dioxide—SO_2 has been experimentally proven to be a nonlinear molecule in which the distances between the central sulfur atom and the two oxygen atoms are about equal. The formula :O::S:O: is incorrect because it is linear, and the double bond involving sulfur and the oxygen on the left suggests a shorter distance between these atoms than does the single bond between sulfur and the other oxygen atom. This problem is resolved to some extent by the concept of **resonance.**

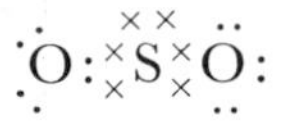

6.6 THE CONCEPT OF RESONANCE

A **resonance hybrid structure** is one which is a blend of the alternative structures that conform to the octet principle. In other words, the resonance concept applies to molecules in which no single electron-dot formula conforms to the octet principle and the dictates of experimental evidence. The resonance hydrid structure for SO_2 is

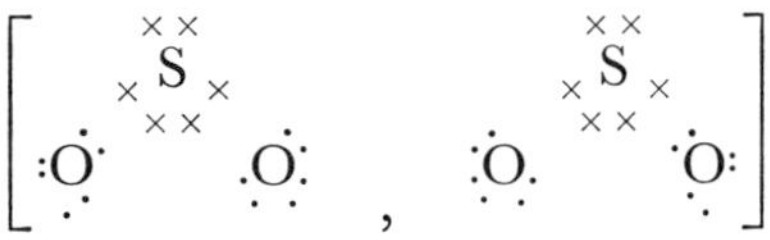

$$\left[\begin{array}{ccc} & S & \\ :O: & & :O: \end{array} , \begin{array}{ccc} & S & \\ :O: & & :O: \end{array}\right]$$

These few selected exceptions (that is, water, borane, and sulfur dioxide) serve to illustrate the fact that the writing of electron-dot formulas is often more than a trial and error manipulation of dots. The section on molecular geometry that appears later in this chapter may help to illuminate this problem further.

6.7 POLYATOMIC IONS

A very special and important aspect of covalent bonding involves the **polyatomic ion**, sometimes referred to by its old name, **radical.** A polyatomic ion is a group of covalently bound atoms that bear an electrical charge because of an imbalance between the total number of nuclear protons and the electrons. While a polyatomic ion and a compound may have the same apparent formula, the compound is electrically neutral and the polyatomic ion is not:

SO_3	SO_3^{2-}
formula for the compound sulfur trioxide	formula for the sulfite ion (polyatomic ion)

Polyatomic ions do not normally exist as independent particles outside of solutions. They are usually found in combination with metal ions that have transferred their electrons. This is what causes the imbalance between the total positive and negative charges. The compounds involving polyatomic ions are often ionic because of the *transfer* of electrons, but the polyatomic ion itself exhibits covalent bonding. Some examples of polyatomic ions are presented in Table 6–4.

TABLE 6–4 EXAMPLES OF POLYATOMIC IONS

Name	Electron-dot Formula	Polyatomic Ion	Ionic Charge
nitrate ion	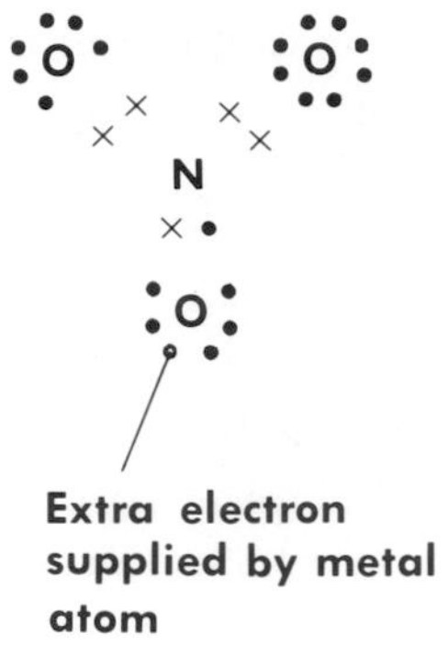O O N O Extra electron supplied by metal atom	NO_3^-	1–
sulfate ion	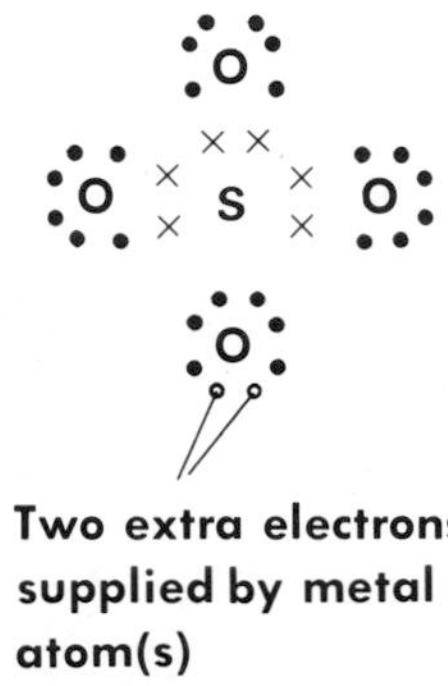O O S O O Two extra electrons supplied by metal atom(s)	SO_4^{2-}	2–
chromate ion	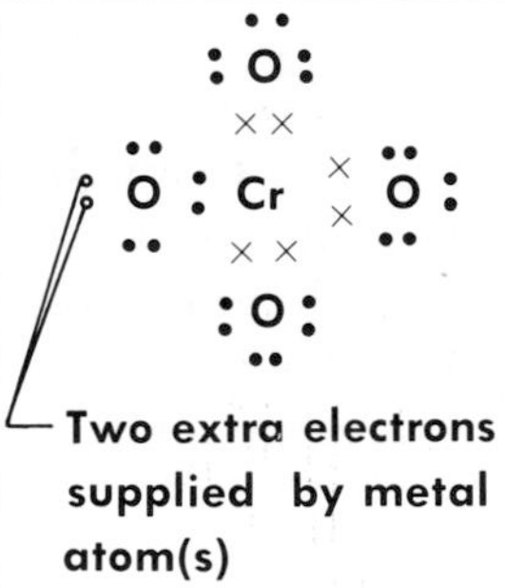O O Cr O O Two extra electrons supplied by metal atom(s)	CrO_4^{2-}	2–
phosphate ion	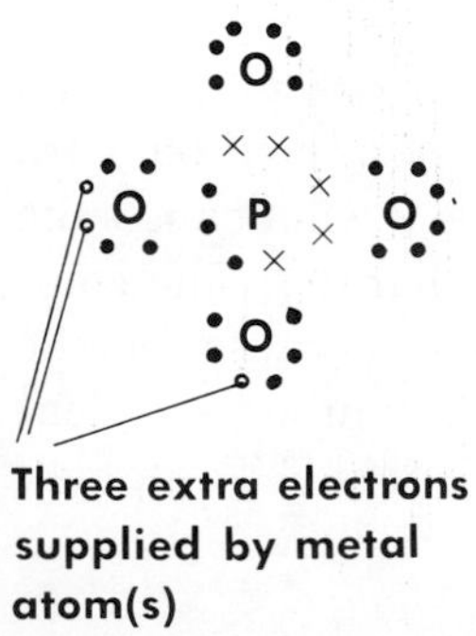O O P O O Three extra electrons supplied by metal atom(s)	PO_4^{3-}	3–

Some examples of compounds involving polyatomic ions are listed in Table 6–5. Note that the sum of the positive and negative charges equals zero in every compound.

TABLE 6–5

Na^{+}, $[NO_3]^{-}$ sodium nitrate	$3K^{+}$, $[PO_4]^{3-}$ potassium phosphate
$2Na^{+}$, $[SO_4]^{2-}$ sodium sulfate	Mg^{2+}, $2[OH]^{-}$ magnesium hydroxide
Ca^{2+}, $2[NO_3]^{-}$ calcium nitrate	Fe^{2+}, $[SO_4]^{2-}$ iron (II) sulfate
$3Ca^{2+}$, $2[PO_4]^{3-}$ calcium phosphate	$2Li^{+}$, $[CO_3]^{2-}$ lithium carbonate

Exercise 6.1

1. Write the electron-dot formulas for the following ionically bonded compounds:

a. potassium bromide
b. strontium chloride
c. cesium fluoride
d. magnesium iodide
e. sodium oxide
f. nickel (II) chloride (·Ni·)
g. lead (II) iodide (·Pb·)
h. lithium nitrate
i. calcium phosphate
j. rubidium sulfate

2. Write the electron-dot formulas for the following covalently bonded compounds:

a. chlorine (Cl_2)
b. hydrogen chloride (HCl)
c. ammonia (NH_3)
d. carbon monoxide (CO)
e. methane (CH_4)
f. silicon dioxide (SiO_2)
g. lead (IV) sulfide (·Ṗḅ·)
h. nitrogen (N_2)
i. hydrogen cyanide (HCN)
j. phosphorus trichloride (PCl_3)

6.8 POLARITY OF MOLECULES

The concept of polarity can be grasped quickly if a bar magnet, our planet Earth, or a simple electrical dry cell is visualized. Each of these examples, illustrated in Figure 6–9, has ends that are described as positive or negative, north or south.

Polarity is a descriptive term that is usefully applied to molecular

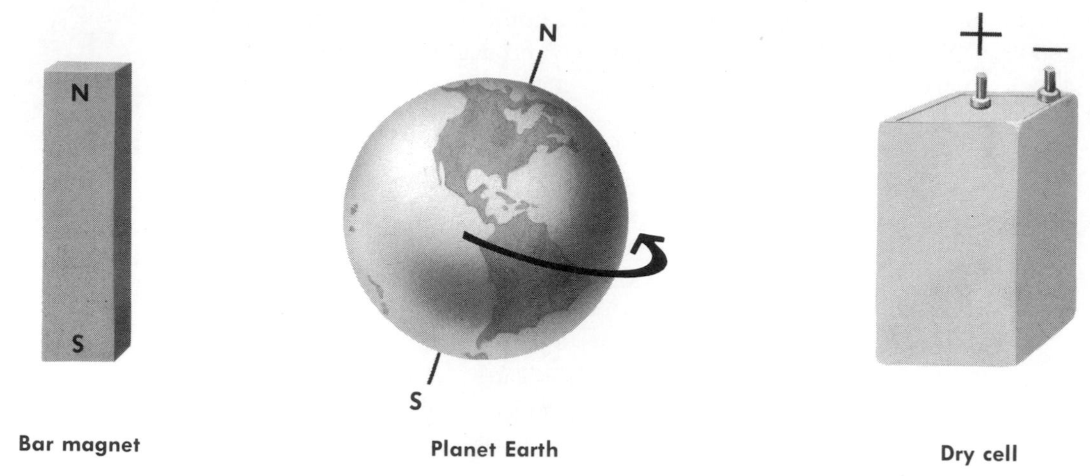

Figure 6–9 The concept of polarity.

structures. A molecule can, and often does, have a physical separation between a region of distinct negative charge and one of relative positive charge. Such a molecule is commonly called a **dipole** (Fig. 6–10). It must be noted that the concept of polarity is only meaningful when applied to molecules and not to ionic crystals. A crystal structure, being composed of many distinctly charged particles (ions), cannot by any stretch of the imagination be thought of as having oppositely charged *ends*.

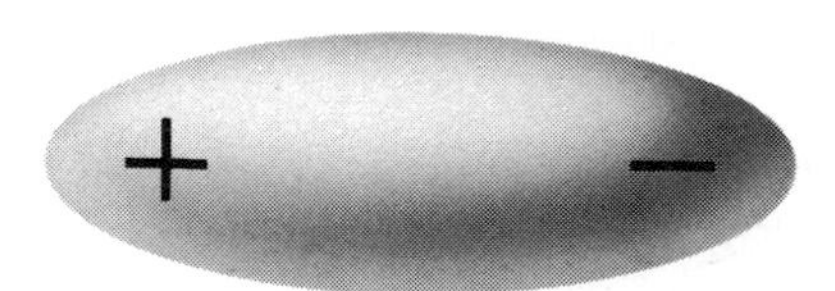

Figure 6–10 A model of a dipole structure.

6.9 DEGREE OF POLARITY

The description of a molecule as highly polar, slightly polar, or nonpolar depends primarily on two factors: *the magnitude of the positive and negative charges* and *the distance* separating the geographical centers of charge. When a molecule has a high degree of polarity, it is also described as having a considerable amount of "ionic character." The usual symbol applied to dipoles to indicate their *partial ionic character* is the small Greek letter delta, δ. The positive region of the molecules is marked δ^+ and the opposite pole, the region of high electron density, is noted as δ^- (Fig. 6–11).

The gradual change in structure observed in compounds with varying degrees of polarity may be visualized by the aid of the models illustrated in Figure 6–12, in which the range of bonds is provided for comparison.

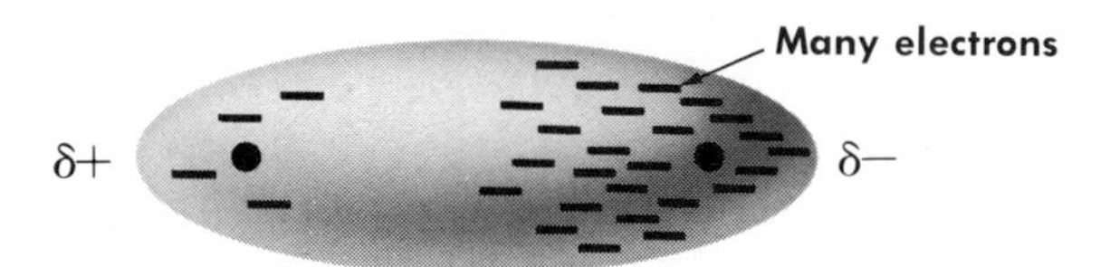

Figure 6–11 Partial ionic character in a molecular model.

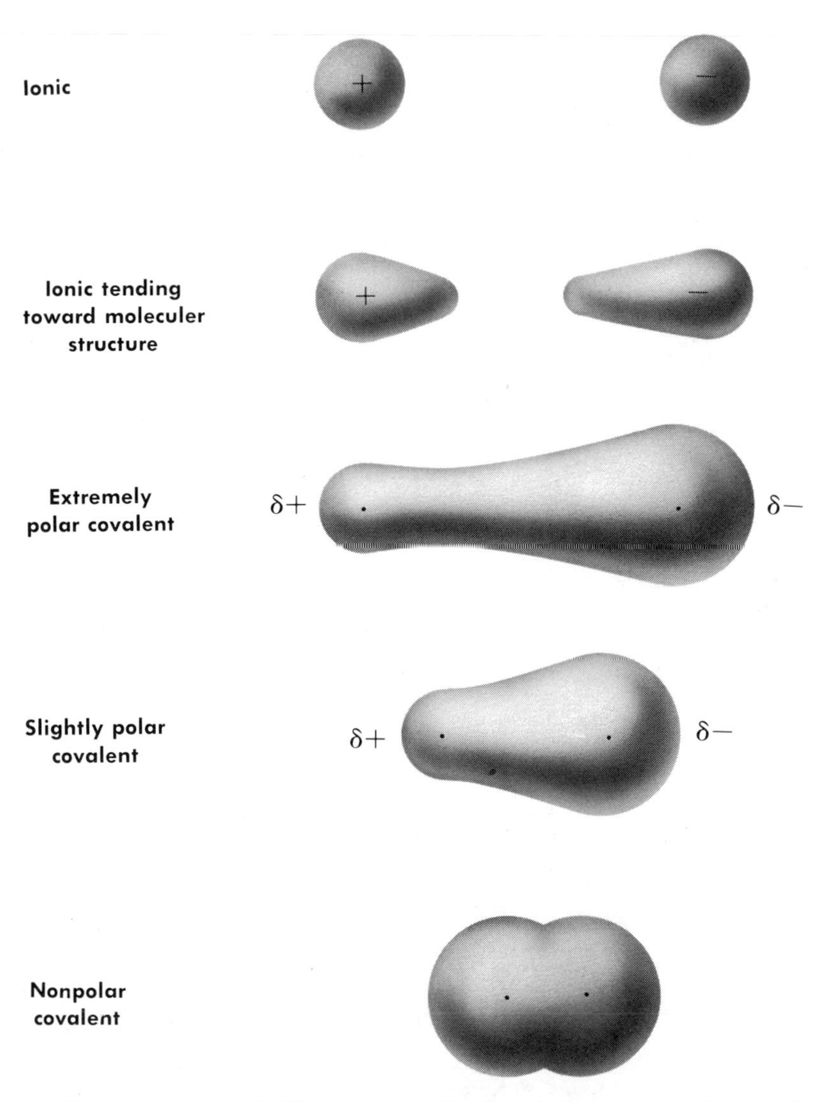

Figure 6–12 Models illustrating structural difference, proceeding from ionic structure to the nonpolar covalent molecule.

6.10 ELECTRONEGATIVITY

A method of determining the polarity and ionic character of covalent molecules is by the use of values assigned to the elements. These values reflect each element's strength of attraction for the pair of electrons that constitutes a chemical bond. This is somewhat similar to electron affinity. However, while electron affinities are measured in terms of the amount of energy released when an electron is "captured" by an atom, **electronegativities** reflect the strength of a chemical bond. The more difficult it is to break a bond, the more ionic its character is likely to be. This is to say that an *unequal* sharing of the pair of electrons forming the bond between two atoms leads to the conclusion that one atom of the pair has a significantly stronger attraction for the electrons than the other one does. The electronegativity values of the elements are predictably higher for the nonmetallic elements which have characteristically high *charge-to-radius ratios*. Observe the comparison between fluorine and lithium in Figure 6–13. From the comparison of lithium and fluorine, both second period elements, it can be seen that fluorine has a very small radius of negative charge that would offer little resistance to the strong pull which the nucleus can exert on a pair of electrons in a chemical bond. By contrast, the slight attraction for an electron pair that a metallic atom such as lithium (with its puny nuclear charge and its wide radius of repelling electrons) has is negligible.

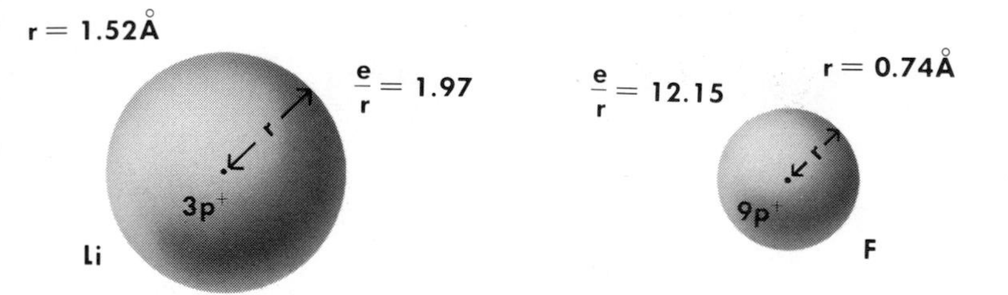

Figure 6–13 Comparison between the charge (e) to radius (r) ratios of Li and F, two atoms in the second period of the periodic table.

The distribution of electronegativity values on the periodic table is generalized in Figure 6–14.

A more precise listing of the values, as developed by Linus Pauling, is given in Figure 6–15. Pauling assigned the electronegativity value of 4.0 to fluorine (the highest), and all the other values represent comparisons to fluorine. The rule of thumb that is most usefully derived from the table of electronegativity values is the following: *the greater the difference between the electronegativity values of the two atoms sharing a pair of electrons, the more ionic is the character of the bond.* Some examples are illustrated in Figure 6–16.

A short summary of the relationship between electronegativity difference and degree of ionic character is given in Table 6–6. From

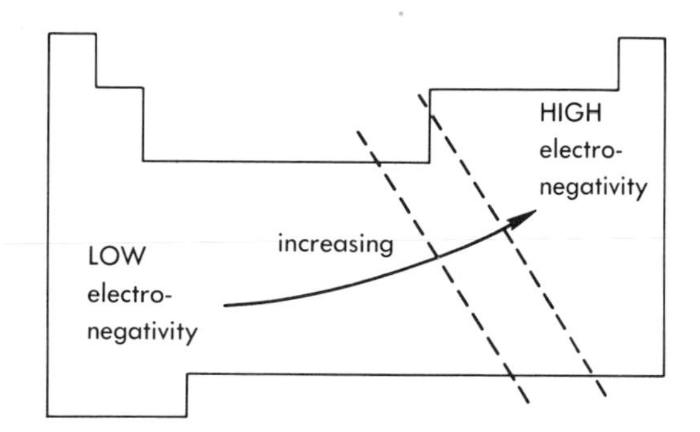

Figure 6–14 Electronegativity trend on the periodic table.

H 2.1																He 2.1
Li 1.0	Be 1.5											B 2.0	C 2.5	N 3.0	O 3.5	F 4.0
Na 0.9	Mg 1.2											Al 1.5	Si 1.8	P 2.1	S 2.5	Cl 3.0
K 0.8	Ca 1.0	Sc 1.3	Ti 1.5	V 1.6	Cr 1.5	Mn 1.8	Fe 1.8	Co 1.8	Ni 1.8	Cu 1.9	Zn 1.6	Ga 1.6	Ge 1.8	As 2.0	Se 2.4	Br 2.8
Rb 0.8	Sr 1.0	Y 1.2	Zr 1.4	Nb 1.6	Mo 1.8	Tc 1.9	Ru 2.2	Rh 2.2	Pd 2.2	Ag 2.4	Cd 1.7	In 1.7	Sn 1.8	Sb 1.9	Te 2.1	I 2.5
Cs 0.7	Ba 0.9	La 1.1	Hf 1.3	Ta 1.5	W 1.7	Re 1.9	Os 2.2	Ir 2.2	Pt 2.2	Au 2.4	Hg 1.9	Tl 1.8	Pb 1.8	Bi 1.9	Po 2.0	At 2.2

Figure 6–15 Electronegativity values of the elements. (After L. Pauling.)

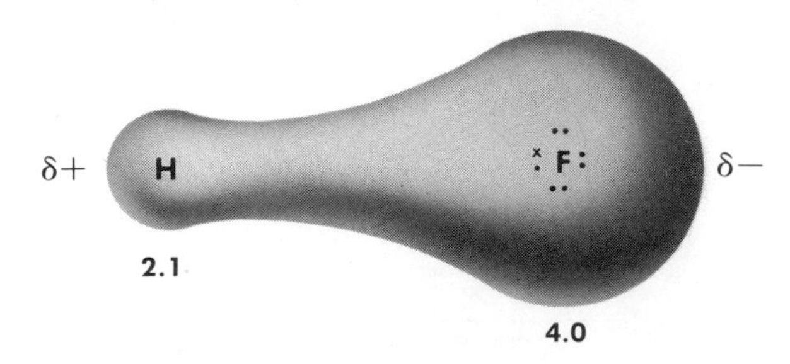

Electronegativity difference is 1.9. HF has a great deal of ionic character.

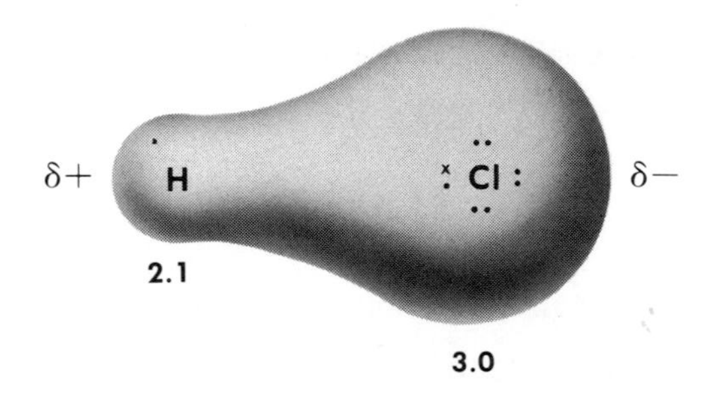

Electronegativity difference is 0.9. HCl is moderately polar.

Figure 6–16 The relationship between electronegativity and the degree of ionic character.

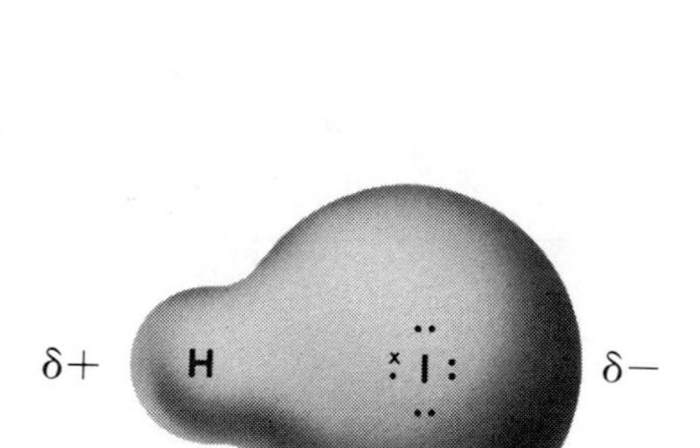

Electronegativity difference is 0.3. HI is very slightly polar and therefore has very little ionic character.

TABLE 6–6

Difference in Electronegativity	Percent of Ionic Character
0.2 to 1.0	1 to 22
1.2 to 1.8	30 to 55
2.0 to 3.2	63 to 92

the information in Table 6–6, it can be concluded that another useful rule of thumb may be applied to the classification of compounds as ionic or covalent. The rule is that *if the electronegativity difference is 1.8 or more, the compound usually may be described as ionically bound.* However, it must be cautioned that rules of thumb are not without exception. For example, while the electronegativity difference between H and F in the hydrogen fluoride molecule is 1.9, the fact is that hydrogen does not form an ionic bond. This apparent deviation from the rule is understood if one recalls the very high ionization energy of hydrogen. Furthermore, some metallic elements in the 1+ or 2+ oxidation state will form ionic bonds with nonmetals even though the electronegativity differences are less than 1.8.

6.11 POLARITY IN MULTIATOM MOLECULES

The fact that molecules composed of more than two atoms exhibit varying degrees of polarity can be explained in terms of the **symmetrical** or **asymmetrical** distribution of all the bonds between pairs of atoms in a polyatomic molecule. For example, methane is composed of 1 carbon atom bonded to 4 hydrogen atoms. The hydrogen atoms are distributed at the four points of a regular tetrahedron, with the carbon atom in the center (Fig. 6–17). Carbon has an electronegativity value that is slightly higher than that of hydrogen (H = 2.1, C = 2.5). This means that the pair of electrons, constituting the shared-pair bond between a carbon atom and a hydrogen atom, will be closer to the carbon atom. Each of the four C—H bonds will be just the same. The result is that the carbon atom will be a center of negative charge, while each of the hydrogen atoms will be an equal-strength region of positive charge. However, the *center* of positive charge will be equidistant from each hydrogen atom, which places this center right at the carbon atom.

Figure 6–17 The structure of the methane molecule.

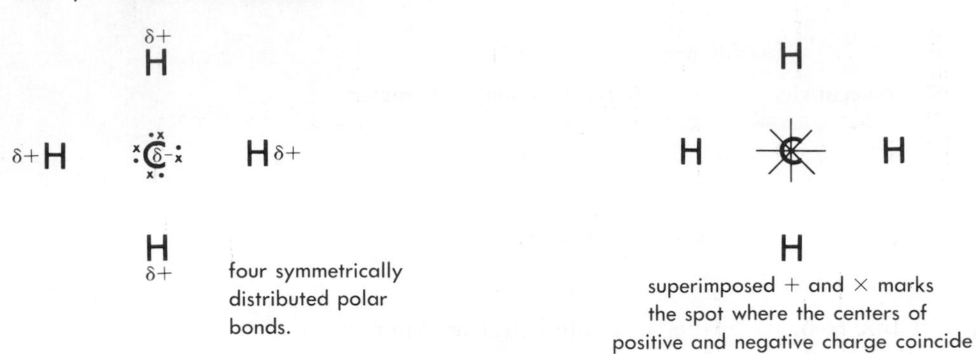

Figure 6–18 The nonpolar methane molecule.

It may be said that *in a molecule* where there is a symmetrical (geometrically balanced) distribution of polar bonds, the centers of positive and negative charge *coincide*. Any molecule fitting such a description is described as *nonpolar*. See Figure 6–18 for a visualization of the methane molecule symmetry. The single-plane model is not meant to suggest that molecules are "flat" any more than a wall map means that the world is flat. It is simply easier to illustrate polarity in this way.

The water molecule is known to have an angle of about 105° between the hydrogen atoms. This information, expressed in the format of an electron-dot formula, serves as the basis for the explanation of why water is a decidedly polar molecule. Notice how the electronegativity differences (indicated in Fig. 6–19) are related to the unequal sharing of the electron pairs that causes the oxygen atom to be the center of negative charge. The resulting polarity of the water molecule is clearly indicated in Figure 6–19. The asymmetrical (geometric imbalance) of the polar bond distribution leads to a *separation* of the centers of charge—an absolute requirement of any molecule that has some degree of polarity.

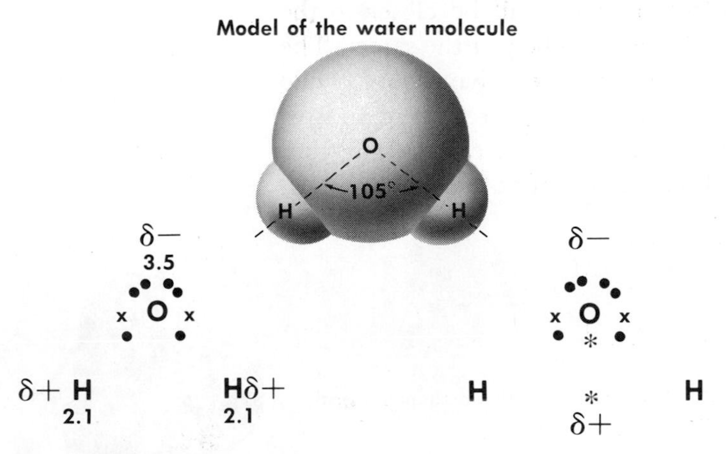

Figure 6–19 The polarity of the water molecule.

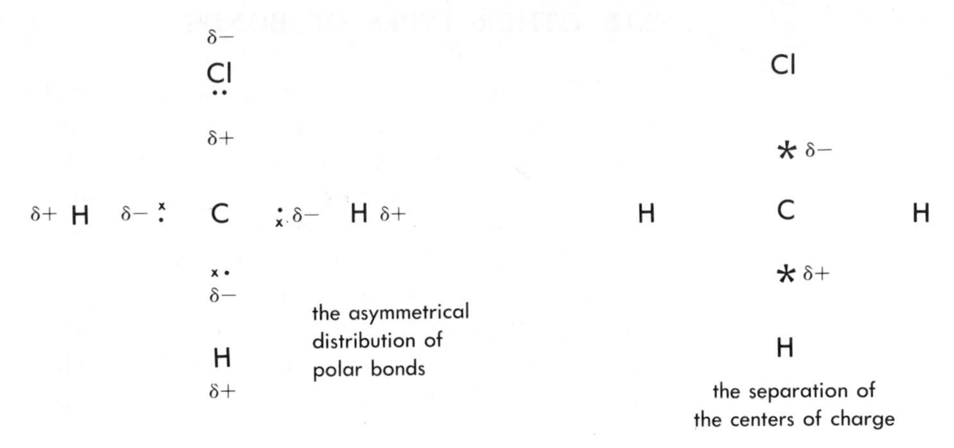

Figure 6–20 The polarity of the CH_3Cl molecule.

Another example of how the polarity of a molecule can be explained by an electron-dot formula that reflects differences in electronegativity is the molecule *methyl chloride* (or, chloromethane). The asymmetrical distribution of the individual polar bonds is due to the fact that atoms of three different electronegativities are present in the same molecule: Cl = 3.0, C = 2.5, and H = 2.1. Figure 6–20 illustrates the net effect of these values.

Exercise 6.2

Use the electronegativity values given in Figure 6–15 to draw illustrative electron-dot formulas of the following molecules. Label each molecule as polar or nonpolar:

1. Carbon dioxide (CO_2), a linear molecule
2. Ammonia (NH_3)
3. Silicon tetrafluoride (SiF_4)
4. Nitrogen (N_2)
5. Cis-dichloroethene
6. Trans-dichloroethene

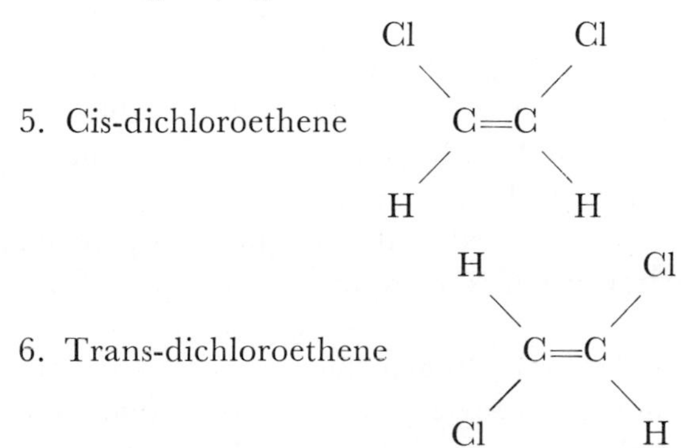

6.12 OTHER TYPES OF BONDS

While the majority of compounds may be described in terms of ionic and covalent bonds, other classes of substances demonstrate bonding characteristics that are sufficiently unique so that chemists assign them to special categories. Some examples of types of bonds other than ionic and covalent are as follows:

1. **Metallic Bond**—This is the bonding among atoms in metals. It is usually described as positive atomic kernels (nucleus plus lower energy level electrons) in a "sea" of valence electrons. Figure 6–21 diagrammatically pictures metallic bonding. This model explains the electrical conductivity of metals, and their strength and heat conductivity.

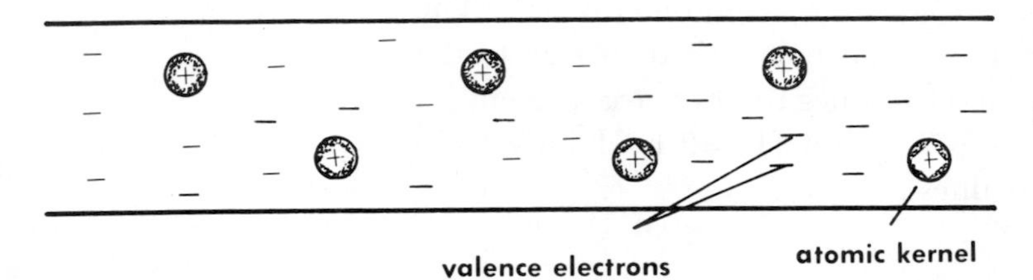

Figure 6–21 Metallic bonding.

2. **Network Bond**—The network bond is a case of complex sharing of electron pairs. The covalent bonding is so extensive that huge crystals result rather than molecules. Combinations (compounds formed by network bonding are usually very hard) of carbon and silicon with each other, or with nitrogen and oxygen, often result in network bonding. Typical examples are diamond (a network of carbon atom to carbon atom bonds) and quartz (silicon dioxide), illustrated in Figure 6–22.

3. **Hydrogen Bond**—This type of bond describes the tendency of small nonmetallic atoms of high electronegativity (notably *fluorine*, *oxygen*, and *nitrogen*) to "share" hydrogen atoms. The small nonmetallic atoms exert a powerful attraction on the hydrogen atom because its electron is so far removed that it is almost a hydrogen ion. Remember that the hydrogen ion is a proton—a tiny unit of positive charge.

The empirical formula for hydrogen fluoride is HF. In reality, HF forms a chain because of hydrogen bonding. Its actual structure might be H_6F_6, or some other number of atoms in a 1 to 1 ratio. Figure 6–23 illustrates hydrogen bonding in hydrogen fluoride.

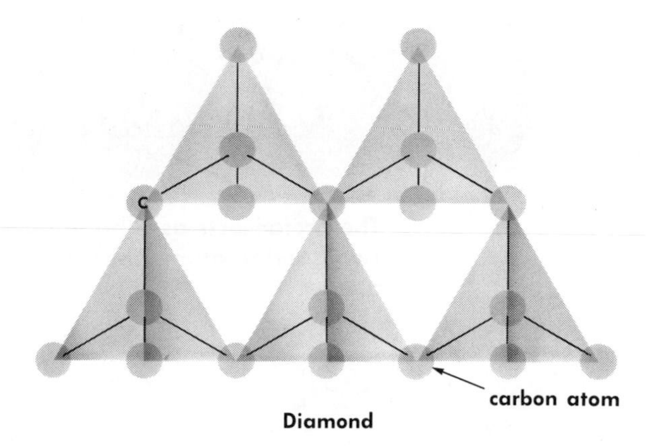

Figure 6–22 Examples of network bonding in diamond and quartz.

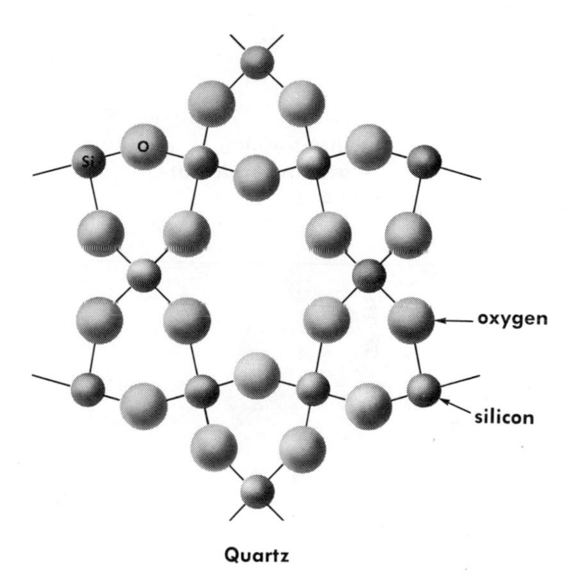

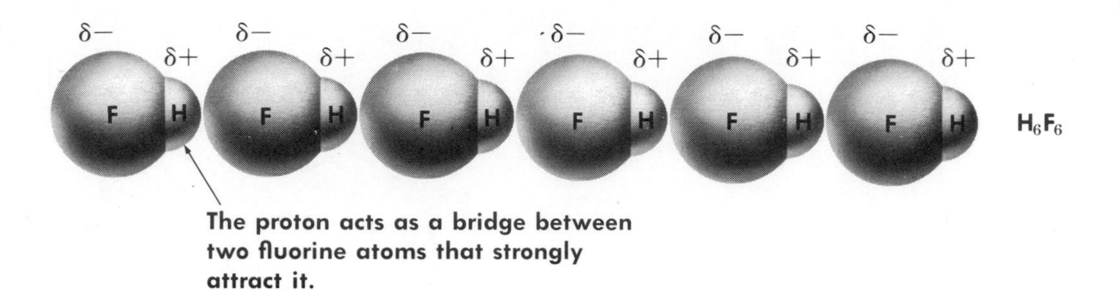

Figure 6–23 Hydrogen bonding in HF.

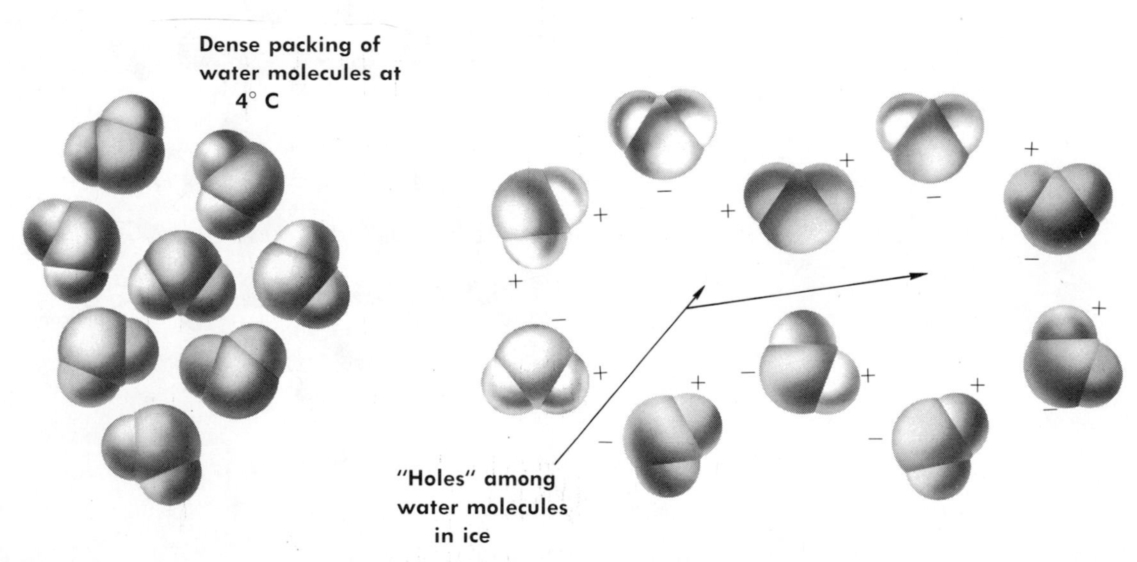

Figure 6–24 Hydrogen bonding among water molecules.

Ice is another example of hydrogen bonding. The fact that ice floats in water is due to a reduction in density resulting from "hole" formation as hydrogen bonding spreads the molecules. Figure 6–24 represents the effect that hydrogen bonding has on the arrangement of water molecules.

A final example of hydrogen bonding is found in proteins. Amino acids join to form peptides, and polypeptides link together through hydrogen bonding to form proteins as illustrated in Figure 6–25.

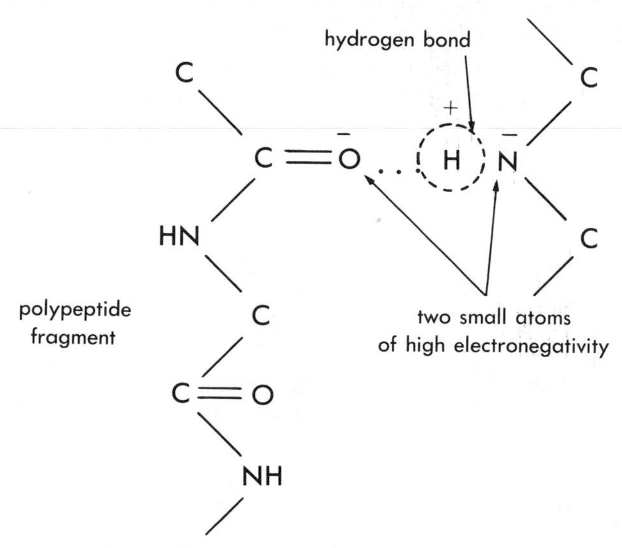

Figure 6–25 Hydrogen bonding in proteins.

6.13 THE GEOMETRY OF SELECTED MOLECULES*

It has been pointed out that the polarity of molecules is related to their physical structure. In a larger sense, it may be inferred that many of the physical and chemical properties of molecules depend on the geometric arrangement of valence electrons and the participating atoms.

The actual shapes of many molecules are known as a result of measurements made in the laboratory by modern instrumental methods. This is as it should be. It is not necessarily the primary function of the electron-dot method, or of any theoretical method for that matter, to predict what the shapes of molecules *ought* to be. When a chemist talks about the overlapping of bonding orbitals or orbital hybridization, he is attempting to present a rational explanation of why the molecule is shaped as it is.

Experience permits the chemist to make predictions about the shapes of molecules on the basis of some guiding principles that work most of the time. However, these predicted shapes must be regarded as attractive hypotheses until supporting empirical evidence is available.

The guiding principles regarding the prediction of molecular shapes may be summarized as follows:

1. The central atom in a molecule will most often have the full complement of electron pairs resulting from covalent bonding.
2. A central atom may have more or less than four bonds that determine the molecular shape, because the octet principle is not an inviolate law.
3. The pairs of bonding electrons will be expected to be distributed at maximum distances apart, since the forces of repulsion between pairs of electrons will tend to produce this arrangement.
4. When the measured angles between atoms in a molecule deviate from what is expected, the explanation may be found in the electron pair repulsion effect due to unbonded pairs of electrons. This interpretation is often called the **electron pair repulsion theory.**

The application of these guidelines will be illustrated by the following examples.

Water

The measured angle between the hydrogen atoms in the water molecule is about 105°, which is contrary to the 90° angle suggested by the electron-dot formula,

```
  ××
 ×O× H
  ×·
  H
```

An examination of the electron configuration of oxygen, $_8O\ 1s^2\ 2s^2\ 2p^4$, indicates a total of six valence electrons. The two unpaired electrons can form covalent bonds with hydrogen, while the remaining two pairs of electrons remain unbonded. The maximum distance between the electron pairs, both bonded and unbonded, is obtained by distributing them at the four corners of a regular tetrahedron as illustrated in Figure 6–26.

The incorrect model in Figure 6–26 is rectified by application of the electron pair repulsion theory, which, specifically related to the water

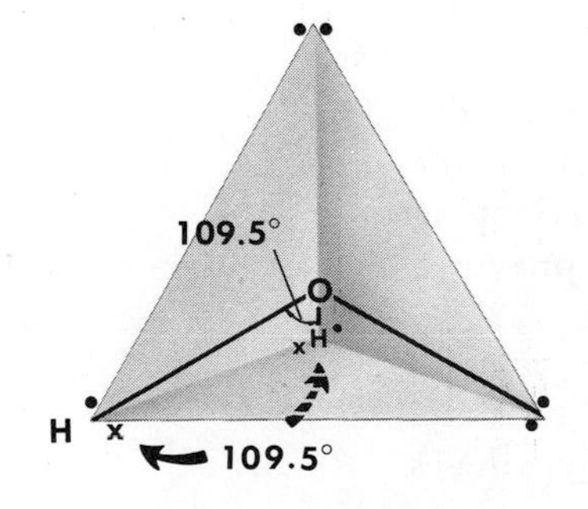

Figure 6–26 A model of the water molecule that does not agree with its actual shape.

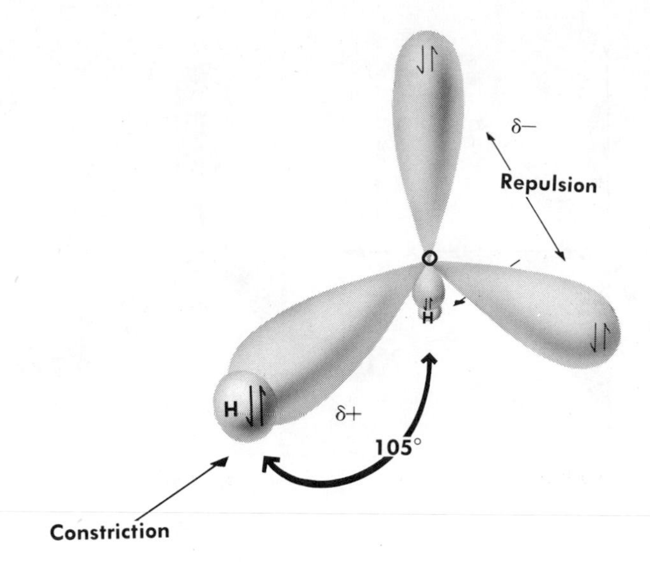

Figure 6–27 Electron pair repulsion in the water molecule, resulting in an angle of 105° between the hydrogen atoms.

molecule, says that the two unbonded pairs of electrons would repel each other to the point of distorting the tetrahedral structure (Fig. 6–27).

6.14 ORBITAL HYBRIDIZATION*

The formation of the tetrahedral arrangement of the valence electrons in oxygen is explained by the concept of **orbital hybridization.** Orbital hybridization means that a new probability distribution of electrons may result from the absorption of energy that allows the electronic configuration of an atom to achieve an **excited state.** The normal configuration is called the **ground state.** In the case of oxygen, the tetrahedral distribution of electrons results from the $2s$ and $2p$ electrons forming what is called sp^3 **hybrid.** An sp^3 hybrid orbital is a combination of one s orbital electron and three p orbital electrons in the excited state. The sp^3 hybrid is *not* a simple composite of an s and three p orbitals; rather, it is an entirely different structure (Fig. 6–28). The sp^3 hybridization can be illustrated by altering the ground state pictorial representation of oxygen:

		$1s^2$	$2s^2$	$2p_x^2$	$2p_y^1$	$2p_z^1$
ground state	$_8$O	↑↓	↑↓	↑↓	↑	↑
excited state		↑↓	↑↓ sp^3	↑↓ sp^3	↑ sp^3	↑ sp^3

The most useful generalization that can be made at this point is that sp³ *orbital hybridization is always equated with a tetrahedral structure.*

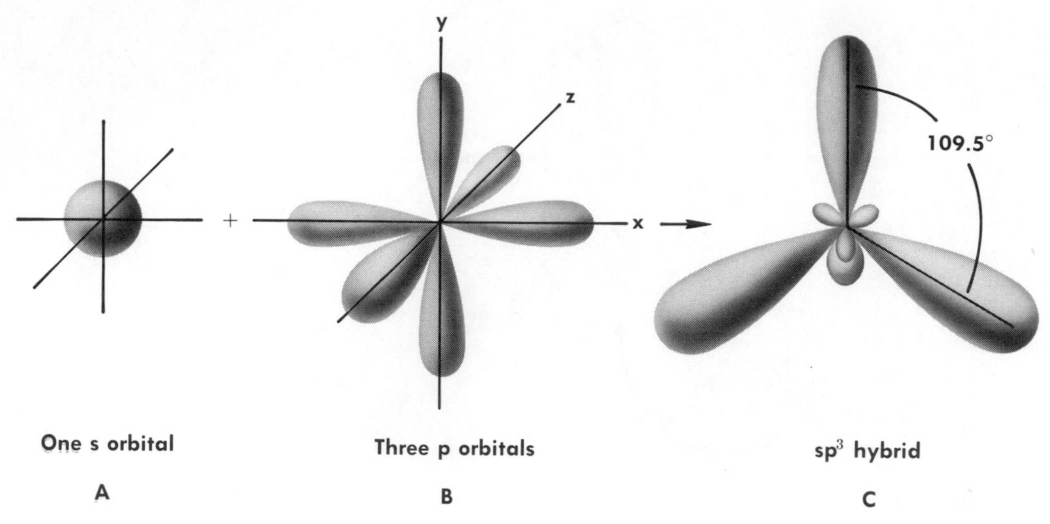

Figure 6–28 Formation of an sp^3 hybrid orbital from one s orbital and three p orbitals.

Ammonia

Ammonia is another example of sp^3 orbital hybridization. The electron-dot formula

$$\begin{array}{ccc} & \times\times & \\ H & \overset{\times}{\underset{\cdot}{}}\,N\,\overset{\cdot}{\underset{\times}{\cdot}} & H \\ & \times\cdot & \\ & H & \end{array}$$

indicates four pairs of electrons (3 bonded and 1 unbonded) that deviate slightly from the regular tetrahedral arrangement, 107° instead of 109.5°, because of the electron pair repulsion. The electronic configuration and pictorial representation of nitrogen in the ground and excited states as illustrated in Figure 6–29 demonstrates the orbital hybridization:

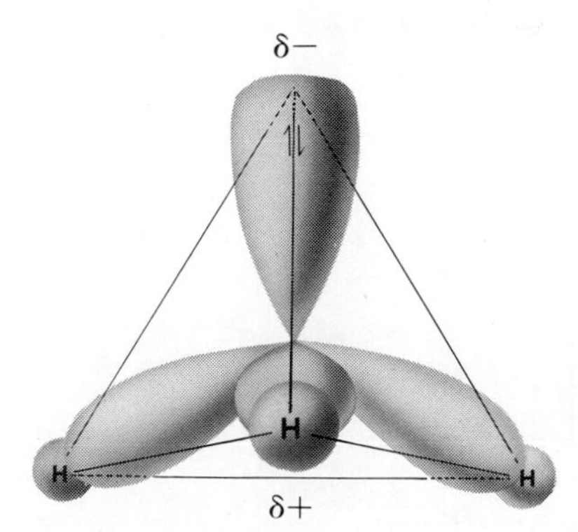

Figure 6–29 The pyramidal structure of the NH_3 molecule due to sp^3 hybridization of the orbitals.

$_7$N	$1s^2$	$2s^2$	$2p_x^1$	$2p_y^1$	$2p_z^1$
ground state	↑↓	↑↓	↑	↑	↑
excited state	↑↓	↑↓	↑	↑	↑
		sp^3	sp^3	sp^3	sp^3

Boron Trifluoride

An explanation of the fact that boron can form 3 bonds is found in the hybridization of the orbitals in the excited state. The electronic configuration and pictorial representation of boron is

$_5$B	$1s^2$	$2s^2$	$2p_x^1$	$2p_y^0$	$2p_z^0$
ground state	↑↓	↑↓	↑	—	—
excited state	↑↓	↑	↑	↑	—
		sp^2	sp^2	sp^2	

Notice how the energy absorption related to the excited state permits an **uncoupling** of the 2*s* electrons. The three unpaired electrons that constitute the sp^2 (one orbital electron and two orbital electrons) hybrid will be arranged at the three points of a triangle, each bond being 120° apart. *The* sp^2 *hybrid is always a trigonal planar distribution* (Fig. 6–30).

The fluorine atom has the following electronic configuration and pictorial representation:

$_9$F	$1s^2$	$2s^2$	$2p_x^2$	$2p_y^2$	$2p_z^1$
	↑↓	↑↓	↑↓	↑↓	↑

The single *p* orbital electron will form the covalent bond with boron. Remembering that a *p* orbital probability distribution resembles the shape shown

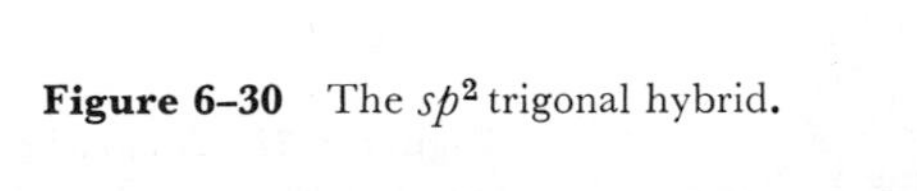

Figure 6–30 The sp^2 trigonal hybrid.

120°

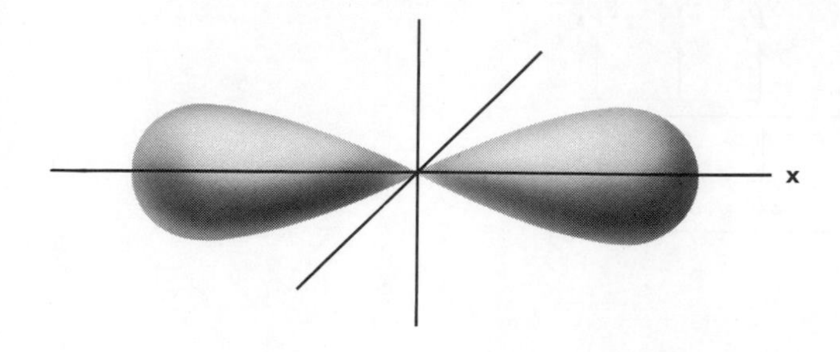

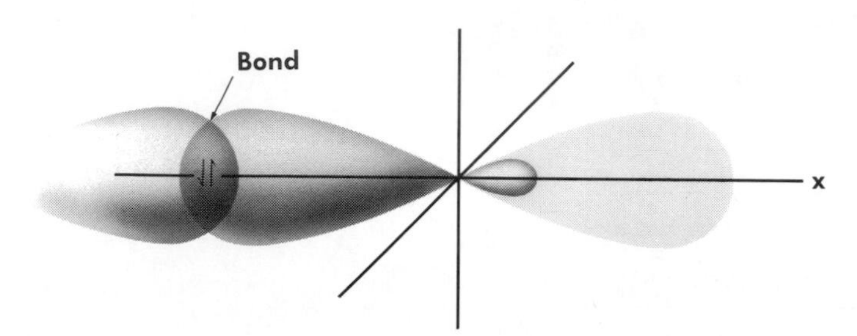

Figure 6–31 Normal and bonded p orbital electron in the fluorine atom.
A, A normal p orbital distribution.
B, "Distorted" p orbital due to bonding.

in Figure 6–31, the "distorted" shape can be said to result from the much higher probability of the fluorine electron being at the site of the covalent bond. The combined diagrammatic representation of the boron and fluorine atoms is illustrated in Figure 6–32.

Sulfur Hexafluoride

Another example of the relationship of orbital hybridization to molecular structure is the sulfur hexafluoride molecule. When a central atom

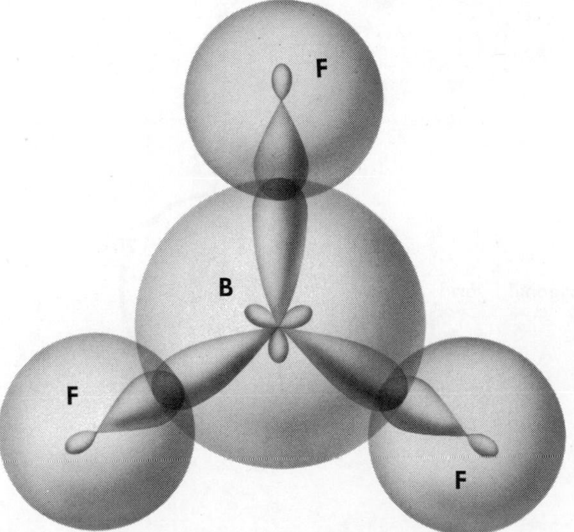

Figure 6–32 A model of sp^2 orbital hybridization in the BF_3 molecule.

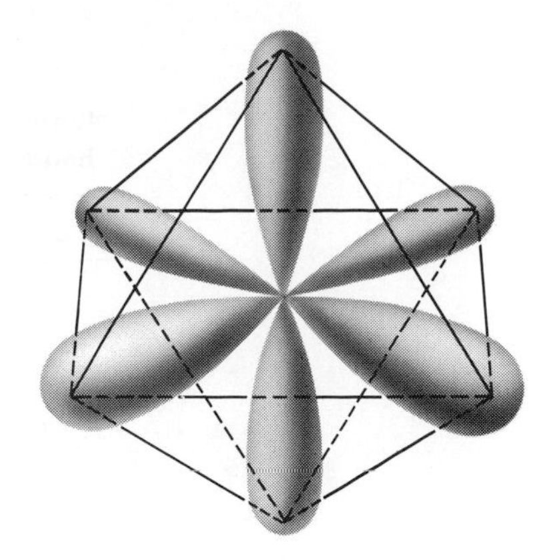

Figure 6–33 The octahedral structure.

is bonded symmetrically to six other atoms, the maximum distances between the pairs of bonding electrons can be obtained by an **octahedral** arrangement. This means that the bonded atoms must be at the six corners of an eight-sided figure, which is a base-to-base arrangement of two four-sided pyramids (Fig. 6–33). The orbital hybridization that produces the octahedral structure is illustrated by sulfur in the form of an *sp^3d^2* **hybrid.**

$_{16}S$	$1s^2$	$2s^2$		$2p^6$		$3s^2$		$3p^4$			$3d^0$			
ground state	↑↓	↑↓	↑↓	↑↓	↑↓	↑↓	↑↓	↑	↑	—	—	—	—	—
excited state	↑↓	↑↓	↑↓	↑↓	↑↓	↑	↑	↑	↑	↑	↑			
						sp^3d^2	sp^3d^2	sp^3d^2	sp^3d^2	sp^3d^2	sp^3d^2			

The shape of the SF_6 molecule is illustrated in Figure 6–34.

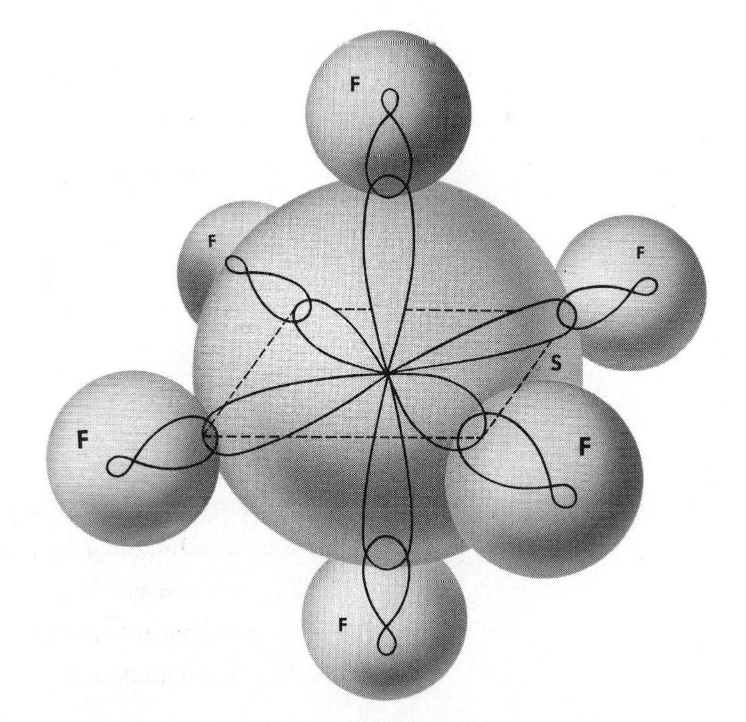

Figure 6–34 The octahedral structure of the SF_6 molecule.

Table 6–7 presents a partial summary of molecular shapes.

TABLE 6–7

Number of bonds to central atom	Type of geometry around a central atom	Possible orbital rotation	Model	Example
2	Linear	sp		CO_2
3	Trigonal planar	sp^2		BF_3
4	Tetrahedral	sp^3		CH_4
5	Octahedral	sp^3d^2		XeF_6

Exercise 6.3

Draw the probable geometric shapes of the following molecules, and label the type of orbital hybridization:

1. CS_2, carbon disulfide
2. PI_3, phosphorus triiodide
3. NH_4^+, ammonium ion

4. SO_4^{2-}, sulfate ion
5. CCl_4, carbon tetrachloride
6. XeF_6, xenon hexafluoride

QUESTIONS AND PROBLEMS

6.1 Define a chemical bond.

6.2 What is the essential difference between an ionic bond and a covalent bond?

6.3 How are electron affinity and electronegativity similar? On what basis do they differ?

6.4 Write the resonance hybrid structures for the nitrate, NO_3^-, ion.

6.5 Write the electron-dot formulas for the following:
a. sodium bromide
b. potassium chlorate
c. iron (II) chloride (Fe:)
d. silicon tetrafluoride
e. hydrogen bromide

6.6 What is the significance of high charge-to-radius ratio in an atom with regard to ionization energy?

6.7 Describe the molecules CH_4 (methane) and water in terms of polarity, symmetry, and separation of the centers of charge.

6.8 What is hydrogen bonding?

6.9 Predict the most probable shapes of the following molecules:
a. CH_3F
b. SO_2
c. PCl_3
d. BH_3

6.10 What is the difference between a polar bond and a polar molecule?

6.11 Place the appropriate code letters (A, B, C, D, E) in the related section of the periodic table.
A. highest ionization energy
B. highest electron affinity
C. highest electronegativity
D. smallest charge-to-radius ratio
E. elements most easily oxidized

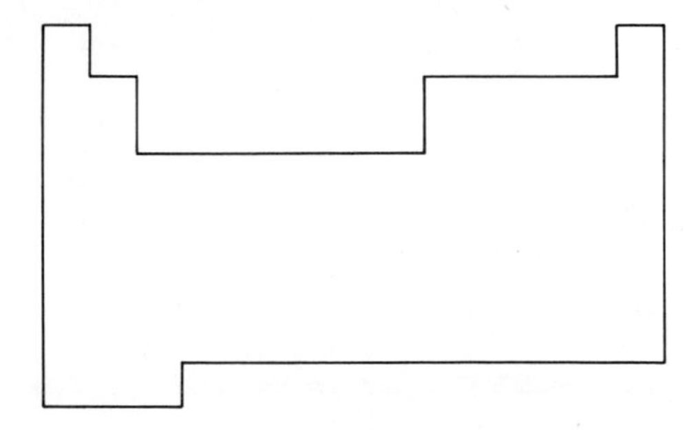

CHAPTER 7

Therefore, the two processes, that of science and that of art, are not very different. Both science and art form in the course of the centuries a human language by which we can speak about the more remote parts of reality, and the coherent sets of concepts as well as the different styles of art are different words or groups of words in this language.

Werner Heisenberg, *Physics and Philosophy*

Behavioral Objectives

At the completion of this chapter, the student should be able to:

1. Name binary and ternary compounds when the formula is given.
2. Use the classical, trivial, and Stock systems of nomenclature.
3. Name hydrated compounds.
4. Write formulas from given names of compounds.
5. Use tables of common oxidation numbers or ionic charges in writing formulas.
6. Know the rules for correct formula writing.
7. Use the special nomenclature of acids to write and name their formulas.
8. List the prefixes used in the descriptive nomenclature of nonmetallic compounds.
9. Use the descriptive and Stock systems of nomenclature in naming and writing the formulas of nonmetallic compounds.
10. Calculate the percentage composition of compounds.
11. Determine the empirical formulas of compounds from experimental data.
12. Find the true molecular formulas of compounds from empirical formulas and molecular weights.

---

* New York; Harper & Bros., 1958.

FORMULAS AND NOMENCLATURE

7.1 INTRODUCTION: THE STOCK AND CLASSICAL SYSTEMS

The ability to write formulas of compounds and to know the names of formulas that are encountered is a critical skill in chemistry. The efficiency with which formulas are handled depends on a thorough knowledge of the rules involved, and a great deal of practice. The formula is more than a shorthand method of writing the names of compounds. The formula of a compound very specifically indicates the number of each type of atom in a molecule, and the ratios of the ions in an ionic compound. Indirectly, the atomic weights of the constituent atoms in a compound allow the weight ratios to be calculated, as was described in Chapter 3 dealing with the law of definite composition. For example, the formula for water, H_2O, says that one mole of water is composed of two moles of hydrogen atoms and one mole of oxygen atoms. Furthermore, an 18 gram sample of water is composed of 2 grams of hydrogen and 16 grams of oxygen.

Two related types of calculations in which the mole concept is applied to compounds are **percentage composition** and **empirical formulas.** The percentage composition of a compound expresses the weight ratios of the elements and attached water (called *water of hydration*) in compounds. Empirical formulas are expressions of the simplest ratios of the numbers of each type of atom in a compound. The term *empirical* suggests that these weight data are obtained as a result of direct experimentation and observation. The topics of percentage composition and empirical formulas will be discussed in the latter part of this chapter, after a systematic study of nomenclature.

Since the balancing of chemical equations is critically dependent on the correctness of the formulas for all the reactants and products, there is no such thing as a trivial error in formula writing. For example, $FeCl_2$ is the formula of a compound that is very different from $FeCl_3$, although the error in the subscript appears to be small. If $Ca(OH)_2$ is written, the formula states that one mole of the compound is composed of one mole of calcium, and two moles each of oxygen and hydrogen. The weight of a mole of $Ca(OH)_2$ is quite different from the weight suggested by the formula $CaOH_2$, where the absence of the parenthesis indicates that only one mole of oxygen is in a mole of compound. The rules governing the writing and naming of formulas that chemists prefer to use was developed by the German inorganic chemist Alfred Stock, and is known as the **Stock system of nomenclature.** While some chemists and many commercial chemical supply companies still use

classical and **trivial** (common) names for compounds, the Stock system is preferred. A table of trivial names is presented in Table 7–1.

TABLE 7–1 SOME TRIVIAL NAMES OF COMMON COMPOUNDS

Formula	Trivial Name
$NaHCO_3$	baking soda
$Na_2B_4O_7 \cdot 10H_2O$	borax
$MgSO_4 \cdot 7H_2O$	epsom salts
CH_3CH_2OH	grain alcohol
CH_3OH	wood alcohol
N_2O	nitrous oxide (laughing gas)
CaO	quick lime
$Ca(OH)_2$	slaked lime
NaOH	lye
$NaNO_3$	saltpeter
$C_{12}H_{22}O_{11}$	table sugar

Some trivial names are so deeply ingrained in chemical language that their continued use is generally accepted. A table of some of these compounds is listed in Table 7–2.

TABLE 7–2 SOME GENERALLY ACCEPTED COMMON NAMES ARE USED CONVENTIONALLY AND PROPERLY AS PART OF THE STOCK SYSTEM

Formula	Accepted Name
H_2O	water
NH_3	ammonia
H_2S	hydrogen sulfide
PH_3	phosphine
C_6H_6	benzene
CH_4	methane

7.2 THE BASIS OF FORMULA WRITING

The writing of a formula of a compound depends on knowledge of, or access to, a table of common oxidation numbers, and on the proper application of the rules that compose the Stock system. It has been discussed previously that the common oxidation number of an atom is often equivalent to the charge on the ion. The tendency of metals to form ions having a positive charge (by virtue of their low ionization energies) and the tendency of nonmetallic atoms to form negatively charged ions (due to their relatively high electron affinities) have been established.

$$\underset{\text{metal sodium atom}}{Na\cdot} \rightarrow \underset{\text{ion}}{Na^+} + e^- \text{ (oxidation)}$$

The common oxidation number of sodium is 1+.

$$\underset{\text{nonmetal chlorine atom}}{:\ddot{\underset{..}{Cl}}.} + e^- \rightarrow \underset{\text{chloride ion}}{[:\ddot{\underset{..}{Cl}}:]^-} \text{ (reduction)}$$

The common oxidation number of chlorine is 1−. The negative sign indicates that its formation was due to a process (reduction) which is the reverse of oxidation.

The common oxidation numbers or ionic charges of many frequently used atoms and polyatomic ions are shown in Tables 7–3, 7–4, 7–5, 7–6, 7–7, 7–8, and 7–9. Notice how the Stock system uses Roman numerals in the naming of those metal ions that have more than one

TABLE 7–3 NAMES AND FORMULAS OF COMMON 1+ OXIDATION NUMBER SPECIES

Ion Name	Formula
hydrogen	H^+
lithium	Li^+
sodium	Na^+
potassium	K^+
ammonium	NH_4^+
mercury (I) (mercurous)*	Hg_2^{2+}
copper (I) (cuprous)	Cu^+
silver	Ag^+
rubidium	Rb^+
cesium	Cs^+

* The mercury (I) ion is an unusual species. It occurs naturally as a diatomic ion having a total charge of 2+.

common oxidation number. The older classical method of applying a distinguishing suffix, *-ous* or *-ic* to a positively charged ion **(cation)*** is shown in parentheses. Alternate common names of negatively charged ions **(anions)*** are also included parenthetically.

* The names cation and anion arise from the observation that, in solution or liquid state, positively charged ions will migrate toward a negatively charged pole, called the *cathode*. Negatively charged ions move toward a positively charged pole, called the *anode*.

TABLE 7–4 NAMES AND FORMULAS OF COMMON 2+ OXIDATION NUMBER SPECIES

Ion Name	Formula
magnesium	Mg^{2+}
calcium	Ca^{2+}
barium	Ba^{2+}
strontium	Sr^{2+}
cadmium	Cd^{2+}
zinc	Zn^{2+}
cobalt (II) (cobaltous)	Co^{2+}
mercury (II) (mercuric)	Hg^{2+}
nickel (II) (nickelous)	Ni^{2+}
iron (II) (ferrous)	Fe^{2+}
copper (II) (cupric)	Cu^{2+}
lead (II) (plumbous)	Pb^{2+}
manganese (II) (manganous)	Mn^{2+}
tin (II) (stannous)	Sn^{2+}
chromium (II) (chromous)	Cr^{2+}

TABLE 7–5 NAMES AND FORMULAS OF COMMON 3+ OXIDATION NUMBER SPECIES

Ion Name	Formula
aluminum	Al^{3+}
cobalt (III) (cobaltic)	Co^{3+}
nickel (III) (nickelic)	Ni^{3+}
iron (III) (ferric)	Fe^{3+}
chromium (III) (chromic)	Cr^{3+}
bismuth (III)*	Bi^{3+}
antimony (III)*	Sb^{3+}
arsenic (III)*	As^{3+}

* The trivial names of bismuth, antimony, and arsenic compounds are commonly expressed according to the rules applied to the nomenclature of nonmetallic compounds.

TABLE 7–6 NAMES AND FORMULAS OF COMMON 4+ OXIDATION NUMBER SPECIES

Ion Name	Formula
carbon	C^{4+}
silicon	Si^{4+}
lead (IV) (plumbic)	Pb^{4+}
manganese (IV) (manganic)	Mn^{4+}
tin (IV) (stannic)	Sn^{4+}

TABLE 7–7 NAMES AND FORMULAS OF COMMON 3− OXIDATION NUMBER (OR IONIC CHARGE) SPECIES

Ion Name	Formula
nitride	N^{3-}
phosphide	P^{3-}
phosphite	PO_3^{3-}
phosphate	PO_4^{3-}
arsenite	AsO_3^{3-}
arsenate	AsO_4^{3-}

TABLE 7–8 NAMES AND FORMULAS OF COMMON 2− OXIDATION NUMBER (OR IONIC CHARGE) SPECIES

Ion Name	Formula
oxide	O^{2-}
sulfide	S^{2-}
carbonate	CO_3^{2-}
sulfite	SO_3^{2-}
sulfate	SO_4^{2-}
thiosulfate	$S_2O_3^{2-}$
hydrosulfite (dithionite)	$S_2O_4^{2-}$
chromite	CrO_3^{2-}
chromate	CrO_4^{2-}
dichromate (bichromate)	$Cr_2O_7^{2-}$
monohydrogen phosphate (dibasic phosphate)	HPO_4^{2-}
oxalate	$C_2O_4^{2-}$ $^{-}(OOCCOO)^{-}$

TABLE 7–9 NAMES AND FORMULAS OF COMMON 1— OXIDATION NUMBER (OR IONIC CHARGE) SPECIES

Ion Name	Formula
hydride	H^-
hydroxide	OH^-
fluoride	F^-
chloride	Cl^-
bromide	Br^-
iodide	I^-
cyanide	CN^-
cyanate	OCN^-
thiocyanate	SCN^-
hydrogen sulfide (bisulfide)	HS^-
hypochlorite	ClO^-
chlorite	ClO_2^-
chlorate	ClO_3^-
perchlorate	ClO_4^-
nitrite	NO_2^-
nitrate	NO_3^-
hydrogen carbonate (bicarbonate)	HCO_3^-
permanganate	MnO_4^-
dihydrogen phosphate (monobasic phosphate)	$H_2PO_4^-$
hydrogen sulfite (bisulfite)	HSO_3^-
hydrogen sulfate (bisulfate)	HSO_4^-
peroxide	O_2^{2-}
acetate	OAc* (CH_3COO^-) ($C_2H_3O_2^-$)

* A contrived formula used by many inorganic chemists.

TABLE 7–10 A SUMMARY OF COMMON OXIDATION NUMBERS (OR IONIC CHARGES) FOR EASY REFERENCE

Cations

1+	*2+*	*3+*	*4+*
H^{+}	Mg^{2+}	Al^{3+}	C^{4+}
Li^{+}	Ca^{2+}	Co^{3+}	Si^{4+}
Na^{+}	Ba^{2+}	Ni^{3+}	Pb^{4+}
K^{+}	Sr^{2+}	Fe^{3+}	Mn^{4+}
NH_4^{+}	Cd^{2+}	Cr^{3+}	Sn^{4+}
Hg_2^{2+}	Zn^{2+}	Bi^{3+}	
Cu^{+}	Co^{2+}	Sb^{3+}	
Ag^{+}	Hg^{2+}	As^{3+}	
Rb^{+}	Ni^{2+}		
Cs^{+}	Fe^{2+}		
	Cu^{2+}		
	Pb^{2+}		
	Mn^{2+}		
	Sn^{2+}		
	Cr^{2+}		

Anions

3−	*2−*	*1−*
N^{3-}	O^{2-}	H^{-}
P^{3-}	S^{2-}	OH^{-}
PO_3^{3-}	CO_3^{2-}	F^{-}
PO_4^{3-}	SO_3^{2-}	Cl^{-}
AsO_3^{3-}	SO_4^{2-}	Br^{-}
AsO_4^{3-}	$S_2O_3^{2-}$	I^{-}
	$S_2O_4^{2-}$	CN^{-}
	CrO_3^{2-}	OCN^{-}
	CrO_4^{2-}	SCN^{-}
	$Cr_2O_7^{2-}$	ClO^{-}
	HPO_4^{2-}	ClO_2^{-}
	$C_2O_4^{2-}$	ClO_3^{-}
		ClO_4^{-}
		NO_2^{-}
		NO_3^{-}
		HCO_3^{-}
		MnO_4^{-}
		$H_2PO_4^{-}$
		HSO_3^{-}
		HSO_4^{-}
		O_2^{2-}
		OAc^{-}
		HS^{-}

7.3 RULES FOR WRITING FORMULAS

The rules for the writing of formulas are summarized as follows:

1. Careful attention must be paid to capital and lowercase letters in the symbols used:

   Co means the element cobalt.

   CO means the compound carbon monoxide.

2. Roman numerals are not used in formulas:

   $FeCl_2$ is iron(II) chloride.

   $Fe(II)Cl_2$ is unconventional and therefore incorrect.

3. Parentheses are used when *more* than one polyatomic ion is indicated in the formula. Parentheses around a single polyatomic ion are not usually necessary:

   $Ca(OH)_2$ 1 calcium atom is calcium hydroxide.
   ② oxygen atoms
   2 hydrogen atoms

   $CaOH_2$ 1 calcium atom is wrong.
   ① oxygen atom
   2 hydrogen atoms

   KOH is potassium hydroxide.
   K(OH) contains unnecessary parentheses and is therefore considered wrong by convention.

4. In the writing of **binary** (two element) compounds the symbol for the metal is usually written first, followed by the symbol for the nonmetal:

   NaCl is correct.
   ↗ ↖
   metal nonmetal
   ClNa is incorrect.

   The same rule for the ordering of symbols applies to **ternary** (three element) compounds.

5. When a compound is composed of nonmetals exclusively, the symbol of the more metallic (less electronegative) element is written first. This is the element furthest to the left in a period or, as a secondary choice, lower in a group when both elements belong to the same group.* Consider examples taken from a small section of the periodic table:

	Group		
	IVA	**VA**	**VIA**
second period	C	N	O
third period	Si	P	S

* Oxygen-halogen compounds are exceptions to the rule. A common oxide of chlorines is Cl_2O_7. This comes about because oxygen and chlorine have almost the same electronegativity.

The following examples will serve to illustrate this rule.

SiC — below carbon in group IVA
NO_2 — to the left of oxygen in the second period
SO_3 — below oxygen
P_4O_{10} — to the left of and below oxygen
SiO_2 — far to the left of and below oxygen
CS_2 — far to the left of sulfur

6. The most important rule in formula writing is that *the sum of the oxidation numbers in a compound must equal zero.*

The rules for formula writing will now be applied and emphasized in a series of examples.

EXAMPLE 7.1

Write the formula for potassium hydroxide.

1. Set down the components with their respective ionic charges:

$$K^+ \qquad OH^-$$

2. See if the sum of the electrical charges equals zero:

$$\underset{\text{potassium}}{1+} \quad \text{and} \quad \underset{\text{hydroxide}}{1-} \quad = 0$$

3. The original number of potassium and hydroxide units is accepted as correct. The formula is KOH.

EXAMPLE 7.2

Write the formula for copper(I) chloride.

1. Set down the components with their respective oxidation numbers:

$$Cu^+ \qquad Cl^-$$

2. See if the sum of the electrical charges equals zero:

$$\underset{\text{copper}}{1+} \quad \text{and} \quad \underset{\text{chloride}}{1-} \quad = 0$$

3. The original number of ions is accepted. The formula is CuCl.

EXAMPLE 7.3

Write the formula for copper(II) chloride.

1. Set down the appropriate symbols for the elements involved:

$$Cu^{2+} \quad Cl^-$$

2. In order to balance the electrical charges, two chloride ions are required:

$$Cu^{2+} \quad Cl_2^-$$

3. Balancing formulas *always* involves *subscripts*. The formula is $CuCl_2$.

EXAMPLE 7.4

Write the formula for potassium sulfate.

1. $K^+ \quad SO_4^{2-}$
2. Two potassium ions are needed for balancing:

$$K_2^+ \quad SO_4^{2-}$$

3. The formula is K_2SO_4
4. **Note:** There are *no* parentheses used in the case of a single polyatomic ion in the formula.

EXAMPLE 7.5

Write the formula for iron(III) oxalate.

1. $Fe^{3+} \quad C_2O_4^{2-}$
2. The balancing of ionic charges requires a total of six positive and six negative charges (the lowest common denominator):

$$Fe_2^{3+} \quad (C_2O_4)_3^{2-}$$

3. Parentheses *must* be placed around the oxalate ion. The subscript, 3, multiplies the number of both the carbon atoms and the oxygen atoms composing the polyatomic ion.
4. The finished formula is $Fe_2(C_2O_4)_3$.

EXAMPLE 7.6

Write the formula for mercury(I) chloride.

1. Set down the appropriate symbols, remembering that the mercury (I) ion exists as a diatomic unit:

$$Hg_2^{2+} \quad Cl^-$$

2. The balancing of the charges requires two chloride ions:

$$Hg_2^{2+} \quad Cl_2^-$$

3. The charges *appear* to be unbalanced. However, the Hg_2 notation means that there is in reality only one unit of (Hg-Hg). If it were customary to write the formula as

$$(Hg\text{-}Hg)^{2+} \quad Cl_2^-$$

the appearance of the formula might be less confusing.
4. The finished formula is Hg_2Cl_2, and the subscripts should *not* be reduced to make the formula HgCl.

EXAMPLE 7.7

Write the formula for hydrogen peroxide.

1. Set down the components, bearing in mind that the peroxide ion is a diatomic species:

$$H^+ \quad O_2^{2-}$$

2. The peroxide ion may be somewhat misleading, as in the case of the Hg_2^{2+} ion. The formation of the hydrogen peroxide molecule may be easier to visualize by using electron-dot formulas:

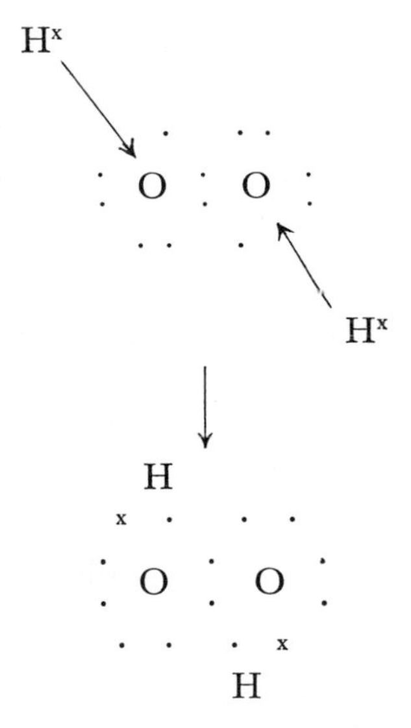

3. The electron-dot formula clearly indicates the need for two hydrogen atoms in the molecule:

$$H_2^+ \quad O_2^{2-}$$

4. While the formula might be interpreted as $H_2^+(OO)^{2-}$, the conventional form is H_2O_2.
5. Notice once again that the formula is *not* reduced to the simplest ratio, HO.

EXAMPLE 7.8

Write the formula for lead(IV) sulfide.

1. Set down the symbols for the lead(IV) and sulfide ions:

$$Pb^{4+} \qquad S^{2-}$$

2. Notice that two sulfide ions are sufficient to balance the electrical charges:

$$Pb^{4+} \qquad S_2^{2-}$$

3. The correct formula then is PbS_2, and *not* Pb_2S_4.

The possibility of writing Pb_2S_4 for lead(IV) sulfide, as indicated in Example 7.8, is a more likely mistake when a short cut method is used for the writing of formulas. This use of the quicker method that will now be described can be justified only after a fundamental understanding of formula writing has been developed and applied in the manner of the previous examples.

The Short Cut Method of Formula Writing

The rapid technique for writing formulas consists of "switching" the ionic charges of the anion and cation in a compound. When the numerical values, *without* the positive and negative signs, are exchanged between the cation and anion and then written as *subscripts*, the formula is usually completed.

EXAMPLE 7.9

Use the short cut method to write the formula for sodium chromate.

1. Set down the appropriate symbols:

$$Na^{+} \qquad CrO_4^{2-}$$

2. Switch the ionic charge values, ignoring the positive and negative signs:

$$\underset{2}{Na^{+}} \qquad \underset{1}{CrO_4^{2-}}$$

3. The final form of the formula is Na_2CrO_4.

EXAMPLE 7.10

Use the short cut method to write the following formulas:

1. cobalt(III) sulfide
2. tin(IV) oxide
3. ammonium monohydrogen phosphate
4. nickel(III) hydroxide

1. Co^{3+} S^{2-} $= Co_2S_3$

2. Sn^{4+} O^{2-} $= Sn_2O_4$

The ratio 2:4 must be reduced:

$$Sn_2O_4 = SnO_2$$

3. NH_4^+ HPO_4^{2-} $= (NH_4)_2HPO_4$

4. Ni^{3+} OH^- $= Ni(OH)_3$

Note: In samples 3 and 4, parentheses are needed to indicate the presence of more than one polyatomic ion in the formula.

Exercise 7.1

Write the formulas for the following compounds:

1. copper (II) sulfate
2. magnesium hydroxide
3. aluminum arsenite
4. cobalt (III) bromide
5. barium chlorate
6. sodium peroxide
7. mercury (I) nitrate
8. potassium permanganate
9. calcium oxalate
10. nickel (III) sulfate
11. tin (IV) nitrite
12. cadmium nitride
13. chromium (II) hydroxide
14. silver acetate
15. iron (III) monohydrogen phosphate
16. ammonium sulfide
17. potassium thiocyanate
18. strontium hydrogen carbonate
19. lead (IV) chromate
20. mercury (II) hydroxide

7.4 NOMENCLATURE OF BINARY AND TERNARY COMPOUNDS

Naming formulas correctly is most easily accomplished by starting with the subscripts and switching them back to their original ionic charge values. This process is exactly the reverse of the short cut method previously described to write formulas.

EXAMPLE 7.11

What is the name of the formula $Mg(CN)_2$?

1. Since magnesium has only one common oxidation number, there is no need to make any special distinction, such as calling it magnesium (I) or magnesium (II) (said as, "magnesium-one" or "magnesium-two"), nor does the older system of nomenclature have the forms magnes*ious* or magnes*ic*. The inclusion of a Roman numeral when it is unnecessary is wrong by convention.
2. The name of the formula is simply the addition of the names of the ions, magnesium cyan*ide*.

EXAMPLE 7.12

Write the name of the formula $PbCl_2$.

1. Lead has more than one possible common oxidation number. The particular oxidation state of the lead *must* be included in the name of compound:

   Stock: lead (II) or lead (IV)

   classical: plumb*ous* or plumb*ic*

2. By switching the subscripts to their oxidation number positions, the lead is observed to be in the $2+$ oxidation state:

   $$Pb_1 \; Cl_2 = Pb^{2+} \; Cl^-$$

3. The name of the compound is lead (II) chlor*ide* (plumbous chloride).

EXAMPLE 7.13

What is the name of the compound having the formula $Fe_3(AsO_4)_2$?

1. Separating the formula into two identifiable components, the name will be

   Stock: iron (II) or iron (III)

   classical: ferr*ous* or ferr*ic*

2. The AsO_4 is the arsen*ate* ion.
3. Switching the subscripts to their ionic charge positions will indicate the ion charges:

$$Fe_3^{2+} \quad (AsO_4)_2^{3-}$$

4. The name of the compound is iron (II) arsenate (ferrous arsenate).

EXAMPLE 7.14

Write the name of the formula $Sn(SO_3)_2$.

1. The metallic part of this ternary compound will be

   Stock: tin (II) or tin (IV)

   classical: stannous or stannic

2. The SO_3 is the sulf*ite* ion.
3. Switching the subscripts to the ionic charge positions creates a problem!

$$Sn^{2+} \quad (SO_3)_2^{1-} \qquad \text{wrong!}$$

4. The net charge on the sulfite ion is *not* 1 —, it is 2 —.
5. If the sulfite ion is correctly expressed as SO_3^{2-}, then the charge on the tin ion must also be doubled:

$$Sn^{4+}(SO_3)_2^{2-}$$

6. The name of the compound is tin (IV) sulfite (stannic sulfite). The correct formula for tin (II) sulfite (stannous sulfite) is $SnSO_3$.

A Summary of Binary and Ternary Compound Nomenclature

1. Lowercase letters are used throughout. Names of compounds should not be capitalized.
2. When the metallic component of a formula has more than one common oxidation number, it must be noted by a Roman numeral in parentheses, or by the appropriate *-ous* or *-ic* suffix if the classical system is used. However, it is emphasized that the Stock system is preferred.
3. The suffix on the nonmetallic member of the formula depends on the formal name of the ion or **oxyanion** (negative ions containing oxygen).

4. It is difficult to generalize oxyanion nomenclature except to say that the ones with the *lower* number of oxygen atoms per ion are likely to be linked to the *-ite* suffix, while the ones with more oxygen atoms have the *-ate* suffix.
5. The nonoxyanions usually have the suffix, *-ide*. Table 7–11 illustrates some sample oxyanion nomenclature.

TABLE 7–11 EXAMPLES OF ANION AND OXYANION NOMENCLATURE

MEANS LESS	Cl^-	chlor*ide*	lower number of oxygens
THAN USUAL	ClO^-	*hypo*chlor*ite*	
	ClO_2^-	chlor*ite*	
FROM "HYPER,"	ClO_3^-	chlor*ate*	higher number of oxygens
MEANING MORE	ClO_4^-	*per*chlor*ate*	
THAN USUAL			

***Exercise** 7.2*

Write the names of the following compounds according to the Stock system:

1. KF
2. $NaHCO_3$
3. $FeCl_2$
4. $MgSO_3$
5. $Ca(OH)_2$
6. Li_3AsO_4
7. NH_4Cl
8. HgO
9. K_2HPO_4
10. PbS_2
11. $K_2Cr_2O_7$
12. AlN
13. CsCN
14. $KMnO_4$
15. CdI_2
16. Hg_2I_2
17. $Ni(OCN)_2$
18. $Sn(CO_3)_2$
19. $Co_2(C_2O_4)_3$
20. $Mn(NO_2)_2$

7.5 THE NOMENCLATURE OF ACIDS

The special nomenclature of acids is derived from the standard nomenclature of binary and ternary compounds and is expressed according to a firm set of rules.

Although acids as a class of compounds will be more systematically investigated in a later chapter, they can usually be recognized. They generally have a formula beginning with *hydrogen*, such as HCl, HNO_3, and H_2SO_4 (hydrochloric acid, nitric acid, and sulfuric acid, respectively, and are dissolved in water, usually).

First rule: When the compound has the suffix, *-ide* in the standard nomenclature, the corresponding acid name begins with the prefix *hydro-*, then the *root* word, and finally the suffix *-ic*.

EXAMPLE 7.15

Name the formula HBr as an acid.

1. The formula begins with hydrogen, indicating an acid.
2. The standard nomenclature would be hydrogen brom*ide*.
3. Since the suffix is *-ide*, the root, *brom*, is preceded by *hydro-* and followed by *-ic*. The name is

hydro *brom* *ic* acid = hydrobromic acid
↑ ↑ ↑
prefix root suffix

Table 7–12 lists common roots of some of the most common nonmetals that are included in acids.

TABLE 7–12 ROOT NAMES OF SOME NONMETALS

Nonmetal	Root
fluorine	*fluor*
chlorine	*chlor*
bromine	*brom*
iodine	*iod*
sulfur	*sulfur* or *sulf*
phosphorus	*phosph* or *phosphor*
nitrogen	*nitr*

Examples of the application of the first rule are listed below in Table 7–13.

TABLE 7–13 APPLYING THE FIRST RULE IN ACID NOMENCLATURE

Formula	Standard Nomenclature	Acid Nomenclature
HCl	hydrogen chlor*ide*	*hydro*chlor*ic* acid
HCN	hydrogen cyan*ide*	*hydro*cyan*ic* acid
HI	hydrogen iod*ide*	*hydr*iod*ic* acid
HF	hydrogen fluor*ide*	*hydro*fluor*ic* acid

Second rule: When the suffix of the compound is *-ite* or *-ate*, the corresponding name of the **oxyacid** begins with the root and is followed by the suffix *-ous* in place of *-ite*, and *-ic* in place of *-ate*.

EXAMPLE 7.16

Name the formula HNO_2 as an acid.

1. The standard nomenclature is hydrogen nitr*ite*.
2. The root is *nitr* and the corresponding acid suffix is *-ous*.
3. The name of the compound is

nitr *ous* acid = nitrous acid

↗ root ↑ suffix related to *-ite*

EXAMPLE 7.17

Name the compound H_2SO_4.

1. The standard name is hydrogen sulf*ate*.
2. The root is *sulfur* and the corresponding acid suffix is *-ic*.
3. The name of the compound is

sulfur *ic* acid = sulfuric acid

↗ root ↑ suffix related to *-ate*

TABLE 7–14 APPLYING THE SECOND RULE IN ACID NOMENCLATURE

Formula	Standard Nomenclature	Acid Nomenclature
HNO_2	hydrogen nitr*ite*	nitr*ous* acid
H_2SO_3	hydrogen sulf*ite*	sulfur*ous* acid
H_3PO_3	hydrogen phosph*ite*	phosphor*ous* acid
H_3AsO_3	hydrogen arsen*ite*	arsen*ous* acid
HNO_3	hydrogen nitr*ate*	nitr*ic* acid
H_2CrO_4	hydrogen chrom*ate*	chrom*ic* acid
H_2CO_3	hydrogen carbon*ate*	carbon*ic* acid
H_2SO_4	hydrogen sulf*ate*	sulfur*ic* acid

Third rule: When the number of oxygens in the oxyacid is very low or unusually high, such as in ClO^- (hypochlorite) or ClO_4^- (perchlorate), respectively, the prefixes *hypo-* and *per-* are maintained in the acid nomenclature. Table 7–15 shows some selected examples of *hypo-* and *per-*acids.

TABLE 7–15 APPLYING THE THIRD RULE IN ACID NOMENCLATURE

Formula	Standard Nomenclature	Acid Nomenclature
HClO	hydrogen *hypo*chlor*ite*	*hypo*chlor*ous* acid
HBrO	hydrogen *hypo*brom*ite*	*hypo*brom*ous* acid
$HClO_4$	hydrogen *per*chlor*ate*	*per*chlor*ic* acid
$HMnO_4$	hydrogen *per*mangan*ate*	*per*mangan*ic* acid

EXAMPLE 7.18

Write the formula for oxalic acid.

1. Note the suffix, *-ic*. This corresponds to the standard name, hydrogen oxal*ate*.
2. Write the symbols and switch the appropriate ionic charges to subscripts:

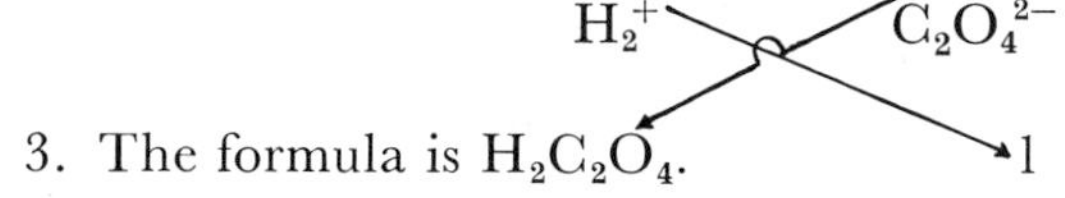

3. The formula is $H_2C_2O_4$.

EXAMPLE 7.19

Write the formula for phosphorous acid.

1. Note the suffix *-ous*. This corresponds to the standard name, hydrogen phosph*ite*.
2. Write the symbols and switch the ionic charges to the subscript position:

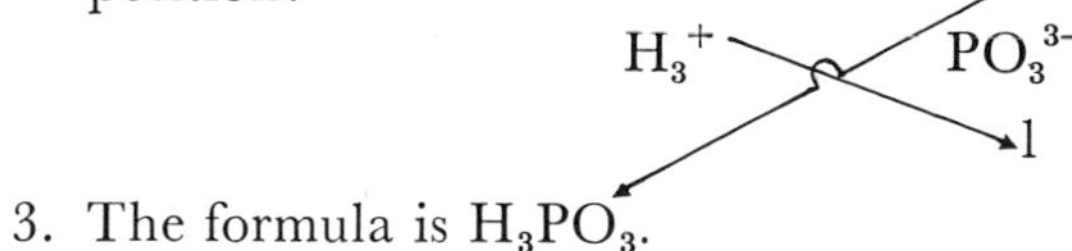

3. The formula is H_3PO_3.

Exercise 7.3

Write the formulas for the following acids:

1. cyanic acid
2. acetic acid
3. hypoiodous acid
4. permanganic acid
5. hydrobromic acid
6. nitrous acid
7. arsenic acid
8. hydrocyanic acid
9. chromous acid
10. hydrofluoric acid

EXAMPLE 7.20

What is the name of $HClO_2$?

1. Identify the compound as a possible classical acid since the formula begins with hydrogen.
2. Note the standard name of the compound as hydrogen chlor*ite*.
3. The suffix, *-ite* corresponds to the acid name, chlor*ous* acid.

EXAMPLE 7.21

What is the name of HPO_4^{2-}?

1. *Caution:* Observe the notation of electrical charge on the formula. This is *not* a neutral compound. It is not likely to function as an acid (not even by a more sophisticated definition).
2. The formula should be identified as the monohydrogen phosphate ion.
3. This example serves to illustrate the importance of being concerned with detail in the writing and naming of formulas. While a hydrogen sulfate ion, HSO_4^-, may behave as an acid, it is not subject to the classical acid nomenclature designed for neutral compounds.

Exercise 7.4

Name the following formulas:

1. HBrO
2. H_2CrO_4
3. HCN
4. HCO_3^-
5. H_3PO_4
6. HOAC
7. $H_2PO_4^-$
8. $HClO_4$
9. HNO_2
10. HSO_3^-

7.6 THE NOMENCLATURE OF NONMETALLIC COMPOUNDS

In the naming of compounds in which no metals are present, or if there is a metal (or metalloid) and it has an oxidation number of 4+ or higher, as especially descriptive system of nomenclature is often used. The method involves the use of Greek or Latin prefixes that describe the number of atoms present in the molecule. The prefixes are illustrated by the examples in Table 7–16. Also shown are the alternative names according to the Stock system. The apparent oxidation number of the more metallic element (the first symbol in the formula) is calculated on the basis of the common oxidation number of the more nonmetallic (electronegative) element (the second symbol in the formula).

EXAMPLE 7.22

Name the formula SO_2 by the Stock system.

1. Calculate the total *negative* charge on the oxide on the basis of one oxide ion having a charge of 2−:

$$SO_2^{2-}$$

2. The total negative charge is 4−, since there are two oxide atoms.
3. Axiomatically, the charges must balance in the formula of a compound. Therefore, the single sulfur atom must have an oxidation number of 4+:

$$S_1^{4+}\ O_2^{2-}$$

4. The Stock name of the compound is sulfur (IV) oxide.

TABLE 7–16 DESCRIPTIVE AND STOCK NOMENCLATURE FOR SOME SELECTED COMPOUNDS

Prefix	Example	Descriptive Name	Stock Name
mono = 1	CO	carbon ***mono***oxide	carbon (II) oxide
di = 2	SiO_2	silicon ***di***oxide	silicon (IV) oxide
tri = 3	SO_3	sulfur ***tri***oxide	sulfur (VI) oxide
tetra = 4	CCl_4	carbon ***tetra***chloride	carbon (IV) chloride
penta = 5	N_2O_5	***di***nitrogen ***pent***oxide	nitrogen (V) oxide
hexa = 6	XeF_6	xenon ***hexa***fluoride	xenon (VI) fluoride
hepta = 7	Cl_2O_7	***di***chlorine ***hept***oxide	chlorine (VII) oxide
octa = 8	Cl_2O_8	***di***chlorine ***oct***oxide	chlorine (IV) oxide
nona = 9	I_4O_9	***tetra***iodine ***non***oxide	iodine (V) oxide (dimer)*
deca = 10	P_4O_{10}	***tetra***phosphorus ***dec***oxide	phosphorus (V) oxide (dimer)*

* A **DIMER** is a double structure. The I_4O_9 is the result of two molecules of I_2O_5 sharing a common oxygen atom. Similarly, P_4O_{10} is a dimer of P_2O_5 (diphosphorus pentoxide).

EXAMPLE 7.23

Name the following compounds according to the descriptive and Stock systems.

1. N_2O $(N_2^{+}\ O_1^{2-})$

descriptive: dinitrogen monoxide
Stock: nitrogen(I) oxide

2. MnO_2 $(M_1^{4+} O_2^{2-})$

descriptive: manganese dioxide
Stock: manganese(IV) oxide

3. P_4O_6 $(P_4^{3+} O_6^{2-})$

descriptive: tetraphosphorus hexoxide
Stock: phosphorus(III) oxide dimer

4. Cl_2O $(Cl_2^{+} O_1^{2-})$

descriptive: dichlorine monoxide
Stock: chlorine(I) oxide

5. V_2O_5 $(V_2^{5+} O_5^{2-})$

descriptive: divanadium pentoxide
Stock: vanadium(V) oxide

EXAMPLE 7.24

Write the formulas for the following compounds:

1. phosphorus pentachloride

PCl_5

2. disulfur dichloride

S_2Cl_2

3. nitrogen(III) oxide

N^{3+} $O^{2-} = N_2O_3$

4. silicon(IV) bromide

Si^{4+} $Br^{-} = SiBr_4$

5. diarsenic pentasulfide

As_2S_5

Exercise 7.5

1. Write the Stock or descriptive names of the following compounds:

a. NO_2
b. CF_4
c. I_4O_9
d. N_2O_5
e. VCl_4
f. BF_3
g. I_2Cl_7
h. PbO_2
i. CrO_3
j. Sb_2O_5

2. Write formulas for the following compounds:
   a. diarsenic trisulfide
   b. uranium hexachloride
   c. selenium dioxide
   d. dinitrogen trioxide
   e. carbon(II) oxide
   f. sulfur(VI) fluoride
   g. xenon hexaiodide
   h. carbon(IV) fluoride
   i. diantimony trioxide
   j. bismuth(V) oxide

7.7 THE NOMENCLATURE OF HYDRATES

A hydrated compound is one in which a specific number of moles of water are bound to the crystal structure of a compound in a definite ratio of water molecules to compound molecules. When the water is removed (a dehydration process) the compound is called an **anhydride,** or it is described as being **anhydrous.** The nomenclature is quite descriptive. For example, $CuSO_4 \cdot 5H_2O$ means that there are 5 moles of water per mole of salt in the crystal.

EXAMPLE 7.25

Write the alternate names of $CuSO_4 \cdot 5H_2O$.
1. copper(II) sulfate 5-water
2. copper(II) sulfate 5-hydrate
3. copper(II) sulfate pentahydrate
4. cupric sulfate pentahydrate

EXAMPLE 7.26

Write the alternate names of $MgSO_4 \cdot 7H_2O$.
1. magnesium sulfate 7-water
2. magnesium sulfate 7-hydrate
3. magnesium sulfate heptahydrate

EXAMPLE 7.27

Write the formulas for the following hydrates.
1. calcium chloride 6-water

$$CaCl_2 \cdot 6H_2O$$

2. ferrous sulfate heptahydrate

$$FeSO_4 \cdot 7H_2O$$

3. calcium sulfate 1-hydrate

$$CaSO_4 \cdot H_2O$$

4. iron(II) phosphate 8-water

$$Fe_3(PO_4)_2 \cdot 8H_2O$$

Exercise 7.6

1. Write two alternate names for the following hydrates:
   a. $Ca(NO_3)_2 \cdot 3H_2O$
   b. $Sn(NO_3)_2 \cdot 20H_2O$
   c. $Na_2SO_3 \cdot 7H_2O$
   d. $U(SO_4)_3 \cdot 9H_2O$
   e. $ZnSO_4 \cdot 7H_2O$
   f. $NaI \cdot 2H_2O$
2. Write formulas for the following hydrates:
   a. aluminum oxide 3-hydrate
   b. cobalt(II) perchlorate 5-water
   c. thallium(III) sulfate heptahydrate
   d. sodium sulfate decahydrate
   e. sodium acetate 3-water
   f. cadmium chloride 3/2-water
   (Hint: The nomenclature becomes simpler by doubling the formula to eliminate the fraction. The name then becomes dicadmium chloride 3-water.)

7.8 PERCENTAGE COMPOSITION

The calculation of the mass composition of a compound by percentage is a very useful skill in analytical chemistry. Since the formula weight of a compound in grams equals the weight of a mole of the compound, the individual mole weights of the elements in the compound can be compared to the total weight and expressed as a percent.

EXAMPLE 7.28

Calculate the percentage composition of ammonium sulfide.

1. Write the formula:

$$(NH_4)_2S$$

2. Calculate the weight of a mole of $(NH_4)_2S$, which is composed of

a. 2 moles of nitrogen atoms:

$$(NH_4)_2S$$

$$2 \text{ mols } (14.00 \text{ g mol}^{-1}) = 28.00 \text{ g}$$

b. 8 moles of hydrogen atoms:

$$(NH_4)_2S$$

$$8 \text{ mols } (1.01 \text{ g mol}^{-1}) = 8.08 \text{ g}$$

c. 1 mole of sulfur atoms = 32.06 g

d. 1 mole of $(NH_4)_2S$ units = 68.14 g

3. The percentage of each element is determined by calculating its fraction of the whole:

$$\text{percent N} = \frac{28.00}{68.14} = 0.412 = 41.2\%$$

$$\text{percent } H = \frac{8.08}{68.14} = 0.118 = 11.8\%$$

$$\text{percent S} = \frac{32.06}{68.14} = 0.470 = 47.0\%$$

$$100.0\%$$

4. The sum of the percentages should equal 100%, allowing 0.6% for slide rule estimation.

EXAMPLE 7.29

If 593.0 mg of an unknown compound composed of carbon, hydrogen, and chlorine is chemically combined with oxygen, 1060.0 mg of carbon dioxide and 145.0 mg of water are formed. Calculate the percentage composition.

1. In the formation of carbon dioxide, all of the carbon came from the unknown compound.
   a. Find the fraction by weight of carbon in carbon dioxide. This can be used to yield the original weight of carbon (in the unknown):

$$CO_2$$

$$(2 \text{ mols}) \times (16.00 \text{ g mol}^{-1}) = 32.00 \text{ g}$$

$$(1 \text{ mol}) \times (12.01 \text{ g mol}^{-1}) = 12.01 \text{ g}$$

$$\text{Total} \quad 44.01 \text{ g}$$

$$\text{weight fraction of C} = \frac{12.01}{44.01} = \sim\frac{3}{11}$$

b. Since any size sample of CO_2 always contains 3/11 carbon by weight (law of definite composition), then 3/11 of 1060.0 mg of CO_2 will be equal to the original weight of carbon in the unknown:

$$\frac{3}{11} \times 1060.0 \text{ mg} = 289.0 \text{ mg}$$

c. Find the weight fraction of hydrogen in the water that is produced. This can be used to yield the weight of hydrogen in the unknown:

H_2O

$$(1 \text{ mol}) \times (16.00 \text{ g mol}^{-1}) = 16.00 \text{ g}$$

$$(2 \text{ mols}) \times (1.01 \text{ g mol}^{-1}) = 2.01 \text{ g}$$

$$\text{Total} \quad 18.01 \text{ g}$$

$$\text{weight fraction of H} = \frac{2.01}{18.01} = \sim\frac{1}{9}$$

d. The weight of hydrogen in water will always be equal to 1/9 of the total weight:

$$\frac{1}{9} \times 145.0 \text{ mg} = 16.1 \text{ mg}$$

e. By computing the percentages of C and H in the unknown, the percentage of Cl will be the difference between the sum of the C and H and 100%

$$C = \frac{289.0 \text{ mg}}{593.0 \text{ mg total}} = 0.487 = 48.7\%$$

$$H = \frac{16.1 \text{ mg}}{593.0 \text{ mg total}} = 0.027 = 2.7\%$$

$$\text{C and H total} = 51.4\%$$

$$Cl = 100\% - 51.4\% = 48.6\%$$

Exercise 7.7

1. Calculate the percentage composition of the following compounds:
   a. KBr
   b. $Ca(OH)_2$
   c. H_2SO_4
   d. $Na_2S_2O_3$
   e. $Fe_2(HPO_4)_3$
2. Find the percentage of water in the compound $MgSO_4{\cdot}7H_2O$.
3. What is the percentage composition of indium chloride if 0.50 g in solution is added to enough silver nitrate solution so that 0.99 g of silver chloride is produced?

7.9 EMPIRICAL FORMULAS

The empirical formula of a compound is primarily obtained by direct experimentation. It expresses the simplest whole-number ratios of the atoms in a molecule or of the ions in a crystal.

An empirical formula is not necessarily the true formula of a compound. For example, mercury(I)chloride is written Hg_2Cl_2 although the simplest (empirical) formula is HgCl. However, if data are available so that the true molecular weight of a substance can be determined, the true molecular formula can be easily calculated, since its weight will usually be a small whole-number multiple of the empirical formula weight. Some examples follow:

***

EXAMPLE 7.30

Find the empirical formula of a compound that was analyzed to be 32.4% sodium, 22.5% sulfur, and 45.1% oxygen.

1. The first step is to express the percentages as grams. This is immediately accomplished if 100 g of compound is assumed. The numerical values are unchanged:

$$32.4\,\%\text{ of }100\text{ g} = 32.4\text{ g of Na}$$

$$22.5\,\%\text{ of }100\text{ g} = 22.5\text{ g of S}$$

$$45.1\,\%\text{ of }100\text{ g} = 45.1\text{ g of 0}$$

2. The second step is to calculate the comparative number of moles of Na, S, and O atoms. There are

$$(n = \text{moles})\ n_{Na} = \frac{32.4\text{ g}}{23.0\text{ g mol}^{-1}} = 1.41\text{ mols}$$

$$n_S = \frac{22.5\text{ g}}{32.1\text{ g mol}^{-1}} = 0.70\text{ mol}$$

$$n_O = \frac{45.1\text{ g}}{16.0\text{ g mol}^{-1}} = 2.82\text{ mols}$$

3. The formula $Na_{1.41}S_{0.70}O_{2.82}$ is not acceptable. The ratios of the atoms in a formula must be in terms of small *whole* numbers. This is done by dividing all the mole values by the smallest number present. The resulting formula will then have no number less than 1.

$$\text{Na}\,\frac{1.41}{0.70} = 2$$

$$\text{S}\,\frac{0.70}{0.70} = 1$$

$$\text{O}\,\frac{2.82}{0.70} = 4$$

(smallest number of moles → S)

4. The simplest whole-number ratio of the atoms in the compound is observed to be

$$Na_2S_1O_4$$

which is finally written as:

$$Na_2SO_4$$

(sodium sulfate)

EXAMPLE 7.31

Determine the formula of *hydrated* copper(II) sulfate if the residue weighs 0.89 g after the water is driven off by heating a 1.39 g sample.

1. Determine the weight of the water by calculating the difference between the weight of the hydrate and the dehydrated salt (residue):

$$\begin{array}{r} 1.39 \text{ g hydrate} \\ -0.89 \text{ g residue} \\ \hline 0.50 \text{ g water} \end{array}$$

2. Calculate the comparative number of moles of $CuSO_4$ and H_2O by dividing the grams of each by their respective mole weights:

$$CuSO_4 = 63.5 + 32.0 + (4 \times 16.0) = 159.5 \text{ g mole}^{-1}$$

$$H_2O = (2 \times 1.0) + 16.0 = 18.0 \text{ g mole}^{-1}$$

$$CuSO_4 = \frac{0.89 \cancel{g}}{159.5 \cancel{g} \text{ mol}^{-1}} = 0.00558 \text{ mole}$$

$$H_2O = \frac{0.50 \cancel{g}}{18.0 \cancel{g} \text{ mol}^{-1}} = 0.0278 \text{ mole}$$

Hint: The arithmetic could be made simpler and less liable to computational error if the mole numbers are made larger without changing their ratios. This may be done by multiplying both mole numbers by 100 or 1000 or whatever is most convenient. For example, the calculated ratio 0.00558/0.0278 is exactly the same as 55.8/278, since both numbers were multiplied by 10^4.

3. Use the simplified mole ratios to find the whole-number values. Divide both numbers by the smaller value:

$$CuSO_4 = \frac{55.8}{55.8} = 1$$

$$H_2O = \frac{278}{55.8} = 5$$

4. The formula of the hydrate is

$$CuSO_4 \cdot 5H_2O$$

(copper(II) sulfate 5-water)

Occasionally, the calculation of the empirical formula needs some extra refinement before it works out to a small whole number answer.

EXAMPLE 7.32

200.0 mg of a K, Cr, and O compound was analyzed and found to contain 70.8 mg of chromium and 76.0 mg of oxygen. Find the empirical formula.

1. Find the weight of potassium in the sample by subtracting the Cr and O sum from the total:

$$\begin{aligned} \text{total weight of sample} &= 200.0 \text{ mg} \\ \text{Cr and O weight} &= \underline{146.8 \text{ mg}} \\ \text{K} &= 53.2 \text{ mg} \end{aligned}$$

2. Calculate the comparative number of moles of each element after converting the weights to grams for convenience. Divide all the mole numbers by the smallest value.

$$\text{K} = \frac{53.2 \not{g}}{39.1 \not{g} \text{ mol}^{-1}} = 1.36 \text{ mols}, \frac{1.36}{1.36} = 1$$

$$\text{Cr} = \frac{70.8 \not{g}}{52.0 \not{g} \text{ mol}^{-1}} = 1.36 \text{ mols}, \frac{1.36}{1.36} = 1$$

$$\text{O} = \frac{76.0 \not{g}}{16.0 \not{g} \text{ mol}^{-1}} = 4.75 \text{ mols}, \frac{4.75}{1.36} = 3.5$$

3. The formula $K_1Cr_1O_{3.5}$ is not acceptable. The entire formula must be multiplied by the smallest number possible to produce a whole-number ratio. Obviously, 3.5 moles of oxygen must be *doubled*. The empirical formula becomes

$$K_2Cr_2O_7$$

(potassium dichromate)

It was mentioned previously that the empirical formula is not necessarily the true molecular formula. This is especially true in the case of **hydrocarbons,** which are compounds composed of C and H. Example 7.32 illustrates a method for determining the true molecular formula if data are available for determining the molecular weight.

EXAMPLE 7.32

A gaseous hydrocarbon is found to be 82.9% carbon. If the density of gas is 2.59 g liter^{-1} at STP, what is the true molecular formula?

1. The first step is to find the empirical formula of the hydrocarbon. Since the carbon is 82.9%, the hydrogen must be 17.1% (82.9% + 17.1% = 100.0%):

	100.0 *grams*	*Moles*	*Whole-Number Ratio*	*Corrected Whole-Number Ratio*
C	$= \frac{82.9\ \text{g}}{12.0\ \text{g mol}^{-1}}$	$= 6.9$	$\frac{6.9}{6.9} = 1$	2
H	$= \frac{17.1\ \text{g}}{1.0\ \text{g mol}^{-1}}$	$= 17.1$	$\frac{17.1}{6.9} = 2.5$	5

2. The empirical formula, C_2H_5, has an apparent molecular weight of 29 g mole^{-1}.
3. Use the mole concept as it relates to gases at STP to calculate the true molecular weight from the density:

$$(2.59\ \text{g liter}^{-1})(22.4\ \text{liter mol}^{-1}) = 58\ \text{g mol}^{-1}$$

$$\text{true molecular weight} = 58\ \text{g mol}^{-1}$$

4. Divide the empirical formula weight into the true molecular weight to find if the true formula is double, triple, or a larger multiple of the empirical formula:

$$\frac{58\ \text{g mol}^{-1}\ \text{(true)}}{29\ \text{g mol}^{-1}\ \text{(empirical)}} = 2$$

5. The true molecular formula is observed to be *double* the empirical formula:

$$\underset{\text{empirical}}{C_2H_5} \quad \text{becomes} \quad \underset{\text{true}}{C_4H_{10}}$$

6. As a final check, it can be seen that the formula weight of C_4H_{10} is indeed 58.

Exercise 7.8

1. Find the empirical formulas of compounds having the following percentage compositions:
   a. 75.0% C, 25.0% H
   b. 72.4% Fe, 27.6% O
   c. 21.8% Mg, 27.9% P, 50.3% O
   d. 70.0% Fe, 30.0% O
   e. 59.30% C, 4.55% H, 23.00% N, 13.15% O
2. If 10.0 g of a compound is analyzed and found to be composed of 4.21 g sodium, 1.89 g phosphorus, and 3.90 g oxygen, find the empirical formula.
3. A hydrocarbon is 85.7% C and 14.3 % hydrogen. If the vapor density at 1.0 atm and −20°C is 4.05 g $liter^{-1}$, what is the true molecular formula?

QUESTIONS AND PROBLEMS

7.1 Write the names of the following ions:

a. S^{2-}
b. SO_3^{2-}
c. $S_2O_3^{2-}$
d. ClO^-
e. ClO_2^-
f. HPO_4^{2-}
g. HSO_3^-
h. SCN^-
i. $H_2PO_4^-$
j. HCO_3^-

7.2 Write the formulas for the following ions:

a. phosphide
b. iodide
c. cyanide
d. phosphite
e. arsenate
f. hydrogen sulfate
g. bicarbonate
h. ferrous
i. ammonium
j. cyanate

7.3 Explain what is *wrong* with each of the following formulas:

a. $Cr(III)Cl_3$
b. $FE(NO_3)_2$
c. $MgOH_2$
d. KCn
e. $H_3PO_4^-$

7.4 Name the following compounds according to the Stock system:

a. Na_2SO_3
b. $ZnHPO_4$
c. $Fe(OAc)_3$
d. $NaNO_2$
e. $Hg_2(NO_3)_2$
f. $Ba(OH)_2$
g. $AgHSO_4$
h. $FeBr_2$
i. Na_3N
j. K_2O_2
k. $CuSO_4 \cdot 3H_2O$
l. Rb_3AsO_3
m. $Bi(OCN)_3$
n. $Ba(IO_3)_2$
o. $KMnO_4$
p. Be_3N_2
q. ZnS
r. $Co_2(Cr_2O_7)_3$
s. $Pb(HCO_3)_2$
t. $Sn(SO_4)_2$

7.5 Name the following compounds according to the classical (ous, ic) system:

a. $CuCl_2$
b. HgO
c. $Fe_2(C_2O_4)_3$
d. PbS_2
e. $CoHPO_4$
f. $Fe(BrO_3)_3$
g. $Ni(OH)_2$
h. Cu_2S
i. $Sn(CO_3)_2$
j. $Sb(ClO)_2$

7.6 Name the following nonmetallic compounds by the Stock or the descriptive (mono, di, tri . . .) methods:

a. PCl_3
b. SiF_4
c. Bi_2O_5
d. NO_2
e. P_4S_7
f. S_2Cl_2
g. P_2O_5
h. Br_3O_8
i. VF_6
j. As_4O_6

7.7 Write the names of the following acids:

a. HBr
b. HNO_2
c. H_3PO_3
d. HIO_3
e. $HClO_4$
f. HCN
g. $HOCN$
h. H_2S
i. $HOAc$
j. $H_2C_2O_4$

7.8 Write formulas for the following compounds:

a. copper(II) sulfate
b. carbon disulfide
c. magnesium hydroxide
d. aluminum arsenate
e. phosphoric acid
f. cobalt(II) bromide
g. barium chlorate
h. zinc iodide
i. acetic acid
j. mercurous iodide
k. diarsenic pentoxide
l. potassium permanganate
m. sodium peroxide
n. nickelic sulfate
o. stannic nitrate
p. cadmium hydrogen carbonate
q. carbon(II) oxide
r. mercuric cyanide
s. chromium(II) hydroxide
t. ammonium carbonate

7.9 Name the following compounds according to the Stock system:

a. NH_4OH
b. HCN
c. As_2S_3
d. SiC
e. $NaMnO_4$
f. CdI_2
g. $Ag_2Cr_2O_7$
h. $Ba(ClO_3)_2$
i. SiF_4
j. $Na_2SO_4 \cdot 10H_2O$
k. $Ni(OCN)_2$
l. $XeCl_6$
m. $CoCl_3 \cdot 3H_2O$
n. HI
o. $BiCl_3$
p. AlN
q. Li_3AsO_4
r. $H_2C_2O_4$
s. $Fe(SCN)_3$
t. $CuHPO_4$

7.10 Write formulas for the following compounds and ions:

a. sulfur trioxide
b. diiodine pentoxide
c. iron (II) hydrogen carbonate
d. boron trifluoride
e. potassium arsenide
f. stannic oxide
g. hydrocyanic acid
h. sulfite ion
i. cuprous thiosulfate
j. ammonia
k. cobalt(III) ion
l. hydrosulfuric acid
m. antimony(V) perchlorate
n. carbon monoxide
o. zinc sulfate 20-water
p. dichlorine heptoxide
q. strontium hydroxide
r. barium hypochlorite 2-hydrate
s. ammonium ion
t. cupric phosphite

7.11 Complete the following table by writing the possible names and formulas.

	Br^-	CO_3^{2-}	PO_4^{3-}
K^+	KBr potassium bromide		
Fe^{2+}			
Ni^{3+}			
As^{5+}			
NH_4^+			

7.12 Calculate the percentage composition of the following:
- a. KOCN
- b. $Fe(NO_3)_2$
- c. Fe_3O_4
- d. NH_4HCO_3

7.13 If 3.20 g of $Na_2SO_4 \cdot 10H_2O$ is heated to complete dehydration, what weight of residue remains?

7.14 What is the empirical formula of a compound that is analyzed to contain 75.8% As and 24.2% oxygen?

7.15 Find the true molecular formula of a gaseous hydrocarbon that is 92.3% carbon if 1.1 g occupies 0.316 liter at STP.

CHAPTER 8

My business is to teach my aspirations to conform themselves to fact, not to try to make facts harmonise with my aspirations. Sit down before fact as a little child, be prepared to give up every preconceived notion, follow humbly wherever nature leads, or you will learn nothing.

Thomas Huxley, quoted in *The Art of Scientific Investigation**

At the completion of this chapter, the student should be able to:

1. Define the term stoichiometry and describe a stoichiometric reaction.
2. List four experimental observations that identify a reaction as stoichiometric.
3. Balance equations by inspection.
4. State the information that a balanced equation conveys.
5. Classify chemical equations as representative of the following types: decomposition, single replacement, and double replacement.
6. Use the activity series to predict the products of single replacement reactions.
7. Perform stoichiometric calculations from balanced equations when given data in the form of moles, masses, and gas volumes.
8. Use the factor-label method to perform calculations.
9. Apply the equation of state to the solution of stoichiometric problems involving gases at nonstandard conditions.
10. Distinguish between theoretical and actual product yields in reactions.★
11. Identify reaction-limiting compounds in the solution of stoichiometric★ problems.

---

* By W. I. B. Beveridge. W. W. Norton & Company Inc., 1957.

CHEMICAL EQUATIONS AND STOICHIOMETRY

8.1 INTRODUCTION

The chemical equation is a very special shorthand that chemists use to describe the **reactants** and the **products** of a chemical change. The reactants, conventionally written to the left of a **yield** sign (an arrow, $\rightarrow$), state what substances interact to produce new substances. The new substances, the products, are the end results, after the reaction has gone to completion. A reaction is said to be complete when any one of the reactants has been completely converted to end products. A reaction that goes to completion (often evidenced by the production of a *gas*, the formation of a *precipitate*, a large *energy change*, or possibly a *color change*) is called a **stoichiometric** reaction. A reaction that does not go to completion—i.e., reaches a point where both the original reactants and the products coexist in a dynamic balance, is called an **equilibrium** reaction. An equilibrium reaction is distinguished from a stoichiometric reaction by the use of a double-arrow yield sign ($\rightleftharpoons$), in which the arrow pointing to the right ($\rightarrow$) indicates the *forward reaction*, while the arrow pointing to the left ($\leftarrow$) indicates the *reverse reaction.*

The factor that makes a chemical equation a special kind of shorthand is its **quantitative** nature. An equation not only describes what reacts and what is produced, but also *how much* of each participating species is involved. The law of conservation is paramount in this regard. When a chemist begins with a given number of grams of reactants, he expects to end up with the same number of grams of products regardless of how dramatic the chemical change may be. In a chemical change, the amount of matter that may be converted to radiant energy is not detectable by ordinary laboratory balances. Stoichiometric reactions are most conveniently described in terms of the mole concept.

Before analyzing chemical equations as mole-to-mole ratios of reactants and products, some useful embellishments should be pointed out. While an equation does not necessarily have to indicate more than the mole ratios of original reactants and final products, special symbols may be used to give a fuller picture of the conditions under which the reactants combined and the products emerged. For example, substances can be annotated to indicate whether they are solids, liquids, gases, insoluble, in solution, or in the presence of a catalyst. The overall reaction may be described quantitatively as **exothermic** (heat producing) or **endothermic** (heat absorbing) by writing plus (+) or minus (−) so many calories for the reaction. Heating reactants to supply the energy of activation may be noted by a capital *delta* (Δ). A summary of these kinds of notations is provided in Table 8–1.

TABLE 8-1 CLARIFYING SYMBOLS IN CHEMICAL EQUATIONS

Symbol	Meaning	Examples
(*s*)	solid reactant or insoluble solid product	$Zn(s)$ $PbCrO_4(s)$ $H_2O(s)$ ice
(ℓ)	reactant in liquid state or liquid product	$H_2O(\ell)$ $Hg(\ell)$ $Br_2(\ell)$
(*g*)	vapor state reactant or product evolved as a gas	$CO_2(g)$ $H_2(g)$ $H_2O(g)$ steam
(*aq*) aqueous	reactants or products in water solutions	$NaCl(aq)$ $KBr(aq)$ $Fe(NO_3)_2(aq)$
$\xrightarrow{\Delta}$	reactants heated to supply energy of activation	$A + B \xrightarrow{\Delta} C + D$
Pt	platinum used as a catalyst	$A + B \xrightarrow{Pt} C + D$
or		
MnO_2	MnO_2 used as a catalyst	$XY \xrightarrow{MnO_2} X + Y$
Equation + kcal (or, + cal)	exothermic reaction	$CH_4 + 2O_2 \rightarrow CO_2 + 2H_2O$ +2.13 kcal
Equation − kcal (or, − cal)	endothermic reaction	$2HCl \rightarrow Cl_2 + H_2 - 44{,}100$ cal

8.2 BALANCING EQUATIONS

The process of balancing an equation is analogous to the handling of algebraic equations. In the case of chemical equations, however, a careful distinction must be made between the subscripts and the **coefficients.** The coefficients in chemical equations are the large numbers written *in front of* the formulas (never in the middle of a formula). While the subscripts are concerned with the *law of constant composition*, the coefficients are governed by the *law of conservation.* For example, the chemical combination of aluminum and oxygen produces aluminum oxide. If conservation were the only factor to deal with, an incorrect equation could be written:

$$Al + O_2 \longrightarrow AlO_2$$

wrong

The formula AlO_2 does balance the equation insofar as it results in the same number of Al and O units on both sides of the equation. However, the equation is wrong! The law of constant composition clearly states that there is only one correct formula for aluminum oxide, Al_2O_3, based on the common oxidation numbers of aluminum and oxide:

$$Al^{3+}\, O^{2-} = Al_2O_3$$

The problem can be most efficiently dealt with by establishing a few rules for balancing equations by **inspection.**

Rules

1. Write the correct formulas for every reactant and product. This is based on the oxidation numbers of the species involved. The result is a **skeleton,** or unbalanced equation:

$$Al + O_2 \rightarrow Al_2O_3$$

2. The subscripts must not be changed (law of constant composition). For example

$$Al_2 + O_3 \rightarrow Al_2O_3$$

cannot be used. There is no evidence to support the existence of aluminum as a diatomic molecule. O_3 is the formula for ozone, not oxygen.

3. A recommended procedure for balancing equations is to follow this order:
   a. Balance the metals first.
   b. Balance the nonmetals and polyatomic ions second.
   c. Balance the hydrogen atoms (if present).
   d. Finally, balance the oxygen atoms (if present).
4. The coefficients used should be the smallest numbers possible. The use of a fractional coefficient is perfectly satisfactory if it speeds up the balancing procedure. For example, in the equation

$$Al + O_2 \rightarrow Al_2O_3$$

the aluminum atoms are balanced by placing a coefficient of 2 before the aluminum:

$$2Al + O_2 \rightarrow Al_2O_3$$

The oxygen atoms can be directly balanced by multiplying the O_2 formula by the fraction 3/2:

$$2Al + \frac{3}{2}O_2 \rightarrow Al_2O_3$$

The equation is balanced.

5. If for some reason whole-number coefficients are preferred, the total equation may be doubled:

$$2\left(2Al + \frac{3}{2}O_2 \rightarrow Al_2O_3\right) = 4Al + 3O_2 \rightarrow 2Al_2O_3$$

6. It must be remembered that the coefficient multiplies *every* atom in the formula:

$2Al_2O_3$ means

$2 \times Al_2 = 4$ aluminum atoms, and

$2 \times O_3 = 6$ oxygen atoms

In reality, the coefficient means that in 2 moles of Al_2O_3, there are 4 moles of aluminum atoms and 6 moles of oxygen atoms.

EXAMPLE 8.1

In the formula $2Fe_2(SO_4)_3$, how many iron, sulfur, oxygen, and sulfate units are there?

1. $2Fe_2(SO_4)_3$
   4 iron atoms
2. $2Fe_2(S_1O_4)_3$
   $2 \times 3 \times 1$ sulfur atoms = 6 sulfur atoms
3. $2Fe_2(S_1O_4)_3$
   $2 \times 3 \times 4$ oxygen atoms = 24 oxygen atoms
4. $2Fe_2(SO_4)_3$
   2×3 sulfate ions = 6 sulfate ions

8.3 THE MEANING OF THE BALANCED EQUATION

The balanced equation indicates the proportionate number of moles of each reactant and product involved. Since a number of moles can be expressed as grams, as a volume of gas, and as a number of particles, a great deal of information is available. For example, the equation $2Al + 3/2O_2 \rightarrow Al_2O_3$ means the following:

1. 2 moles of aluminum atoms + $1\frac{1}{2}$ moles of oxygen molecules produces 1 mole of aluminum oxide.
2. Since 1 mole of aluminum atoms is 27.0 g and 1 mole of O_2 molecules is 32.0 g, then

$$2Al + \frac{3}{2}O_2 \rightarrow Al_2O_3$$

$$2\,\cancel{mols}(27.0\text{ g }\cancel{mol^{-1}}) + \frac{3}{2}\cancel{mols}(32.0\text{ g }\cancel{mol^{-1}}) \rightarrow 1\,\cancel{mol}(102.0\text{ g }\cancel{mol^{-1}})$$

$$54.0\text{ g} + 48.0\text{ g} \longrightarrow 102.0\text{ g}$$

3. At STP, 1 mole of any gas occupies 22.4 liters, $1\frac{1}{2}$ moles will occupy 33.6 liters. Therefore

$$2Al + \frac{3}{2}O_2 \longrightarrow Al_2O_3$$

$$54.0\text{ g} + 33.6\text{ liters} \rightarrow 102.0\text{ g at STP}$$

4. The number of particles (atoms or molecules) also relates to the mole concept:

$$2Al + \frac{3}{2}O_2 \longrightarrow Al_2O_3$$

$$2 \times 6.02 \times 10^{23} + 1\frac{1}{2} \times 6.02 \times 10^{23} \rightarrow 6.02 \times 10^{23}$$

atoms of aluminum + molecules of oxygen → molecules of aluminum oxide

EXAMPLE 8.2

Balance the equation

$$C_3H_8 + O_2 \rightarrow CO_2 + H_2O$$

and list the information it conveys.

1. Check the carbon:

$$C_3H_8 + O_2 \rightarrow CO_2 + H_2O$$

3 carbon   1 carbon

Balance the carbon:

$$C_3H_8 + O_2 \rightarrow \mathbf{3}CO_2 + H_2O$$

2. Check the hydrogen:

$$C_3H_8 + O_2 \rightarrow 3CO_2 + H_2O$$

8 hydrogen   2 hydrogen

Balance the hydrogen:

$$C_3H_8 + O_2 \rightarrow 3CO_2 + \mathbf{4}H_2O$$

3. Check the oxygen:

$$C_3H_8 + O_2 \rightarrow 3CO_2 + 4H_2O$$

$$2 \text{ oxygen} = \underbrace{6 \text{ oxygen} + 4 \text{ oxygen}}_{10 \text{ oxygen}}$$

Balance the oxygen:

$$C_3H_8 + \mathbf{5}O_2 \rightarrow 3CO_2 + 4H_2O$$

4. The equation is balanced:

	reactants			products	
	$C_3H_8(g)$	$+ 5O_2(g)$	$\rightarrow$	$3CO_2(g)$	$+ 4H_2O(g)$
moles of molecules:	1 mole	+ 5 moles	$\rightarrow$	3 moles	+ 4 moles
mass ratios:	44.0 g	+ 160.0 g	$\rightarrow$	132.0 g	+ 72.0 g
volume ratios:	22.4 liters	+ (5 × 22.4) liters	$\rightarrow$	(3 × 22.4) liters	+ (4 × 22.4) liters
particle ratios:	6.02×10^{23} molecules	+ $(5 \times 6.02 \times 10^{23})$ molecules	$\rightarrow$	$(3 \times 6.02 \times 10^{23})$ molecules	+ $(4 \times 6.02 \times 10^{23})$ molecules

EXAMPLE 8.3

Write a balanced equation that describes the reaction between solid aluminum and oxygen gas as it produces solid aluminum oxide in an exothermic reaction (380.0 kcal).

$$2Al(s) + \frac{3}{2}O_2(g) \rightarrow Al_2O_3(s) + 380.0 \text{ kcal}$$

EXAMPLE 8.4

Write and balance the equation for the heating of mercury (II) oxide (solid) with the production of oxygen and mercury.

1. Write skeleton equation:

$$HgO(s) \xrightarrow{\Delta} Hg(l) + O_2(g)$$

2. The oxygen atoms are balanced by placing a coefficient of 1/2 before the O_2:

$$HgO(s) \xrightarrow{\Delta} Hg(l) + \frac{1}{2}O_2(g)$$

EXAMPLE 8.5

Write a balanced equation for the reaction between iron (III) nitrate and sodium sulfide, assuming that iron (III) sulfide is precipitated and sodium nitrate is produced in water solution.

1. Write the correct formulas for all the reactants and products:

$$Fe(NO_3)_3 + Na_2S \rightarrow Fe_2S_3 + NaNO_3$$

2. Use the appropriate symbols to indicate what is in solution and what is not:

$$Fe(NO_3)_3(aq) + Na_2S(aq) \rightarrow Fe_2S_3(s) + NaNO_3(aq)$$

3. Check the iron:

$$Fe(NO)_3 \rightarrow Fe_2S_3$$

1 iron atom   2 iron atoms

Balance the iron:

$$2Fe_1(NO_3)_3(aq) + Na_2S(aq) \rightarrow Fe_2S_3(s) + NaNO_3(aq)$$

4. $2Fe(NO_3)_3$ means a total of 6 nitrate ions. Balance the nitrate ions:

$$2Fe(NO_3)_3(aq) + Na_2S(aq) \rightarrow Fe_2S_3(s) + 6NaNO_3(aq)$$

5. Balance the sulfur and sodium simultaneously by a coefficient of 3 before Na_2S. The equation is checked and found to be balanced:

$$2Fe(NO_3)_3(aq) + 3Na_2S(aq) \rightarrow Fe_2S_3(s) + 6NaNO_2(aq)$$

EXAMPLE 8.6

Balance the descriptive equation zinc + hydrochloric acid → zinc chloride + hydrogen gas.

1. Write the correct formulas for each reactant and product:

$$Zn + HCl \rightarrow ZnCl_2 + H_2$$

2. Notice how a coefficient of 2 before the HCl balances the equation:

$$Zn + 2HCl \rightarrow ZnCl_2 + H_2$$

3. Make the equation more descriptive:

$$Zn(s) + 2HCl(aq) \rightarrow ZnCl_2(aq) + H_2(g)$$

Summary of Errors and Erroneous Concepts in Equations

Some errors and erroneous concepts common to beginning students are as follows:

1. Many equations that can be balanced are not real. They are "paper predictions." An equation is primarily designed to quantitatively describe what does happen in fact. For example, the equations

$$CaO \rightarrow Ca + \frac{1}{2}O_2$$

and

$$Cu + 2HCl \rightarrow CuCl_2 + H_2$$

look good. They are balanced. However, they are not real, CaO is not decomposed by heat, and under ordinary laboratory conditions, copper does not react with hydrochloric acid.

2. The equation is not meant to suggest that the simple addition of reactants will produce a chemical change. When the equation $Mg + 1/2\ O_2 \rightarrow MgO$ is written, it does not imply that a strip of magnesium ribbon will visibly react with oxygen at room temperature. Nor will solid $AgNO_3$ react appreciably with NaCl in the solid state.
3. The equation is not really meant to represent individual atoms and molecules (although this is possible). The formula represents a *mole* of the substance. The use of fractional coefficients would not be feasible if the formulas represented individual particles.
4. Coefficients are always placed in front of the formula and not in the middle. For example, the equation $2Al + O_2 \rightarrow Al_2\ 2/3\ O_3$ is quite incorrect, despite the observation that it seems to be balanced.

Exercise 8.1

1. Balance the following skeleton equations, assuming that the reactions do occur:

a. $KClO_3 \xrightarrow{\Delta} KCl + O_2$
b. $Na + H_2O \rightarrow NaOH + H_2$
c. $CuO + H_2 \rightarrow Cu + H_2O$
d. $SO_3 + H_2O \rightarrow H_2SO_4$
e. $H_2SO_4 + Al(OH)_3 \rightarrow Al_2(SO_4)_3 + H_2O$
f. $Pb(OAc)_2 + K_2Cr_2O_7 \rightarrow PbCr_2O_7 + KOAc$
g. $(NH_4)_2S + Fe(NO_3)_3 \rightarrow NH_4NO_3 + Fe_2S_3$
h. $Ca + H_2O \rightarrow Ca(OH)_2 + H_2$
i. $(NH_4)_2CO_3 \rightarrow NH_3 + CO_2 + H_2O$
j. $Al + H_2C_2O_4 \rightarrow Al_2(C_2O_4)_3 + H_2$

2. Write balanced equations for the following reactions:
   a. sodium + water → sodium hydroxide + hydrogen
   b. calcium oxide + water → calcium hydroxide
   c. potassium cyanide + nitric acid → potassium nitrate + hydrocyanic acid
   d. silver oxide $\xrightarrow{\Delta}$ silver + oxygen
   e. hydrogen + oxygen → water
   f. hydrogen peroxide → oxygen + water
   g. iron + sulfur → iron (II) sulfide
   h. gold (III) chloride + zinc → zinc chloride + gold
   i. hydrogen + nitrogen → ammonia
   j. phosphoric acid + strontium hydroxide → strontium phosphate + water

3. Write descriptive equations for the following reactions:
   a. Xenon gas reacts with fluorine gas (F_2) to produce crystals of xenon tetrafluoride while giving off 62.5 kcal of heat in the exothermic change.
   b. Zinc metal reacts with a water solution of copper (II) sulfate to produce a solution of zinc sulfate and copper granules.
   c. When chlorine gas (Cl_2) is bubbled into a water solution of potassium bromide, a potassium chloride solution and liquid bromine are produced.

4. Given the skeleton equation $C_2H_4 + O_2 \rightarrow CO_2 + H_2O$, answer the following questions after balancing:
   a. How many moles of oxygen are needed to produce 2 moles of CO_2?
   b. How many grams of C_2H_4 are required to produce 2 moles of water?
   c. What volume of carbon dioxide (STP) is produced from 1 mole of C_2H_4?

8.4 TYPES OF CHEMICAL REACTIONS

Although it has been previously mentioned that the writing of chemical equations is not a matter of magically manipulating symbols and numbers until a balance is achieved, it is possible to predict the products of some reactions because of their clear identity as a particular type of reaction. The ultimate test of the reality of a chemical reaction, however, is made by experimentation.

Reactions are usually classified as **decomposition** (analysis), **combination** (synthesis), **single replacement** (substitution), and **double replacement** (metathesis). More specialized titles include **neutralization** and **hydrolysis** reactions, which in one sense are types of double replacement reactions. The special cases of **oxidation-reduction** (redox) and **ionic** equations will be taken up in another chapter.

Decomposition Equations

Decomposition reactions are ones in which a single unstable compound is treated in such a way that it decomposes. The decomposition products of binary compounds are obviously the two elements making up the binary compound. The decomposition products of ternary compounds, however, are not predictable without additional information. Examples of decomposition reactions are as follows:

$$HgO \rightarrow Hg + \frac{1}{2} O_2$$

$$KClO_3 \rightarrow KCl + \frac{3}{2} O_2$$

$$PbO_2 \rightarrow Pb + O_2$$

$$NaNO_3 \rightarrow NaNO_2 + \frac{1}{2} O_2$$

$$(NH_4)_2CO_3 \rightarrow 2NH_3 + CO_2 + H_2O$$

Combination Reactions

Combination, or synthesis, reactions may involve the simple chemical union of two elements to form a compound. Sometimes compounds may be reactants in a combination reaction that produces a more complex compound. Examples of combination reactions are

$$C + O_2 \rightarrow CO_2$$

$$P_4 + 5O_2 \rightarrow P_4O_{10}$$

$$\frac{1}{8} S_8 + O_2 \rightarrow SO_2$$

$$Mg + \frac{1}{2} O_2 \rightarrow MgO$$

$$K_2O + H_2O \rightarrow 2KOH$$

$$SO_3 + H_2O \rightarrow H_2SO_4$$

Single Replacement Reactions

Single replacement, or substitution, reactions occur when a more chemically active, free element displaces a less chemically active element from a compound. In reality, this is an oxidation-reduction reaction because there must be a loss and gain of electrons in such a process. A free element is in the "zero" oxidation state. When it becomes chemically combined as a result of the replacement reaction, it necessarily exhibits some "valence" or oxidation number that is above or below zero. A gain (more positive) in oxidation number is, by definition, *oxidation*, while the converse is *reduction*.

However, for the sake of simplicity, replacement reactions can be predicted with a fair degree of reliability from an empirically obtained **activity series.** The activity series, shown in Table 8–2, indicates that a free element on the list can replace any element *below* it from a compound, but there is no reaction when the free element is added to a water solution of a compound containing an element *above* it in the activity series.

TABLE 8–2 THE ACTIVITY SERIES

	Element	Activity Description
Decreasing metal activity ↓	Li	Li through Na: will replace hydrogen from liquid water
	K	
	Rb	
	Cs	
	Ba	
	Sr	
	Ca	
	Na	
	Mg	Li through Fe: will replace hydrogen from steam
	Al	
	Mn	
	Zn	
	Cr	
	Fe	Li through Pb: will replace hydogen from acids
	Ni	
	Sn	
	Pb	
	H_2	
	Cu	
	I_2	
	Hg	Ag → I_2: will replace
	Ag	
	Br_2	
	O_2	Cl_2 → Br_2: will replace
	Cl_2	
Increasing nonmetal activity	Au	F_2 → Cl_2: will replace
	F_2	

EXAMPLE 8.7

Write the equation for the reaction between $Zn(s)$ and $CuSO_4(aq)$.

1. Write the reactants in the equation format:

$$Zn(s) + CuSO_4(aq) \rightarrow ?$$

Can zinc replace copper?

2. Observe in Table 8–2 that zinc is *above* copper. Therefore, a safe prediction would be that zinc can replace copper:

$$Zn(s) + CuSO_4(aq) \rightarrow ZnSO_4(aq) + Cu(s)$$

Replace

EXAMPLE 8.8

Write the equation for the reaction between copper and hydrochloric acid.

1. Set up the equation:

$$Cu(s) + HCl(aq) \rightarrow ?$$

Can copper replace hydrogen?

2. From Table 8–2, it is seen that copper is *below* hydrogen and therefore will not react. A common symbol that indicates no reaction is **NR**:

$$Cu(s) + HCl(aq) \rightarrow NR$$

Examples of Single Replacement Reactions

1. Active metals replacing hydrogen from liquid water produce a hydroxide and hydrogen gas:

$$Li(s) + H_2O(\ell) \rightarrow LiOH(aq) + \frac{1}{2} H_2(g)$$

Remember that water can be written as HOH (hydrogen hydroxide). A more pictorial replacement is

$$\mathit{Li} + \mathit{H}OH \rightarrow LiOH + \frac{1}{2} H_2$$

Replace

Other examples are

$$Ca(s) + 2H_2O(\ell) \rightarrow Ca(OH)_2(s) + H_2(g)$$
$$Cs(s) + H_2O(\ell) \rightarrow CsOH(aq) + \tfrac{1}{2} H_2(g)$$
$$Ba(s) + 2H_2O(\ell) \rightarrow Ba(OH)_2(s) + H_2(g)$$

2. Moderately active metals replacing hydrogen from steam produce metal oxides and hydrogen gas:

$$Mg(s) + H_2O(g) \rightarrow MgO(s) + H_2(g)$$
$$2Fe(s) + 3H_2O(g) \rightarrow Fe_2O_3(s) + 3H_2(g)$$

3. A metal above hydrogen in the activity series will replace hydrogen from acids:

$$Zn(s) + 2HCl(aq) \rightarrow ZnCl_2(aq) + H_2(g)$$
$$Cd(s) + 2HCl(aq) \rightarrow CdCl_2(aq) + H_2(g)$$
$$2Al(s) + 3H_2SO_4(aq) \rightarrow Al_2(SO_4)_3(aq) + 3H_2(g)$$
$$Ag(s) + HCl(aq) \rightarrow NR$$

4. A metal will probably replace any other metal *below* itself in the activity series:

$$Zn(s) + CuSO_4(aq) \rightarrow ZnSO_4(aq) + Cu(s)$$
$$Fe(s) + Pb(NO_3)_2(aq) \rightarrow Fe(NO_3)_2(aq) + Pb(s)$$
$$Cu(s) + 2AgNO_3(aq) \rightarrow Cu(NO_3)_2(aq) + 2Ag(s)$$

5. A nonmetal will probably replace any other nonmetal *above* itself in the activity series:

$$F_2(g) + 2NaCl(aq) \rightarrow 2NaF(aq) + Cl_2(g)$$
$$Cl_2(g) + 2KBr(aq) \rightarrow 2KCl(aq) + Br_2(\ell)$$
$$Br_2(\ell) + 2NaI(aq) \rightarrow 2NaBr(aq) + I_2(s)$$
$$I_2(s) + NaBr(aq) \rightarrow NR$$

Double Replacement Reactions

A double replacement, or metathesis, reaction occurs between two compounds in which there is a "switch" of positive and negative ions. It can be thought of as an exchange of partners in which the metal ion of compound A becomes the positive ion member of compound B, as the original metal ion of compound B switches over and becomes the positive ion member of compound A:

compound A compound B

M A + M B ⟶ M A + M B

It is very important to bear in mind that the double replacement reaction must result in compounds that have net oxidation numbers of zero. This cannot be accomplished if the formulas of supposed products result from the combination of ions that have the same charge. For example, in the reaction between sodium chloride and silver nitrate, the sodium and silver ions have positive oxidation numbers while the chloride and nitrate ions are negative:

$$\overset{+\ -}{\mathrm{NaCl}} + \overset{+\ -}{\mathrm{AgNO_3}} \rightarrow$$

The only reasonable double replacement would be for the Na^+ to combine with the NO_3^- and the Ag^+ to combine with the Cl^-:

$$\overset{+\ -}{\mathrm{NaCl}} + \overset{+\ -}{\mathrm{AgNO_3}} \rightarrow \overset{+\ -}{\mathrm{NaNO_3}} + \overset{+\ -}{\mathrm{AgCl}}$$

The only other change that might be tried tentatively would result in a grossly incorrect equation:

$$\overset{+\ -}{\mathrm{NaCl}} + \overset{+\ -}{\mathrm{AgNO_3}} \rightarrow \overset{+\ +}{\mathrm{NaAg}} + \overset{-\ -}{\mathrm{ClNO_3}}$$

obviously incorrect

If the student is at all thoughtful about the products predicted in a double replacement reaction, he will see that formulas such as NaAg and $ClNO_3$ violate almost every rule and law thus far encountered that apply to the behavior of matter. The names sodium silveride or argentide and chlorine nitrite are discordant noises to the ear. Of course, there are exceptions to many "sacred" rules and laws, but that will be a subject of future concern.

With regard to the question of why a double replacement reaction occurs, the answer will become more apparent as the topics of electrolyte strength and solubility rules are taken up in Chapter 9. However, it is useful in the meantime to categorize double replacement reactions as ones in which more stable products emerge. The more stable compounds are most notably *precipitates* and *water*.

Examples of Double Replacement Reactions

1. When two soluble compounds interact to form an insoluble precipitate, the reaction usually occurs:

$$\overset{+\ -}{\mathrm{AgNO_3}}(aq) + \overset{+\ -}{\mathrm{KCl}}(aq) \rightarrow \overset{+\ -}{\mathrm{AgCl}}(s) + \overset{+\ -}{\mathrm{KNO_3}}(aq)$$

$$3\overset{+\ -}{\mathrm{NaOH}}(aq) + \overset{+\ -}{\mathrm{FeCl_3}}(aq) \rightarrow 3\overset{+\ -}{\mathrm{NaCl}}(aq) + \overset{+\ -}{\mathrm{Fe(OH)_3}}(s)$$

$$\overset{+\ -}{\mathrm{Pb(NO_3)_2}}(aq) + \overset{+\ -}{\mathrm{K_2CrO_4}}(aq) \rightarrow \overset{+\ -}{\mathrm{PbCrO_4}}(s) + 2\overset{+\ -}{\mathrm{KNO_3}}(aq)$$

$$\mathrm{BaCl_2}(aq) + \mathrm{ZnSO_4}(aq) \rightarrow \mathrm{BaSO_4}(s) + \mathrm{ZnCl_2}(aq)$$

$$2\mathrm{Na_3PO_4}(aq) + 3\mathrm{CaBr_2}(aq) \rightarrow \mathrm{Ca_3(PO_4)_2}(s) + 6\mathrm{NaBr}(aq)$$

2. When an acid and base react to form a salt and water, the reaction is described as an acid-base neutralization. It is also a double replacement reaction. Sometimes the base is not in the obvious form of a hydroxide. It may be a metallic oxide, which is known as a **basic anhydride.** The term basic anhydride means a dehydrated base, which is a way of saying that the compound would become a typical hydroxide if water were added and the metallic oxide were water soluble. For example, $Na_2O(s) + H_2O(\ell) \rightarrow 2NaOH(aq)$ illustrates the reaction between the basic anhydride Na_2O and water, which forms the typically alkaline solution of NaOH. The effect of this observation is to provide the basis for understanding how a metallic oxide behaves similarly to metal hydroxides in reacting with acids in acid-base neutralization:

$$NaOH(aq) + HCl(aq) \rightarrow NaCl(aq) + H_2O(\ell)$$

$$2KOH(aq) + H_2SO_4(aq) \rightarrow K_2SO_4(aq) + 2H_2O(\ell)$$

$$2HNO_3(aq) + CaO(s) \rightarrow Ca(NO_3)_2(aq) + H_2O(\ell)$$

$$2Al(OH)_3(s) + 3H_2SO_4(aq) \rightarrow Al_2(SO_4)_3(aq) + 6H_2O(\ell)$$

$$SrO(s) + 2HBr(aq) \rightarrow SrBr_2(aq) + H_2O(\ell)$$

Exercise 8.2

1. Complete and balance the following reactions and label each one as *decomposition*, *combination*, *single replacement*, *double replacement*, or *no reaction:*

   a. $Ca + O_2 \rightarrow$
   b. $Al + S_8 \rightarrow$
   c. $Au_2O_3 \xrightarrow{\Delta}$
   d. $P_4 + Cl_2 \rightarrow$ phosphorus trichloride
   e. $Cs + H_2O(\ell) \rightarrow$
   f. $Al + H_2C_2O_4 \rightarrow$
   g. $N_2O_5 + H_2O \rightarrow$
   h. $NiCl_2 + KOH \rightarrow$
   i. $Fe_2O_3 + C \rightarrow$
   j. $CuO + HNO_3 \rightarrow$

2. Complete and balance the following equations:

   a. strontium + water →
   b. cadmium + copper (II) sulfate →
   c. copper (II) oxide + hydrogen →
   d. sulfur trioxide + water →
   e. antimony (III) sulfide + hydrochloric acid →
   f. platinum (IV) oxide $\xrightarrow{\Delta}$
   g. sulfuric acid + chromium (III) oxide →

h. benzene (C_6H_6) + oxygen →
i. aluminum + bromine →
j. lead (II) nitrate + potassium chromate →

8.5 MOLES, MASS, AND VOLUME RELATIONSHIPS IN EQUATIONS

The process of performing useful, sometimes essential, calculations on the basis of balanced equations is called stoichiometry. It has been emphasized that the correctly balanced equation is an absolute necessity if one is to know the mole-to-mole ratios of reactants and products. The importance of calculating the weights or the volumes, or both, of the constituents of a reaction is both an economic and a safety issue. For example, silver nitrate *is* very expensive. It would be irresponsible to use a large excess if the precise amount needed in a reaction can be calculated. If everyone in a freshman chemistry laboratory used more silver nitrate than was necessary, there would be a waste of many dollars.

Suppose, as another example, chlorine gas is to be produced and collected in bottles. Unless the expected volume of chlorine gas is calculated, there is no sure way of knowing how many gas-collecting bottles are needed. If several people were to produce more chlorine than could be collected in the available bottles, many students in the laboratory could suffer from the irritating and toxic overflow of chlorine gas. While the empirical (trial and error) approach has its place, the use of reference texts for data on physical and chemical properties and the performance of stoichiometric calculations describe the practice of chemistry. The try-it-and-see approach is a last resort for the *experienced* chemist, and then only with appropriate safety equipment and fume exhaust hoods.

In effect, all stoichiometric problems are **mole-to-mole** calculations. However, since a mole can be expressed as a formula weight, a gas volume, or a number of particles, stoichiometric problems may be classified as *moles-to-mass*, *mass-to-moles*, *mass-to-mass*, *mass-to-volume*, *volume-to-mass*, and *volume-to-volume*.

Review for a moment the equation $KClO_3 \rightarrow KCl + 3/2\ O_2$. This equation clearly says that the decomposition of 1 mole of $KClO_3$ yields 1 mole of KCl and $1\frac{1}{2}$ moles of O_2. One mole of $KClO_3$ weighs 122.5 g. It is not reasonable, on the basis of economy and safety, to use more than a fraction of a mole of $KClO_3$ to produce oxygen and KCl. The actual amount of $KClO_3$ required should be based on the volume of oxygen that one is prepared to collect. Furthermore, the calculated volume of oxygen would have to be corrected for the effects of temperature and barometric pressure. If the oxygen is to be collected over water—i.e., by displacing water from collecting bottles—the vapor pressure of water, at its specific temperature, would have to be accounted for as described in the discussion of Dalton's law of partial pressures (p. 99). The point is, however, that once the stoichiometry of a reaction is established via the balanced equation, the calculations of various amounts of reactants and products can be determined by dimensional analysis in what is commonly called the *factor-label* method. In order to keep the following examples in proper perspective, it should be remembered that the balanced equation contains all of the stoichiometric information—everything else is arithmetic.

8.6 THE FACTOR-LABEL METHOD

The factor-label method for solving stoichiometric problems is really a lovely wedding of the mole concept to dimensional analysis. The following examples of the types of stoichiometric problems will be generalized in terms of substance A and substance B. In other words, given a specific mass, gas volume, or number of moles of substance A, what mass, gas volume, or number of moles of substance B will be produced or will be needed to react completely? The applications of the generalized procedures to specific problems will be illustrated with examples. Some examples of specific types of problems will deal with the slightly complicating factors of gases at nonstandard conditions and of reactions involving impurities.

Moles-to-Moles Problems

If a balanced equation indicates that 2 moles of A *produces* 3 moles of B in a particular reaction

$$2A \rightarrow 3B$$

the question might be to find the number of moles of B that would be produced by 0.24 moles of A.

The same mathematics (i.e., the same sequence for the dimensional analysis of the labeled factors) applies when the equation indicates that 2 moles of A *reacts with* 3 moles of B to produce some other product:

$$2A + 3B \rightarrow \text{a product}$$

The mole ratio, $\dfrac{2 \text{ moles A}}{3 \text{ moles B}}$, is true in either case.

The technique is to write the number of moles of A and multiply by the mole ratio in order to find moles of B:

1. moles A → moles B
2. (actual number of moles of A) × (mole ratio) → (actual number of moles of B)

3. $(0.24 \text{ mol } \not{A}) \times \left(\frac{3 \text{ mols B}}{2 \text{ mols } \not{A}}\right) = 0.36 \text{ moles B}$

EXAMPLE 8.9

How many moles of $KClO_3$ are needed to produce 0.5 moles of oxygen?

1. Write the balanced equation:

$$KClO_3 \rightarrow KCl + \frac{3}{2} O_2$$

2. The essential data are

$$KClO_3 \rightarrow \frac{3}{2} O_2$$

$$1 \text{ mole} \rightarrow 1.5 \text{ moles}$$

3. Identify the problem as a *moles-to-moles* type.
4. The sequence is

$$\text{moles } O_2 \rightarrow \text{moles } KClO_3$$

5. The steps in this sequence are

$$\text{moles } O_2 \times \text{mole ratio} \rightarrow \text{moles } KClO_3$$

6. Substitute the labeled factors and perform the dimensional analysis:

$$(0.5 \text{ mol } \not{O}_2) \times \left(\frac{1.0 \text{ mol } KClO_3}{1.5 \text{ mols } \not{O}_2}\right) = \chi \text{ moles of } KClO_3$$

$$\chi = 0.33 \text{ mole of } KClO_3$$

Moles-to-Mass-Problems

The sequence of operations in this type of problem is to change moles of A to moles of B and then to grams of B:

moles A → moles B → mass B

1. Use the mole ratio to find the actual number of moles of B produced by the given number of moles of A:

$$\left(\begin{matrix}\text{given number}\\ \text{of moles A}\end{matrix}\right) \times \left(\frac{3 \text{ moles B}}{2 \text{ moles A}}\right) \rightarrow \left(\begin{matrix}\text{actual number}\\ \text{of moles of B}\end{matrix}\right)$$

2. Multiply the actual number of moles of B by the mole weight of B to obtain grams:

$$(\text{moles B}) \times (\text{g mole}^{-1}) = \text{grams B}$$

3. In sequence:

$$\begin{pmatrix}\text{actual number of}\\ \text{moles of A}\end{pmatrix} \times (\text{mole ratio}) \times \begin{pmatrix}\text{mole weight}\\ \text{of B}\end{pmatrix} = \text{grams of B}$$

EXAMPLE 8.10

What mass of $KClO_3$ is needed to produce 0.02 moles of KCl?

1. Write the balanced equation:

$$KClO_3 \rightarrow KCl + \frac{3}{2} O_2$$

2. The balanced equation indicates that 1 mole of $KClO_3$ yields 1 mole of KCl:

$$KClO_3 \rightarrow KCl$$

$$1 \text{ mole} \rightarrow 1 \text{ mole}$$

3. Identify the problem as a *moles-to-mass* type.
4. The sequence is

$$\text{moles KCl} \rightarrow \text{moles } KClO_3 \rightarrow \text{mass } KClO_3$$

5. The steps are

$$(\text{moles KCl}) \times (\text{mole ratio}) \times (\text{mole weight of } KClO_3) = \text{grams of } KClO_3$$

6. Perform the calculation after establishing the mole weight of $KClO_3$ as 122.5 g mole^{-1}:

$$(0.02 \text{ mol KCl}) \times \left(\frac{1 \text{ mol } KClO_3}{1 \text{ mol KCl}}\right) \times (122.5 \text{ g mol}^{-1}) = \chi$$

$$\chi = 2.45 \text{ g } KClO_3$$

Mass-to-Moles Problems

Using the same general equation, $2A \rightarrow 3B$, the problem could be to find the number of moles of B produced from a certain mass of A.

The technique is to change grams of A to moles of A, and then find moles of B:

mass A → moles A → moles B

Changing mass to moles is a straightforward mole concept operation:

$$\text{(moles) n} = \frac{\text{g}}{\text{g mole}^{-1}}$$

Therefore, a more detailed sequential scheme is

1. Change mass A to moles A:

$$\frac{\text{g}}{\text{g mole}^{-1}}$$

2. Multiply by the mole ratio to convert moles A to moles B:

$$\left(\frac{\cancel{\text{g}}\text{ of }\cancel{\text{A}}}{\cancel{\text{g}}\text{ m}\cancel{\text{o}}\text{le}^{-1}}\right) \times \left(\frac{\text{3 moles B}}{\text{2 m}\cancel{\text{o}}\text{les A}}\right) = \left(\begin{matrix}\text{actual number of}\\ \text{moles of B}\end{matrix}\right)$$

$$\left(\begin{matrix}\text{actual number}\\ \text{of moles of A}\end{matrix}\right) \times \text{(mole ratio)} = \left(\begin{matrix}\text{actual number}\\ \text{of moles of B}\end{matrix}\right)$$

EXAMPLE 8.11

How many moles of C_2H_2 are needed to produce 4.0 g of CO_2 when C_2H_2 is burned?

1. Write the balanced equation:

$$C_2H_2 + \frac{5}{2}O_2 \rightarrow 2CO_2 + H_2O$$

2. The stoichiometry of the reaction is 1 mole of C_2H_2 produces 2 moles of CO_2:

$$C_2H_2 \rightarrow 2CO_2$$
$$\text{1 mole} \rightarrow \text{2 moles}$$

3. Establish the mole weight of CO_2 (44.0 g mole^{-1}) and follow the sequence:

$$\text{mass } CO_2 \rightarrow \text{moles } CO_2 \rightarrow \text{moles } C_2H_2$$

4. The steps are

$$\left(\frac{\text{g } CO_2}{\text{g mole}^{-1}}\right) \times \text{(mole ratio)} = \text{moles of } C_2H_2$$

5. Perform the operations:

$$\left(\frac{4.0 \text{ g } CO_2}{44.0 \text{ g mol}^{-1}}\right) \times \left(\frac{1 \text{ mol } C_2H_2}{2 \text{ mols } CO_2}\right) = \chi$$

$$\chi = 0.045 \text{ mole } C_2H_2$$

EXAMPLE 8.12

Given the balanced equation

$$P_4 + 6I_2 \rightarrow 4PI_3$$

calculate the number of moles of I_2 used in the formation of 6.0 g of PI_3.

1. The essential stoichiometry in the reaction is

$$6I_2 \rightarrow 4PI_3$$

$$6 \text{ moles} \rightarrow 4 \text{ moles}$$

or a simpler and equal ratio

$$3I_2 \rightarrow 2PI_3$$

$$3 \text{ moles} \rightarrow 2 \text{ moles}$$

2. Write the solution sequence for this *mass-to-moles* problem:

$$\text{mass } PI_3 \rightarrow \text{moles } PI_3 \rightarrow \text{moles } I_2$$

3. Find the mole weight of PI_3 and follow the sequence:

$$PI_3 = 412 \text{ g mole}^{-1}$$

$$\left(\frac{6.0 \text{ g } PI_3}{412 \text{ g mol}^{-1}}\right) \times \left(\frac{3 \text{ mols } I_2}{2 \text{ mols } PI_3}\right) = \chi$$

$$\text{moles} \times \text{mole ratio} = \text{moles}$$

$$\chi = 0.022 \text{ mole } I_2$$

Mass-to-Mass Problems

This is a very common type of problem, often referred to as a "weight-weight" problem. In effect, it asks what mass of B will be produced from a given mass of A? Or, conversely, if a certain mass of B is obtained, what mass of A must have been used?

The sequence of steps is to change mass of A to moles of A, then moles of A to moles of B, and, finally, moles of B to mass of B.

mass A → moles A → moles B → mass B

1. Change mass of A to moles of A by dividing by the mole weight:

$$\frac{\text{g}}{\text{g mole}^{-1}} = \text{moles of A}$$

2. Change moles of A to moles of B by multiplying by the mole ratio. Use the hypothetical reaction, $2A \rightarrow 3B$.
3. Convert moles of B to mass of B by multiplying by the mole weight of B.
4. In sequence:

$$\left(\frac{\not{\text{g}} \text{ of A}}{\not{\text{g}} \text{ m}\not{\text{o}}\text{le}^{-1}}\right) \times \left(\frac{3 \text{ mo}\not{\text{l}}\text{es B}}{2 \text{ mo}\not{\text{l}}\text{es A}}\right) \times (\text{g m}\not{\text{o}}\text{le}^{-1}) = \text{grams of B}$$

$$\left(\begin{matrix}\text{actual number of}\\ \text{moles of A}\end{matrix}\right) \times (\text{mole ratio}) \times (\text{mole weight}) = \text{mass of B}$$

EXAMPLE 8.13

What mass of aluminum would be used in the production of 10.2 g of aluminum oxide if aluminum reacted directly with oxygen?

1. Write the balanced equation to obtain the stoichiometry:

$$2Al + \frac{3}{2} O_2 \rightarrow Al_2O_3$$

2. The mole ratio data are obtained:

$$2Al \rightarrow Al_2O_3$$

$$\text{2 molcs} \rightarrow \text{1 mole}$$

3. Write the sequence of steps for this mass-to-mass problem:

$$\text{mass } Al_2O_3 \rightarrow \text{moles } Al_2O_3 \rightarrow \text{moles Al} \rightarrow \text{mass Al}$$

4. Find the mole weights of Al and Al_2O_3 and follow the sequence:

$$Al = 27.0 \text{ g mole}^{-1}$$

$$Al_2O_3 = 102.0 \text{ g mole}^{-1}$$

$$\left(\frac{10.2 \not{\text{g}} \text{ A}\not{\text{l}}_2\text{O}_3}{102.0 \not{\text{g}} \text{ mol}^{-1}}\right) \times \left(\frac{2 \text{ m}\not{\text{o}}\text{ls Al}}{1 \text{ m}\not{\text{o}}\text{l A}\not{\text{l}}_2\text{O}_3}\right) \times (27.0 \text{ g m}\not{\text{o}}\text{l}^{-1}) = \chi$$

$$\text{moles} \times \text{mole ratio} \times \text{mole weight} = \chi$$

$$\chi = 5.4 \text{ g of Al}$$

Exercise 8.3

Solve the following problems:

1. Given the skeleton equation $C_3H_8 + O_2 \rightarrow CO_2 + H_2O$:
   a. How many moles of C_3H_8 are needed to produce 0.25 moles of water?
   b. What mass of water is produced from 0.05 moles of C_3H_8?
   c. What mass of oxygen is needed to produce 1.4 g of CO_2 in this reaction?
2. Since aluminum reacts with hydrochloric acid in a simple replacement reaction, answer the following questions:
   a. What is the stoichiometry of the reaction?
   b. How many moles of aluminum are needed to produce 0.3 moles of gas?
   c. What mass of aluminum is needed to produce 2.0×10^{-3} g of hydrogen?
   d. How many moles of hydrochloric acid are needed to react completely with 60.0 mg of aluminum?

Mass-to-Volume Problems

Up to this point, the problems involving stoichiometric reactions have focused on mole and mass relationships. However, when a gas is involved in a reaction, it is often more convenient to deal with the volume of the gas rather than its mass. A 0.01 mole sample of chlorine gas at STP is easier to measure as 224 ml than as 0.71 g. Once again, the mole concept forms the basis of calculations concerned with gases.

This type of problem asks what volume of B (if substance B is produced in the gas state) will be obtained from a given mass of A. If the volume of B is measured at STP, the problem is relatively simple and straightforward. If the conditions are not standard, the equation of state is recommended as a method for the solution of such problems.

Consider another general equation:

$$2 \text{ moles } A(s) \rightarrow 1 \text{ mole } B(g)$$

The sequence of steps would be:

mass A → moles A → moles B → volume B

The only new step in this sequence is the change from moles of B to volume of B. This is done by multiplying the actual number of moles of B by the *molar volume*, 22.4 ℓ mole^{-1} at STP:

$$\cancel{\text{moles}} \times 22.4\ \ell\ \cancel{\text{mole}}^{-1} = \text{liters}$$

Or, use the equation of state to convert moles to liters at nonstandard conditions:

$$V = \frac{nRT}{P}$$

1. Change mass of A to moles of A.
2. Multiply moles of A by the mole ratio to obtain moles of B.
3. Multiply moles of B by 22.4 ℓ $mole^{-1}$ at STP, or use the equation of state.
4. In sequence:

$$\left(\frac{\text{g of A}}{\text{g mole}^{-1}}\right) \times \left(\frac{\text{1 mole B}}{\text{2 moles A}}\right) \times \left(\begin{matrix}22.4\ \ell\ \text{mole}^{-1}\\ \text{at STP}\end{matrix}\right) = \begin{matrix}\text{liters of B}\\ \text{at STP}\end{matrix}$$

or,

$$\left(\frac{\text{g of A}}{\text{g mole}^{-1}}\right) \times \left(\frac{\text{1 mole B}}{\text{2 moles A}}\right) = \text{moles of B}$$

then,

$$V = \frac{nRT}{P} \quad \text{for nonstandard conditions}$$

EXAMPLE 8.14

What volume of hydrogen gas can be collected at STP when 2.0 g of zinc react with sufficient hydrochloric acid?

1. Determine the stoichiometry by writing the balanced equation for the reaction:

$$Zn + 2\ HCl \rightarrow ZnCl_2 + H_2$$

2. The stoichiometry indicates the mole ratio:

$$\begin{matrix} Zn \rightarrow H_2 \\ \text{1 mole} \rightarrow \text{1 mole} \end{matrix}$$

3. Since the answer is to be expressed as a volume of hydrogen gas at STP, the molar volume of an ideal gas is required. This is 22.4 liters $mole^{-1}$. The mole weight of Zn is 65.4 g $mole^{-1}$.
4. Write the steps for the sequence of steps:

$$\text{mass Zn} \rightarrow \text{moles Zn} \rightarrow \text{moles } H_2 \rightarrow \text{volume } H_2$$

5. Perform the calculation:

$$\left(\frac{2.0\ \not{g}\ \text{of}\ \not{Zn}}{65.4\ \not{g}\ \not{mol}^{-1}}\right) \times \left(\frac{1\ \not{mol}\ H_2}{1\ \not{mol}\ \not{Zn}}\right) \times (22.4\ \ell\ \text{mol}^{-1}) = \chi$$

moles of zinc × mole ratio × molar volume

$$\chi = 0.69 \text{ liters of } H_2 \text{ at STP}$$

Volume-to-Mass Problems

A volume-to-mass problem might ask what mass of A would be required to produce a given volume of B measured at STP or some other conditions. The sequence of operations would be:

volume B → moles B → moles A → mass A

1. The conversion of the volume of B to moles of B could be done by

$$\frac{\text{liters of A}}{22.4\ \ell\ \text{mole}^{-1}} = \text{moles of A at STP}$$

Otherwise, for moles of gas under *any* conditions

$$n = \frac{PV}{RT}$$

2. The remainder of the sequence is routine:

$$\begin{pmatrix}\text{actual number}\\ \text{of moles of B}\end{pmatrix} \times \left(\frac{2\ \text{moles A}}{1\ \text{mole B}}\right) \times \begin{pmatrix}\text{mole weight}\\ \text{of A}\end{pmatrix} = \text{grams of A}$$

EXAMPLE 8.15

What mass of $KClO_3$ is needed to produce 600.0 ml of oxygen gas when the volume is measured at 27.0°C and 750 torr of pressure?
1. Write the balanced equation for the reaction:

$$KClO_3 \rightarrow KCl + \frac{3}{2}\,O_2$$

2. Determine the mole ratio of the appropriate substances:

$$KClO_3 \rightarrow \frac{3}{2} O_2$$

$$1 \text{ mole} \rightarrow 1.5 \text{ moles}$$

3. Use the equation of state to find the actual number of moles of oxygen gas:

$$PV = nRT \quad \text{or,} \quad n = \frac{PV}{RT}$$

4. Organize the data and solve for the number of moles:

$$P = \frac{750 \text{ torr}}{760 \text{ torr atm}^{-1}} = 0.993 \text{ atm},$$

$$R = 0.082 \text{ liter-atm mole}^{-1} \text{ K}^{-1}$$

$$T = 270°C = 300.0K$$

$$V = 600.0 \text{ ml} = 0.6 \text{ liter}$$

$$n = \frac{(0.993 \text{ atm})(0.6 \text{ liter})}{(0.082 \text{ liter-atm mol}^{-1} \text{ K}^{-1})(300.0K)}$$

$$n = 0.024 \text{ moles of } O_2$$

5. Establish the sequence of steps:

$$\text{moles } O_2 \rightarrow \text{moles } KClO_3 \rightarrow \text{mass } KClO_3$$

6. Perform the calculation after finding the mole weight of $KClO_3 = 122.5 \text{ g mole}^{-1}$:

$$(0.024 \text{ mol } O_2) \times \left(\frac{1 \text{ mol } KClO_3}{1.5 \text{ mols } O_2}\right) \times (122.5 \text{ g mol}^{-1}) = \chi$$

$$\text{moles} \times \text{mole ratio} \times \text{mole weight}$$

$$\chi = 1.96 \text{ g of } KClO_3$$

EXAMPLE 8.16

If 4.2 grams of iron (III) oxide reacts with carbon monoxide to produce iron and carbon dioxide, what volume of carbon dioxide can be collected over water at 0.92 atm and 23.0°C?

1. Write the balanced equation:

$$Fe_2O_3 + 3CO \rightarrow 2Fe + 3CO_2$$

2. Establish the mole ratio of iron (III) oxide to carbon dioxide:

$$Fe_2O_3 \rightarrow 3CO_2$$
$$1 \text{ mole} \rightarrow 3 \text{ moles}$$

3. Calculate the number of moles of CO_2 produced from 4.2 g of Fe_2O_3:

$$\text{mass } Fe_2O_3 \rightarrow \text{moles } Fe_2O_3 \rightarrow \text{moles } CO_2$$

4. Find the mole weight of Fe_2O_3 and perform the calculation:

$$Fe_2O_3 = 159.7 \text{ g mole}^{-1}$$

$$\left(\frac{4.2 \text{ g } \cancel{Fe_2O_3}}{159.7 \cancel{\text{g}} \cancel{\text{mol}}^{-1}}\right) \times \left(\frac{3 \text{ mols } CO_2}{1 \cancel{\text{mol}} \cancel{Fe_2O_3}}\right) = \chi$$

$$\text{moles} \times \text{mole ratio}$$

$$\chi = 0.079 \text{ mole of } CO_2$$

5. Use the equation of state to find the volume of CO_2 gas collected under the specified conditions:

$$PV = nRT, \quad \text{or} \quad V = \frac{nRT}{P}$$

6. Organize the data and solve for volume:

$$n = 0.079 \text{ mole}$$

$$R = 0.082 \text{ liter-atm mole}^{-1} \text{ K}^{-1}$$

$$T = 23.0°C = 296.0K$$

$$P = P_{CO_2} - P_{H_2O}(g) \text{ at } 23.0°C$$

$$P_{H_2O}(g) \text{ at } 23.0°C \text{ (from Table 4–2)} = 21.0 \text{ torr}$$

$$\frac{21.0 \cancel{\text{torr}}}{760 \cancel{\text{torr}} \text{ atm}^{-1}} = 0.028 \text{ atm (rounded off to 0.03 atm)}$$

$$P = 0.92 \text{ atm} - 0.03 \text{ atm} = 0.89 \text{ atm}$$

$$V = \frac{(0.079 \cancel{\text{mol}})(0.082 \text{ liter-}\cancel{\text{atm}} \cancel{\text{mol}}^{-1} \cancel{\text{K}}^{-1})(296\cancel{\text{K}})}{0.89 \cancel{\text{atm}}}$$

$$V = 2.15 \text{ liters of } CO_2$$

Volume-to-Volume Problems

When two or more reactants or products are in the gaseous state in a reaction where the same conditions of temperature and pressure apply to all the substances involved, the volume ratio is the same as the mole ratio. Recall Avogadro's principle, which says that equal numbers of moles of gases at the same temperature and pressure occupy equal volumes.

A generalized equation $2A(g) \rightarrow 5B(g)$ says, in effect, that just as 2 moles of gas A produces 5 moles of gas B, so will 2 volumes (expressed as liters or milliters, for example) of gas A produce 5 volumes of gas B. Therefore, a volume-to-volume problem follows the sequence

volume of A × volume ratio = volume of B

For this general equation, assume that 300 ml of gas A is changed to gas B. What volume of gas B results, as long as temperature and pressure remain constant? The problem is solved by multiplying the 300 ml of gas A by the volume ratio expressed in the balanced equation:

$$(300 \text{ ml of } \cancel{A}) \times \left(\frac{5 \text{ v}\cancel{o}\text{ls B}}{2 \text{ v}\cancel{o}\text{ls } \cancel{A}}\right) = 750 \text{ ml of B}$$

EXAMPLE 8.17

What volume of hydrogen gas is needed to react completely with 4.0 liters of oxygen gas in the formation of water?

1. Balance the equation to obtain the stoichiometry:

$$H_2 + \frac{1}{2} O_2 \rightarrow H_2O$$

2. The mole ratio of H_2 to O_2 is the same as the volume ratio:

$$H_2 + \frac{1}{2} O_2 \rightarrow H_2O$$

$$1 \text{ mole} + \frac{1}{2} \text{ mole, or } 1 \text{ volume} + \frac{1}{2} \text{ volume,}$$

$$\text{or } 1 \text{ liter} + \frac{1}{2} \text{ liter}$$

3. The 1:1/2 ratio obviously means that 8 liters of hydrogen is needed to combine with 4 liters of oxygen.
4. Solving by the factor-label method, the one-step operation is

$$\text{volume of } O_2 \rightarrow \text{volume of } H_2$$

$$(4 \text{ liters } \cancel{O_2}) \times \left(\frac{1 \text{ vol}\cancel{u}\text{me } H_2}{0.5 \text{ vol}\cancel{u}\text{me } \cancel{O_2}}\right) = \chi$$

$$\text{volume} \times \text{volume ratio}$$

$$\chi = 8 \text{ liters of } H_2$$

Exercise 8.4

1. What volume of carbon dioxide can be collected at STP as a result of decomposing 1.7 g of ammonium carbonate by heating?
2. What mass of P_4 is needed to react completely with 450 ml of chlorine gas measured at 32.0°C and 710 torr of pressure? The skeleton equation is $P_4 + Cl_2 \rightarrow PCl_3$.
3. What mass of calcium is required to react with water in the production of 320 ml of hydrogen when collected over water at 29.0°C and 1.22 atm?
4. Given the skeleton equation $C_3H_8 + O_2 \rightarrow CO_2 + H_2O$, what volume of propane (C_3H_8) is needed to produce 0.70 liter of CO_2 at STP?

8.7 THEORETICAL AND ACTUAL YIELDS*

A fact of experimental work is that many reactions produce less than what the stoichiometric calculations suggest. This is due to varying degrees of purity among the reactants, incompleteness of reactions, and the possibility that what is assumed to be a stoichiometric reaction is in fact an equilibrium phenomenon. Furthermore, there may be more than one reaction occurring during the operation, so that various by-products are obtained at the expense of additional reactants. All of these factors combine to produce yields that usually are not as high as those predicted from the ideal equation.

When the **actual yield** is compared to the **theoretical,** or calculated, yield, the resulting fraction is expressed as a percentage value and is called the **percentage yield**.

EXAMPLE 8.18

If 6.0 g of carbon is burned to form CO_2, what volume of CO_2 would be expected at STP? If the actual yield of CO_2 is measured to be 10.8 liters, what is the percentage yield?

1. Write the balanced equation to determine the stoichiometry:

$$\underset{1\text{ mole}}{C} + O_2 \rightarrow \underset{1\text{ mole}}{CO_2}$$

2. Find the theoretical yield of CO_2:

$$\left(\frac{6.0\ \not{g}\ \not{C}}{12.0\ \not{g}\ \text{m}\not{o}\text{l}^{-1}}\right) \times \left(\frac{1\ \text{m}\not{o}\text{l}\ CO_2}{1\ \text{m}\not{o}\text{l}\ \not{C}}\right) \times 22.4\ \text{liters m}\not{o}\text{l}^{-1} = \chi$$

moles of carbon × mole ratio × molar volume

$$\chi = 11.2\text{ liters } CO_2\text{ (theoretical)}$$

3. The percentage yield can be calculated:

$$\% \text{ yield} = \frac{\text{actual}}{\text{theoretical}}$$

$$\% \text{ yield} = \frac{10.8 \text{ liters}}{11.2 \text{ liters}} = 0.964 = 96.4\%$$

8.8 LIMITING FACTOR OF A REACTION*

When the masses of reactants are known, a calculation must be made to determine whether the masses are stoichiometric. For example, if one of two reactants is present in excess, some of the excess reactant will have no part in the reaction. Any calculation based on the excess weight will be incorrect. Before a percentage yield can be determined, it is necessary to find which reactant is completely changed.

EXAMPLE 8.19

A 6.54 gram sample of zinc is added to 17.0 g of $CuSO_4$ in solution. If 5.42 g of copper is produced, what is the percentage yield?

1. Determine the stoichiometry:

$$Zn(s) + CuSO_4(aq) \rightarrow ZnSO_4(aq) + Cu(s)$$

2. Observe that 1 mole of Zn reacts with 1 mole of $CuSO_4$ to form 1 mole of Cu. Are there equal moles of Zn and $CuSO_4$?

$$Zn = \frac{6.54 \not{g}}{65.4 \not{g} \text{ mol}^{-1}} = 0.100 \text{ mole}$$

$$CuSO_4 = \frac{17.0 \not{g}}{159.6 \not{g} \text{ mol}^{-1}} = 0.107 \text{ mole}$$

3. Note the excess of $CuSO_4$. Therefore, the calculation of the weight copper must be based on the amount of zinc, because zinc is the limiting reagent:

$$(0.1 \text{ m}\not{o}\text{l } \not{Zn}) \times \left(\frac{1 \text{ m}\not{o}\text{l Cu}}{1 \text{ m}\not{o}\text{l } \not{Zn}}\right) \times (63.5 \text{ g m}\not{o}\text{l}^{-1}) = \text{mass of Cu}$$

$$\text{moles} \times \text{mole ratio} \quad \times \text{mole weight of copper} = \chi$$

$$\chi = 6.35 \text{ g of Cu (theoretical)}$$

4. The percentage yield of copper can be calculated:

$$\% \text{ yield} = \frac{\text{actual}}{\text{theoretical}}$$

$$\% \text{ yield} = \frac{5.42 \text{ g}}{6.35 \text{ g}} = 0.853 = 85.3\%$$

QUESTIONS AND PROBLEMS

1. What is the meaning of each of the following symbols in equations: (aq), (s), (ℓ), (g) $(\xrightarrow{\Delta})$?
2. Interpret the following equations:
   a. $K(s) + H_2O(\ell) \rightarrow KOH(aq) + 1/2\ H_2(g)$
   b. $CH_3OH(\ell) + 3/2\ O_2(g) \rightarrow CO_2(g) + 2H_2O(\ell) + 173.4\ \text{kcal}$
3. Balance the following equations:
   a. $Ni + HCl \rightarrow NiCl_2 + H_2$
   b. $Al + C \rightarrow Al_4C_3$
   c. $O_2 \rightarrow O_3$
   d. $MnO_2 + HCl \rightarrow MnCl_2 + Cl_2 + H_2O$
   e. $I_2 + Cl_2 \rightarrow ICl_3$
   f. $As_4 + O_2 \rightarrow As_4O_6$
   g. $MnO_2 \rightarrow Mn_3O_4 + O_2$
   h. $NO + Cl_2 \rightarrow NOCl$
   i. $SiO_2 + H_2F_2 \rightarrow SiF_4 + H_2O$
   j. $H_2O + NOCl \rightarrow HCl + HNO_2$
4. Write skeleton equations and then balance the following:
   a. cesium + water → cesium hydroxide + hydrogen
   b. gold (III) oxide → gold + oxygen
   c. iron (II) sulfide + oxygen → iron (III) oxide + sulfur dioxide
   d. bismuth (III) chloride + sodium sulfide → bismuth (III) sulfide + sodium chloride
   e. silicon dioxide + hydrofluoric acid → silicon tetrafluoride + water
   f. carbon + aluminum oxide → aluminum carbide + carbon monoxide
   g. calcium carbonate → calcium oxide + carbon dioxide
   h. carbon dioxide + calcium hydroxide → calcium hydrogen carbonate
   i. copper (II) phosphate + tin (IV) chloride → copper (II) chloride + tin (IV) phosphate
   j. cadmium + mercury (I) nitrate → cadmium nitrate + mercury
5. Complete and balance the following equations:
   a. $Rb_2O + H_2O \rightarrow$
   b. $Mg + HClO_4 \rightarrow$
   c. $HOAc + NaOH \rightarrow$
   d. $KCN + H_2SO_4 \rightarrow$
   e. $Na_2S + Fe_2(Cr_2O_7)_3 \rightarrow$
   f. $P_4O_{10} + H_2O \rightarrow$
   g. $Co(OAc)_3 + H_2S \rightarrow$
   h. $H_2CrO_4 + Ni_2(C_2O_4)_3$
   i. $Al + H_3PO_4 \rightarrow$
   j. $C_5H_{12} + O_2 \rightarrow$
6. How many moles of zinc are required to react completely with 1.6 moles of $CuSO_4$ in solution?
7. Calculate the mass of HgO needed to produce 0.12 moles of mercury after complete decomposition.
8. What volume of hydrogen gas at STP can be collected when 2.2×10^{-2} moles of magnesium is completely reacted with hydrochloric acid?
9. Iron (III) oxide reacts with aluminum to produce iron and aluminum oxide. What mass of aluminum is needed to reduce 4.3 g of iron (III) oxide?
10. What mass of 80.0% pure gold (III) oxide is needed to yield 1.4 g of pure gold after decomposition?

11. In the reaction $P_4 + 6Cl_2 \rightarrow 4PCl_3$, what mass of phosphorus is required to react completely with 560.0 ml of $Cl_2(g)$ at STP?

12. When calcium reacts with water to produce hydrogen, what mass of calcium is needed to yield 75.0 ml of the gas measured at 20.0°C and 0.91 atm of pressure?

13. In the reaction $H_2S(g) + 3/2\ O_2(g) \rightarrow SO_2(g) + H_2O(g)$, what volume of oxygen must react in order to form 750.0 ml of SO_2 at STP?

14. If 16.3 g of silver oxide is decomposed, what volume (milliliters) of oxygen could be collected over water at 27.0°C and 750 torr of pressure?

15. Find the percentage yield in a reaction where the actual yield of product is 1.55 g while the theoretical yield is 4.02 g.★

16. If 1.12 liters of hydrogen is added to a 0.92 liter container of air (80% oxygen), which reactant is the reaction-limiting factor? Why?★

CHAPTER 9

That is science, learning, knowledge, that is the true source of every spiritual endeavor of man. We try to find out as much as we can about spatial and temporal surroundings of the place in which we find ourselves put by birth. And as we try, we delight in it, we find it extremely interesting.

Erwin Schrödinger, *Science of Humanism**

Behavioral Objectives:

At the completion of this chapter, the student should be able to:

1. Describe the essential characteristics of a solution.
2. Draw sketches to illustrate the dissociation of ionic compounds and polar molecules in water.
3. Contrast the terms strong electrolyte, weak electrolyte, and nonelectrolyte.
4. Use a table of solubilities to predict which products of a reaction are soluble and which are insoluble.
5. Write and balance ionic equations, using a table of weak and strong electrolytes and a solubility table.
6. Recognize spectator ions so that net ionic equations can be written.
7. List five rules for writing ionic equations.
8. Define molarity and calculate molar concentrations of solutions from available data.
9. Write, in mathematical form, the relationship among moles, mole weight, molarity, and volume in a given solution.
10. Calculate new molarities or volumes in dilution problems.
11. Define normality, equivalents, and equivalent weights, and calculate normal concentrations of solutions from given data.⋆
12. Interconvert between normality and molarity.⋆
13. Write the mathematical relationship between equivalents, normality, and volume.⋆
14. Calculate the number of equivalents from a given mass of a compound and its formula.⋆
15. Define percent by weight solutions, and perform calculations with them.
16. Define parts per million solutions, and perform calculations with them.
17. Explain the concept of *formality* in solutions.
18. Describe the colligative effect of solute particles on the vapor pressure of a solution.
19. Explain the effect that the lowering of vapor pressure has on the boiling and freezing points of solvents.
20. Define mole fraction. Describe the effect of the mole fraction in terms of Raoult's law.⋆
21. Define and calculate molality.⋆
22. Calculate the actual changes in the boiling and freezing points of solvents for any given molal concentration of a nonelectrolyte.⋆
23. Use tables of boiling and freezing point data in the solution of related problems.⋆
24. Find true molecular formulas of compounds from their empirical formulas and from information on their freezing or boiling points.⋆

SOLUTIONS

9.1 INTRODUCTION

It is useful to define a solution as a homogeneous combination of a dissolved substance, called a **solute,** and a medium in which the solute is homogeneously dispersed, called the **solvent.** While a solvent could be alcohol, ether, or a variety of liquid hydrocarbon compounds, usually referred to as organic solvents, the most common solvent in chemistry is water. A formula written with the symbol (*aq*) designates a water solution of that compound.

The particles of solute that move about randomly in the solvent are submicroscopic in size. In other words, when a clearly visible load of solid material is mixed with a solvent, there must be some kind of interaction between the unit structures (ions or molecules) of the solute with the solvent molecules, so that the forces of attraction in the solid structure of the solute are overcome and the particles of solid dissolve. The unit structures will then **dissociate,** or dissolve, into the submicroscopic units. However, if the size of the dissociated particles is large enough to reflect light, the mixture is called a **colloidal suspension** rather than a solution. In a true solution the dissolved solute particles will not settle to the bottom of the container, nor can they be filtered out by usual laboratory methods of filtration, nor will they reflect light. While the solution process may drastically alter the physical appearance of both the solute and the solvent, it is not commonly described as a chemical change, because the solute and solvent can be recovered in their original forms by the simple evaporation of the solvent. For example, the white crystals of anhydrous $CuSO_4$ form a beautiful blue solution when dissolved in colorless, liquid water. Complete removal of the water restores the original components of the solution:

1. $\underset{\text{white}}{CuSO_4(s)} + \underset{\text{colorless}}{H_2O(\ell)} \rightarrow \underset{\text{blue solution}}{CuSO_4(aq)}$

2. $CuSO_4(aq) \xrightarrow{\text{evaporation}} \underset{\text{blue crystals}}{CuSO_4{\cdot}5H_2O(s)} + H_2O(g)$

3. $CuSO_4{\cdot}5H_2O \xrightarrow[\text{at 175 C}]{\text{heated in oven}} \underset{\text{white}}{CuSO_4(s)} + 5H_2O(g)$

9.2 SOLUBILITY AND MISCIBILITY

The dissociation of $CuSO_4(s)$ in water, as well as the **solubility** (the ability to dissolve) of table salt, sugar, acids, and thousands more ionic and covalent compounds, raises some questions. Why are so many substances soluble in water while so many other ionic and covalent compounds are insoluble or **immiscible?** (Miscibility is the term used to describe how effectively one liquid-phase substance dissolves in another liquid.)

Most of the reasons that help to explain the degrees of solubility and miscibility of solutes in solvents can be found when data on bond strength and polarity are examined. Bearing in mind that the water molecule is distinctly polar, its interaction with several kinds of solutes will be discussed.

9.3 THE SOLUTION PROCESS

Ionic Solute and Water

When NaCl(*s*) is mixed with water, an attractive force between the positively and negatively charged ions and the water dipole molecules is a reasonable assumption. This is indeed what happens. It is described as an **ion-dipole interaction.** But the question that should be raised is why an ion-dipole interaction is apparently more effective than an ion-ion interaction, since the ionic crystal structure of NaCl has great strength and stability. The answer seems to be related to the insulating property of the cage of water molecules as they surround the surface ions of a crystal and partially neutralize the electrostatic force of attraction among the ions. Figure 9–1 illustrates the ion-dipole interaction between the sodium ions, chloride ions, and water molecules.

Another factor bearing on the solution process is the observed tendency of any system to move toward a more random distribution. This loosely described tendency toward an increase in the **entropy** (randomness) of a system appears to be a kind of driving force behind

Figure 9–1 The dissociation of the sodium and chloride ions from the crystal structure because of ion-dipole interaction.

change. The ionic lattice structure of NaCl is highly ordered (the opposite of random), and the interaction with water provides a pathway for the breakdown of the crystal structure, which culminates in the more random solution system. The fact that so many compounds evidence an increase in solubility when heated relates to the tendency toward randomness. While endothermic changes *seem* to be a standing contradiction of the tendency of matter to lose energy and increase stability, it should preferably be thought of as the application of room heat to increase randomness. When most solutions are heated, the tendency toward increasing randomness is favored because of the availability of energy that can be absorbed. For this reason, in the case of most compounds, the application of heat permits more solid to dissociate.*

At a given temperature, a compound has a specific characteristic solubility value. This means that just so many grams can be dissolved in a stated weight of solvent until the solution is described as **saturated.** When the temperature is raised (more heat energy provided), more solute usually can dissolve.* If the temperature is then lowered, the extra solute will usually crystallize out of solution. Occasionally, a special phenomenon is observed where the excess weight of solute remains in solution even though the temperature is lowered. A solution of this type is called **supersaturated,** and it is likely to be so unstable that an added crystal of solute or a sharp jolt will cause the crystallization of the extra mass of solute.

Heat is the main factor that can affect the solubility of solid and liquid compounds. Solubility tables, found in reference texts, describe the solubility of a substance in terms of grams of solute per 100 g of solvent. These data are related to the temperature. For example, the information given for a compound may state the solubility as 140 g per 100 g of water at 0°C, and 180 g per 100 g of water at 100°C.

The physical acts of grinding large crystals into powder and of vigorous stirring do not increase the solubility of substances. These techniques merely speed up the solution process by allowing more surface area of the solute to come in contact with the water dipoles in a given time. If the solubility of a substance is described as 84 g per 100 g of water at room temperature, stirring and grinding will not allow more than 84 g per 100 g of water to dissolve, but 84 g of powdered compound would probably dissolve in a fraction of the time required to dissolve 84 g of large crystals.

Polar Covalent Solute and Water

A typical example of a covalently bonded compound that is distinctly polar is hydrogen chloride, a gas. When HCl(*g*) is mixed with

* The term *most* rather than *all* compounds is used because there are exceptions. For example, calcium hydroxide, which produces considerable heat upon hydration, shows an increased solubility in cold water.

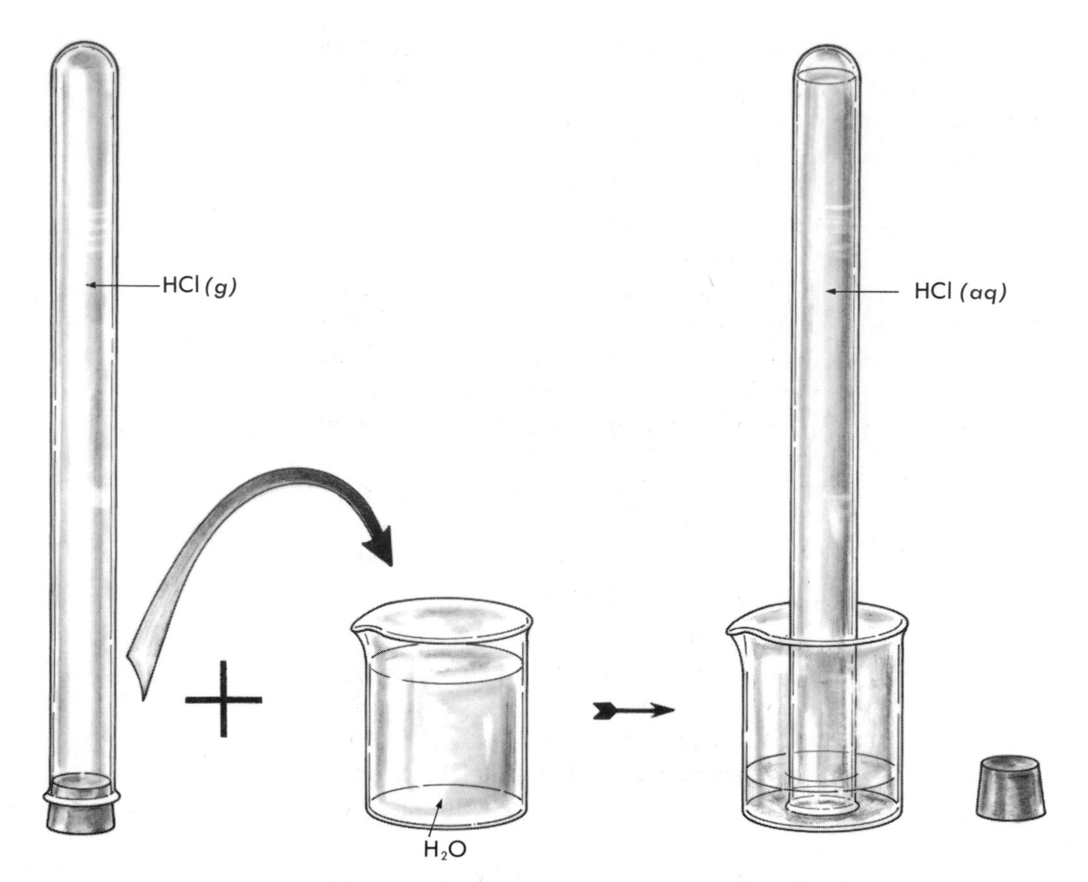

Figure 9–2 The solubility of $HCl(g)$ in water.

water, it is observed to dissolve very rapidly as it forms the solution called hydrochloric acid (Fig. 9–2).

The interaction between the $HCl(g)$ dipoles and the water dipoles is appropriately described as a **dipole-dipole** interaction. Interestingly, the forces of attraction operating between the dipoles actually increase the degree of polarity of each dipole. These are described as **enhanced** dipoles. The hydrogen-to-chlorine bond polarity is actually enhanced to the point where there is a physical separation of the HCl molecule and a resulting formation of hydrated chlorine and hydrogen ions. Figure 9–3 illustrates the enhancement of polarity that may occur in the process of dipole-dipole interaction.

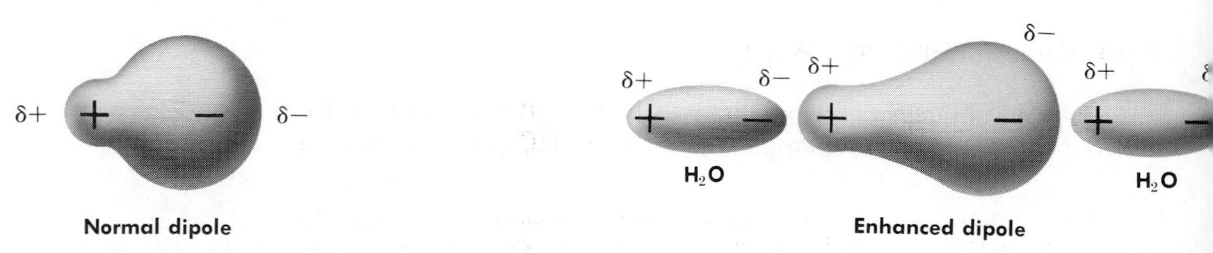

Figure 9–3 The enhancement of polarity.

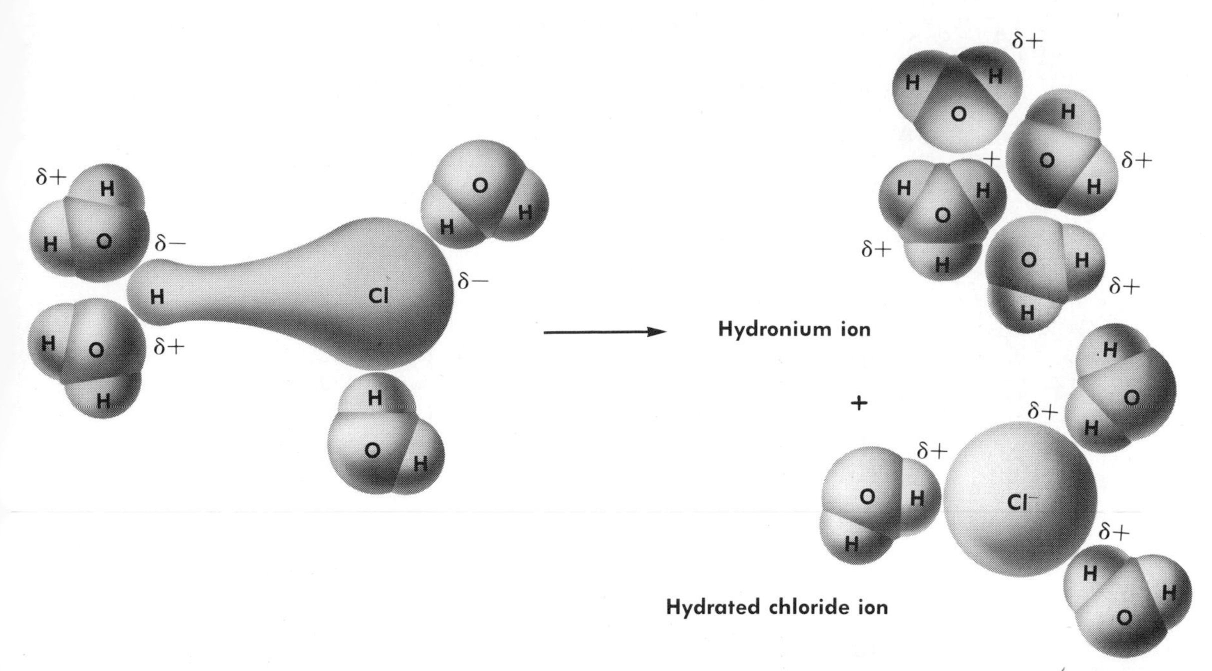

Figure 9–4 The dipole-dipole interaction between H_2O and HCl leading to ionization.

As the HCl molecules separate in water to become hydrated ions, the process is known as **ionization.** The resulting hydrated hydrogen ion, or hydrated proton, as it may be described, is known by the special name, **hydronium ion.** The formula is H_3O^+, and its structure is illustrated in Figure 9–4. A widely used alternative formula for the hydrated proton is $H^+(aq)$.

9.4 ELECTROLYTES AND NONELECTROLYTES

Although acids will be discussed more fully in the next chapter, it is useful to note at this time that the formation of hydrated H^+ ions in solution is a characteristic of acids. In fact, HCl(*aq*) is described as a strong acid because of the vigor and completeness with which it undergoes ionization in the formation of hydrated H^+ ions.

A simple test may be used to determine whether the solution process yields ions in solution. While an ionic crystal only needs to dissociate to produce ions in solution, a molecular compound may or may not ionize. Hydrogen chloride ionizes almost completely in dilute solution (a characteristic of strong acids), but compounds such as sugar and alcohol do not ionize appreciably, although they certainly do dissolve. Some substances, known as weak acids, ionize only slightly because their bond strengths are not easily overcome by dipole-dipole interaction. Substances that ionize extensively, and ionic compounds that dissociate easily, are known as strong **electrolytes.** Weak acids, which ionize slightly, are weak electrolytes. The remaining group of

TABLE 9–1 SOME ELECTROLYTES AND NONELECTROLYTES

Strong Electrolytes		Ions or Molecules in Solution
A. SOLUBLE IONIC COMPOUNDS		
NaCl	~100% dissociation →	$Na^+(aq) + Cl^-(aq)$
KNO_3		$K^+(aq) + NO_3^-(aq)$
NaOAc		$Na^+(aq) + OAc^-(aq)$
$CuSO_4$		$Cu^{2+}(aq) + SO_4^{2-}(aq)$
B. STRONG ACIDS		
HCl	~100% ionization →	$H^+(aq) + Cl^-(aq)$
$HClO_4$		$H^+(aq) + ClO_4^-(aq)$
HNO_3		$H^+(aq) + NO_3^-(aq)$
H_2SO_4		$H^+(aq) + HSO_4^-(aq)$
HBr		$H^+(aq) + Br^-(aq)$
HI		$H^+(aq) + I^-(aq)$
Weak Electrolytes		
HNO_2	slightly less to much less than 1% ionization →	$HNO_2(aq)$ + few $H^+(aq) + NO_2^-(aq)$
H_2SO_3		$H_2SO_3(aq)$ + few $H^+(aq) + HSO_3^-(aq)$
HOAc		$HOAc(aq)$ + few $H^+(aq) + OAc^-(aq)$
HF		$HF(aq)$ + few $H^+(aq) + F^-(aq)$
H_2S		$H_2S(aq)$ + few $H^+(aq) + HS^-(aq)$
Nonelectrolytes		
$C_{12}H_{22}O_{11}(s)$ (table sugar—sucrose)	no appreciable ionization →	$C_{12}H_{22}O_{11}(aq)$
$CH_3OH(\ell)$ (wood alcohol—methanol)		$CH_3OH(aq)$
$CH_3CH_2OH(\ell)$ (grain alcohol—ethanol)		$CH_3CH_2OH(aq)$
$CO(NH_2)_2(\ell)$ (urea)		$CO(NH_2)_2(aq)$
$CH_2OHCH_2OH(\ell)$ (antifreeze—ethylene glycol)		$CH_2OHCH_2OH(aq)$

substances, mostly organic (carbon) compounds, that do not ionize appreciably are appropriately labeled **nonelectrolytes.** Table 9–1 presents a brief summary of some typical electrolytes and nonelectrolytes. An electrolyte may be defined as any substance that produces ions in water solution, and the ions by virtue of their mobility can conduct an electric current between two separated electrodes. The name electrolyte means "carrier of electricity."

The test for ionization is made with an electrolyte tester. This is simply done by placing a light bulb in series with a pair of electrodes, so that the circuit can be completed only when ions act as conductors between the two electrodes. Figure 9–5 illustrates the electrolyte tester being used in several cases.

Investigation of the molecular structures of the organic nonelectrolytes reveals an asymmetrical distribution of the polar bonds and a resulting polarity of the molecules. However, it is not usually correct to describe the polarity of large organic molecules in terms of simple dipole structures. There are often several regions of positive and negative charge, which makes the descriptions of polarity inappropriate.

In the case of sugar, which is obviously very water soluble, the combination of weak forces of attraction among sucrose molecules in the crystal and the attraction that water dipoles have on the hydrogen and oxygen atoms in the sucrose molecule provide an easy pathway for dissolution of the sucrose molecules. The liquid-state organic compounds that dissolve in water are described as being miscible. A solution of 50% pure ethanol and 50% water constitutes 100 proof vodka.

Nonpolar Covalent Solute and Water

Nonpolar compounds are not very soluble in water. For example, iodine crystals dissolve slightly to give water a faint yellow-brown tinge, and carbon dioxide is somewhat soluble (which results in water becoming slightly acidic as it is exposed to the CO_2 in air). The reaction between CO_2 and water may be described as an equilibrium phenomenon, since CO_2 comes out of solution easily. Ordinary soft drinks have CO_2 under pressure in solution to give the drink its "fizz." However, when the cap is removed from a warm bottle, the wet bubbles of carbon dioxide give the drink its layer of foam, or "head." The equation for the equilibrium reaction is

$$CO_2(g) + 2H_2O(\ell) \rightleftharpoons H_3O^+(aq) + HCO_3^-(aq)$$

From another point of view, applicable to both I_2 and CO_2, the slight degree of solubility may be explained by the concept of **induced** polarity. The induction of polarity in molecules that are nonpolar results from the influence of the water dipole on the charge distribution

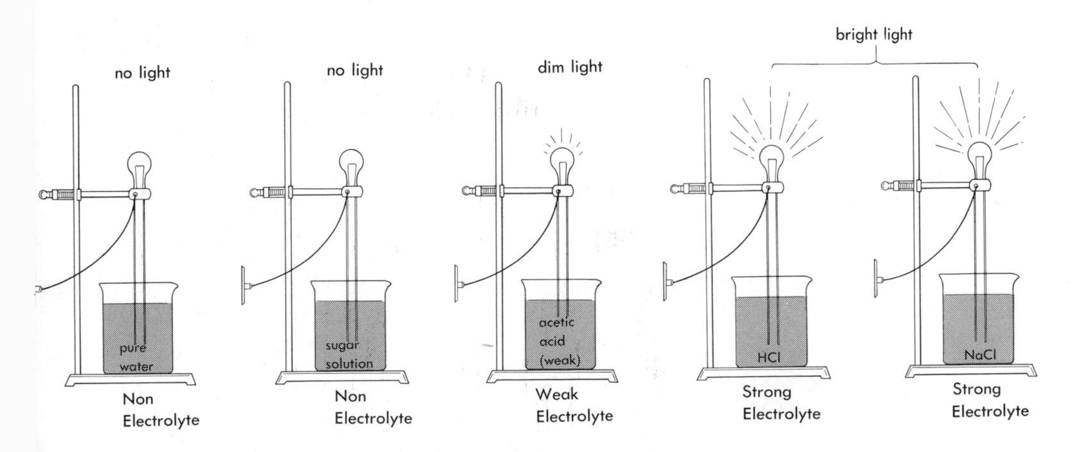

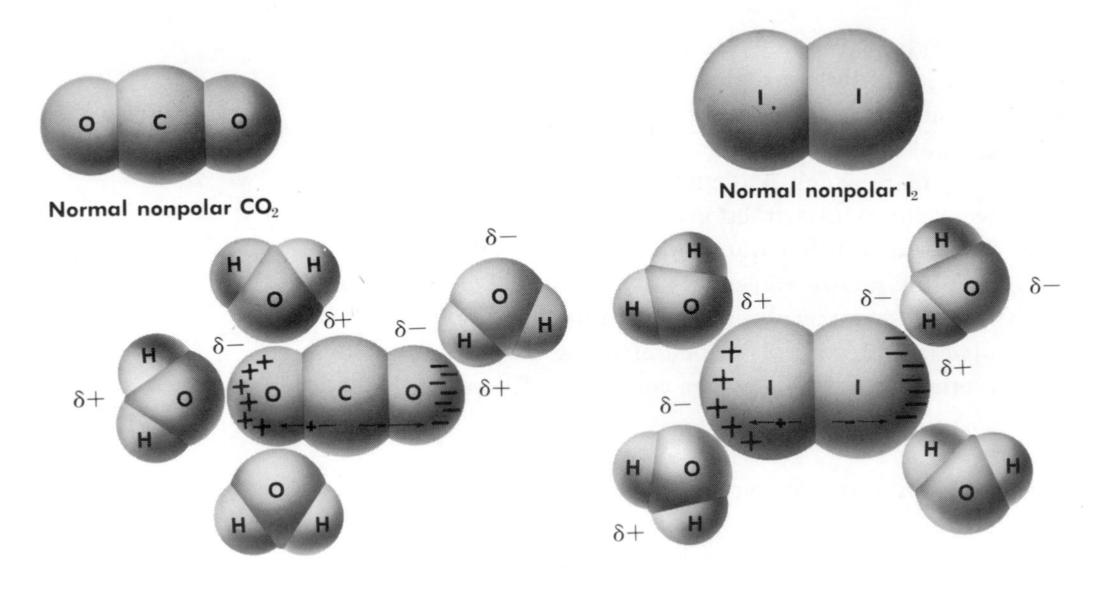

Figure 9–6 The slight water solubility of many nonpolar compounds is due to induced polarity.

in the nonpolar molecule. Normally, the centers of positive and negative charge coincide, as they do in the typical examples of I_2 and CO_2. However, when the polar water molecule interacts with the nonpolar molecules, a slight separation of the centers of charge occurs (Fig. 9–6), and the I_2 and CO_2 molecules temporarily become slight dipoles. This is the phenomenon of induced polarity.

Insoluble and Immiscible Compounds and Water

The term insoluble is a relative term. There is hardly any substance that will not dissolve at all. However, for practical purposes, any substance added to water that does not appear to be dissolving, or one that appears as a nondissolving solid (precipitate) product in a chemical reaction, would be described as being insoluble. The term insoluble is often qualified by saying "slightly" or "very." Or, for example, $PbCl_2$ is described as being insoluble in cold water but soluble in hot water.

Many ionic compounds are soluble. In those instances where an ionic compound is observed to be relatively insoluble, the explanation is found in the exceedingly strong electrostatic force of attraction between the cation and anion. For example, MgF_2 is insoluble because the ion-dipole interaction in water is not sufficient to overcome the great stability of the Mg^{2+} and F^- attraction in the crystal lattice. Other examples of insoluble compounds include other fluorides, AgCl, $PbSO_4$, $Ca(OH)_2$, and many phosphates, carbonates, chromates, and sulfides.

Strong covalent bonding in molecular solids and network bonds are often more powerful than the dipole-dipole interaction and do not allow dissociation. For this reason, many compounds in which the metal or metalloid atom is in the 3+, 4+, 5+, or 6+ oxidation state are called **pseudosalts** ("false" salts), which means that while the compound may have many of the physical properties of salts, it is not an ionic compound.

A useful generalization regarding the solubility of salts and pseudosalts is found in Table 9–2.

When the components of a proposed solution are liquids, there is a necessary concern for the miscibility or degree of "mixability" of the different liquid compounds. Often two liquids, such as oil and water, are seen to be immiscible, and a **phase** separation can be observed (Fig. 9–7). A phase separation between liquids is usually observed as two distinct layers. This is caused by the different light refractive properties of the two liquids. Immiscibility can usually be explained on the basis of polarity. In the case of the oil and water system, the water dipoles interact among themselves much more effectively than they do with the nonpolar oil molecules. Figure 9–8 illustrates the dipole interactions among water molecules leading to the phase separation with oil.

Nonpolar liquids are quite miscible in each other. Nonpolar

TABLE 9–2 SIMPLIFIED TABLE OF THE SOLUBILITY OF COMMON SALTS IN WATER AT 20°C

Anion	Cation	Solubility
acetate, chlorate, nitrate	nearly all	soluble
chloride, bromide, iodide	lead(II), silver, mercury(I)	insoluble
	all others	soluble
hydroxide	group I metals, barium, strontium	soluble
	all others*	insoluble
sulfate	mercury(I) and (II), calcium, barium, strontium, silver, lead(II)	insoluble
	all others	soluble
carbonate, phosphate, chromate	group I metals,** ammonium	soluble
	all others	insoluble
sulfide	group I metals, ammonium, magnesium, calcium, barium	soluble
	all others	insoluble

* $Ca(OH)_2$ slightly soluble. ** Li_3PO_4 insoluble.

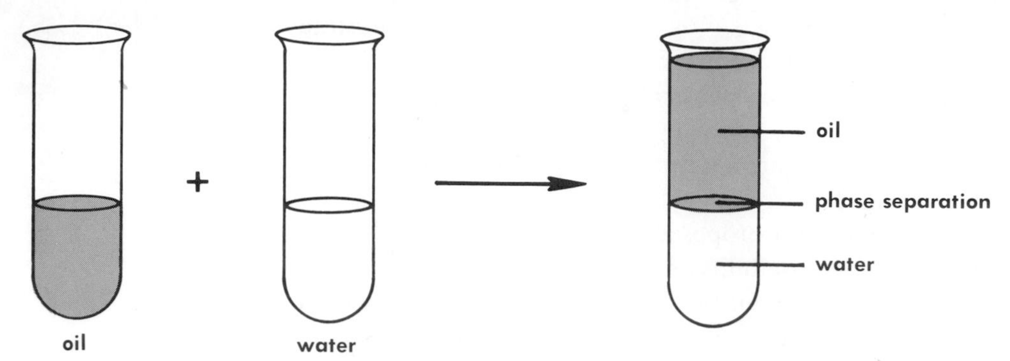

Figure 9–7 The immiscibility of oil and water.

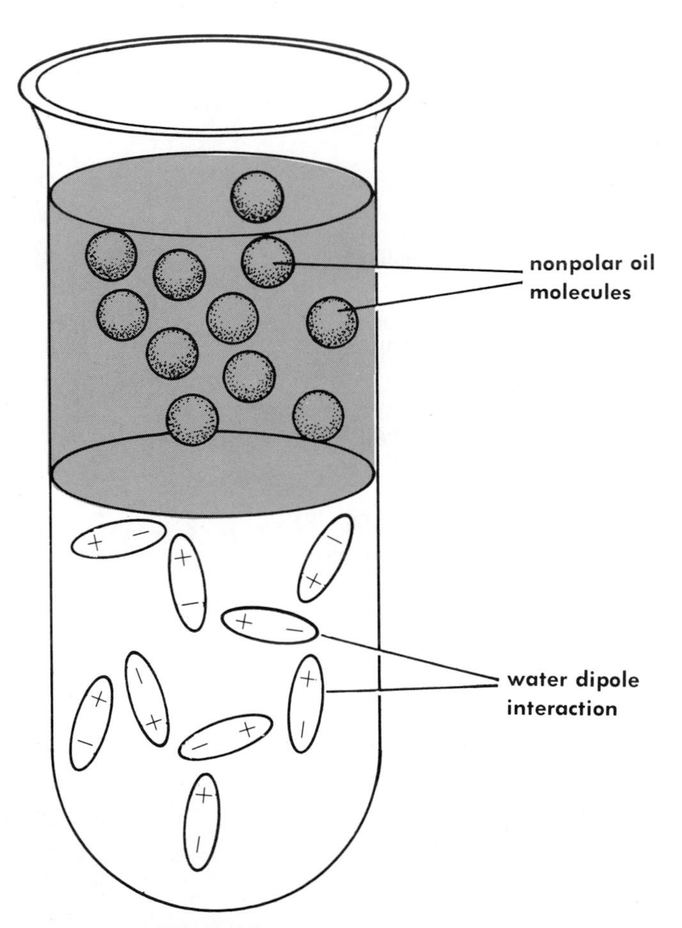

Figure 9–8 The phase separation of oil and water illustrating the effect of polarity.

solids will tend to dissolve readily in nonpolar solvents. For example, iodine crystals dissolve rapidly and easily in CCl_4, C_6H_6 (benzene), CS_2, and other nonpolar solvents.

The liquid phase separation naturally produces a top layer and a lower layer, the order depending on the comparative densities of the two liquids. In the oil and water system, the water will be the lower layer since it is denser than oil. If an even denser nonpolar liquid, such as carbon tetrachloride, were poured into the tube first, a three-phase system would be observed, as illustrated in Figure 9–9.

The previously investigated topic of polarity (Chapter 6) leads us to form several rules of thumb, which are summarized as follows:

Rule 1. Most ionic compounds of groups I and II are water soluble.

Rule 2. Polar covalent compounds tend to be water soluble.

Rule 3. Nonpolar compounds are more soluble in nonpolar solvents.

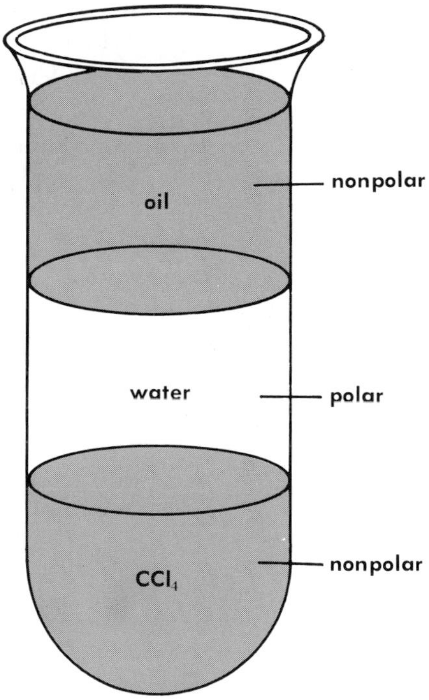

oil — nonpolar

water — polar

CCl_4 — nonpolar

Figure 9–9 A three-phase separation resulting from differences in polarity and density.

9.5 IONIC EQUATIONS

Since aqueous solutions may contain ions (from strong electrolytes), molecules (from nonelectrolytes), and mixtures of both ions and molecules (from weak electrolytes), it is possible, and often preferable, to write equations for reactions that include only what is changed. The purpose of an equation is to represent in a symbolic way the changes that occur. Only those species (ions or molecules) that undergo a change need be included in the equation. For example, in the reaction between $NaCl(aq)$ and $AgNO_3(aq)$, the insoluble precipitate $AgCl(s)$ is formed. Only the silver ions and the chloride ions have participated in the reaction. While the $Ag^+(aq)$ and $Cl^-(aq)$ started out as submicroscopic, invisible ions, they have ended up as the clearly insoluble compound, $AgCl(s)$. The sodium and nitrate ions that were there at the start of the reaction are still there. Ions that are present in a system but which do not undergo a change are appropriately labeled **spectator ions.** As in the case of any "spectator," they are present at the scene of the action but are not involved in the action. The writing of ionic equations takes advantage of the classifications that have been made of the relative solubilities of salts (Table 9–2) and the distinctions made among strong electrolytes, weak electrolytes, and nonelectrolytes (Table 9–1).

Writing an ionic equation for the $AgNO_3$ and NaCl reaction can be shown in three steps. As your experience grows, it will be easy to write a **net ionic equation** for reactions occurring in aqueous solution. A net ionic equation is one in which all of the spectator ions are omitted. When directions call for the addition of chloride ion to silver ion, it will be understood that a variety of soluble ionic metal chloride solutions could be added to a soluble ionic silver compound and produce the same change—the precipitation of $AgCl(s)$.

Step 1. Write the "molecular" equation:

$$AgNO_3(aq) + NaCl(aq) \rightarrow AgCl(s) + NaNO_3(aq)$$

Step 2. Change the "molecular" equation to ionic form for each compound that exists mostly as ions in solution:

$$Ag^+(aq) + NO_3^-(aq) + Na^+(aq) + Cl^-(aq) \rightarrow AgCl(s) + Na^+(aq) + NO_3^-(aq)$$

Step 3. Eliminate the unchanged species (i.e., the spectator ions) and write the net ionic equation:

$$Ag^+(aq) + Cl^-(aq) \rightarrow AgCl(s)$$

Rules for Writing Ionic Equations

1. **Strong electrolytes** are written in **ionic form.**
2. **Nonelectrolotes** and **weak electrolotes** are written in **molecular form.**
3. **Insoluble species, precipitates,** and **gases** are written in **molecular form.**
4. Spectator ions are eliminated.
5. Equations must demonstrate conservation of **mass** and **charge.**

EXAMPLE 9.1

Write a net ionic equation for the reaction between $BaCl_2(aq)$ and $Na_2SO_4(aq)$.

Step 1. Complete and balance the molecular equation:

$$Na_2SO_4(aq) + BaCl_2(aq) \rightarrow BaSO_4(s) + 2NaCl(aq)$$

Step 2. Notice that $BaSO_4(s)$ is a precipitate. This information comes from Table 9–2. Write the ionic equation:

$$2Na^+ + SO_4^{2-} + Ba^{2+} + 2Cl^- \rightarrow BaSO_4(s) + 2Na^+ + 2Cl^-$$

Step 3. Eliminate Na^+ and Cl^- as spectator ions in the writing of the net ionic equation:

$$Ba^{2+}(aq) + SO_4^{2-}(aq) \rightarrow BaSO_4(s)$$

Notice the conservation of charge:

$$(2+) + (2-) = 0$$

EXAMPLE 9.2

Write the net ionic equation for $KOH(aq) + HClO_4(aq)$.

Step 1. Write the molecular equation:

$$KOH(aq) + HClO_4(aq) \rightarrow KClO_4(aq) + H_2O(\ell)$$

Step 2. Note that $H_2O(\ell)$ is a nonelectrolyte (or, a very weak electrolyte at best). Write the ionic equation:

$$K^+ + OH^- + H^+ + ClO_4^- \rightarrow K^+ \quad ClO_4^- + H_2O(\ell)$$

Step 3. Since only the H^+ and OH^- undergo a change, the K^+ and ClO_4^- spectator ions are eliminated in the writing of the net ionic equation:

$$H^+(aq) + OH^-(aq) \rightarrow H_2O(\ell)$$

EXAMPLE 9.3

Write the net ionic equation for $NaBr(aq) + Cl_2(g)$.

Step 1. Write the molecular equation:

$$2NaBr(aq) + Cl_2(g) \rightarrow + 2NaCl(aq)Br_2(\ell)$$

Step 2. Write the ionic equation:

$$2Na^+(aq) + 2Br^-(aq) + Cl_2(g) \rightarrow 2Na^+(aq) + 2Cl^-(aq) + Br_2(\ell)$$

Step 3. The sodium ion is obviously the spectator ion. Both the $Cl_2(g)$ and the $Br^-(aq)$ change. Write the net ionic equation:

$$2Br^-(aq) + Cl_2(g) \rightarrow 2Cl^- + Br_2(\ell)$$

EXAMPLE 9.4

Write the net ionic equation for $Zn(s) + HCl(aq)$.
Step 1. Write the molecular equation:

$$Zn(s) + 2HCl(aq) \rightarrow ZnCl_2(aq) + H_2(g)$$

Step 2. Write the ionic equation:

$$Zn(s) + 2H^+(aq) + 2Cl^-(aq) \rightarrow Zn^{2+}(aq) + 2Cl^-(aq) + H_2(g)$$

Step 3. Eliminate the spectator $Cl^-(aq)$:

$$Zn(s) + 2H^+ \rightarrow Zn^{2+}(aq) + H_2(g)$$

EXAMPLE 9.5

Complete and balance the net ionic equation:

Carbonate ion + hydrogen ion → carbon dioxide + water

1. Write a skeleton ionic equation:

$$CO_3^{2-}(aq) + H^+(aq) \rightarrow CO_2(g) + H_2O(\ell)$$

2. Balance the hydrogens (conservation of mass):

$$CO_3^{2-}(aq) + 2H^+(aq) \rightarrow CO_2(g) + H_2O(\ell)$$

3. Check for conservation of charge:

$$CO_3^{2-} + 2H^+ \rightarrow CO_2 + H_2O$$

$$(2-) + (2+) = 0$$

The ionic equation is complete.

EXAMPLE 9.6

Will $Pb(NO_3)_2(aq)$ and $K_2CrO_4(aq)$ react? If so, write the net ionic equation.

1. Consider the fact that there will be Pb^{2+}, NO_3^-, K^+, and CrO_4^{2-} ions in solution. Referring to Table 9–2, it is seen that CrO_4^{2-} and Pb^{2+} form the insoluble precipitate $PbCrO_4(s)$. Therefore, the reaction should occur.
2. Write the molecular equation:

$$Pb(NO_3)_2(aq) + K_2CrO_4(aq) \rightarrow PbCrO_4(s) + 2KNO_3(aq)$$

3. Since only the Pb^{2+} and CrO_4^{2-} are changed, the net ionic equation can be written directly:

$$Pb^{2+}(aq) + CrO_4^{2-}(aq) \rightarrow PbCrO_4(s)$$

Exercise 9.1

Write net ionic equations for the following reactions:

1. $NaOH(aq) + H_2SO_4(aq) \rightarrow$
2. $Fe(OH)_3(s) + HCl(aq) \rightarrow$
3. $Na_3PO_4(aq) + Pb(NO_3)_2(aq) \rightarrow$
4. $Cu(NO_3)_2(aq) + H_2S(g) \rightarrow$
5. $BaCl_2(aq) + Na_2SO_4(aq) \rightarrow$
6. $HCl(aq) + CaCO_3(s) \rightarrow$
7. $BaO(s) + H_2SO_4(aq) \rightarrow$
8. $Ca(s) + H_2O(\ell) \rightarrow$
9. $Ni(NO_3)_3(aq) + KOH(aq) \rightarrow$
10. $ZnSO_4(aq) + (NH_4)_2S(aq) \rightarrow$

9.6 CONCENTRATIONS OF SOLUTIONS

Any scientific discipline that is involved in the recording and reporting of experimental data in a form suitable for mathematical processing must be concerned with accuracy. It is often not good enough to operate in terms of "dilute," "concentrated," "medium," "small," "hot," and "large." These loosely qualitative terms might be adequate for some occasions, but for quantitative work such as in stoichiometric calculations, the chemist must know precisely how concentrated or how dilute a solution is. One chemist's dilute acid (relative to metal being cleaned) might be deadly concentrated to a colleague working with living organisms.

At the same time, the terms "dilute" and "concentrated" should not be confused with "weak" and "strong." In Chapter 10, the distinction between weak and strong acids and bases will be discussed in detail. The point is, however, that a strong acid, such as hydrochloric acid, can be prepared as a very *dilute* solution by the addition of a great quantity of water. Conversely, the relatively weak acetic acid can be obtained as a highly *concentrated* solution if many moles of it are dissolved in a minimal volume of water.

There are several common methods for expressing the concentrations of solutions.

9.7 MOLAR SOLUTIONS

The most useful way of expressing concentrations of solutions for everyday laboratory activity is in terms of *moles of solute per liter of solution.* Solutions of this type are called **molar,** and are symbolized by the capital letter M. The great advantage of this approach is that it allows for knowing with considerable accuracy the actual number of particles (ions or molecules) per unit of solution volume.

Expressing the definition of molar concentration in the form of an equation is the means by which many practical problems involving solutions can be solved:

$$\text{M} = \text{nV}^{-1}$$

$$\text{molarity} = \text{moles per liter}$$

It must be emphasized that V, the volume in liters, is for *liters of solution, not liters of water.* A liter of 1 molar solution contains 1 mole of solute plus *enough* water to make a final volume (solute + solvent) of 1 liter. This must be done empirically, because even if it were easy to calculate the volume of water needed to yield a liter of solution, it would still not be as fast as pouring water into a volumetric flask.

For the actual preparation of solutions, calculations can be simplified by modifying the basic equation $\text{M} = \text{nV}^{-1}$. Remember that moles can be calculated thus: grams of substance divided by grams per mole:

$$\text{n} = \frac{\text{g}}{\text{g mole}^{-1}}$$

If $\text{g}/\text{g mole}^{-1}$ is substituted for n, and liters (ℓ) is written instead of V, the equation has the form:

$$\text{M} = \frac{\text{g}}{(\text{g mole}^{-1})(\ell)}$$

Call the g mole^{-1} the mole (formula) weight, and use the more descriptive symbol MW. The equation can then be written in a most useful form to solve problems involving the preparation of solutions when the grams of solute per some volume of solution must be calculated:

g =	MW	× M	× ℓ
grams solute =	mole weight	× molarity	× liters

EXAMPLE 9.7

What mass of NaOH(s) is needed to prepare a liter of 1.0 M NaOH(aq)?

1. Calculate the mole weight of NaOH:

$$\text{NaOH} = 40.0 \text{ g mole}^{-1}$$

2. Write the equation needed to find grams of solute:

$$g = MW \times M \times \ell$$

3. Substitute and solve for g:

$$g = (40.0 \text{ g } \cancel{\text{mol}}^{-1})(1.0 \cancel{\text{mol}}\, \cancel{\ell}^{-1})(1.0 \cancel{\ell})$$
$$g = 40.0$$

4. Therefore, place 40.0 g of NaOH(s) in a beaker, and add about 800 ml of water. Stir until dissolved. Finally, add enough water to bring the final volume of solution to 1.0 liter. See Figure 9–10.

STEP 1
Weigh out 40.0 g
of NaOH (s)

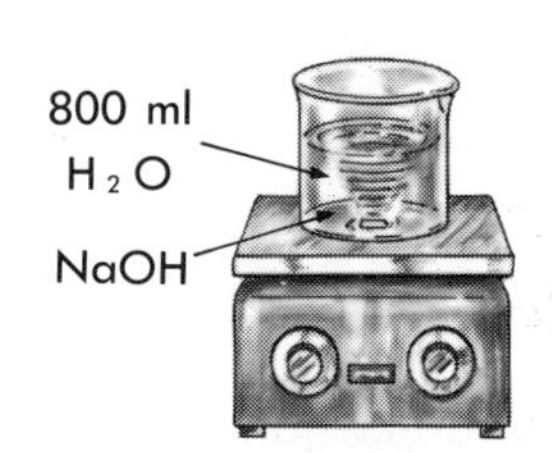

STEP 2
Dissolve the
NaOH in less
than 1 liter
of water

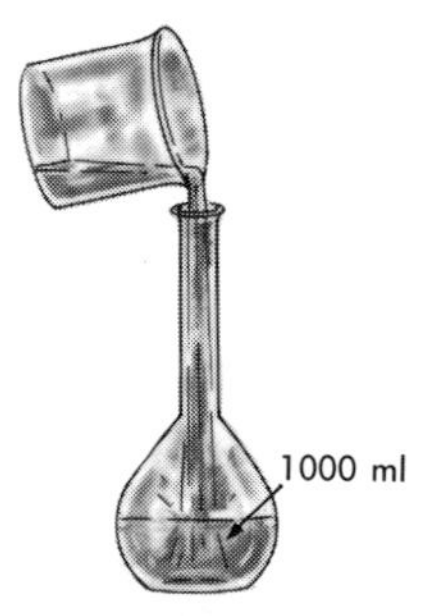

STEP 3
Add water to
bring final volume
of solution to
1 liter

Figure 9–10 Steps in preparing a liter of 1.0 M NaOH(aq).

EXAMPLE 9.8

What mass of NaOH(*s*) is needed to prepare 250.0 ml of a 0.2 M solution of NaOH?

1. Calculate the mole weight of NaOH:

$$NaOH = 40.0 \text{ g mole}^{-1}$$

2. Write the equation to solve for grams:

$$g = MW \times M \times \ell$$

3. Substitute and solve for grams:

$$g = (40.0 \text{ g } \cancel{mol}^{-1})(0.2 \ \cancel{mol}\ \cancel{\ell}^{-1})(0.25\ \cancel{\ell})$$
$$g = 2.0$$

4. Dissolve 2.0 g of NaOH(*s*) in about 200 ml of water and then add enough water to reach the final volume of 250.0 ml of solution.

EXAMPLE 9.9

Calculate the number of moles of KCN in 400.0 ml of 0.5 M solution.

1. Change 400.0 ml to liters:

$$\frac{400.0 \text{ ml}}{10^3 \text{ ml } \ell^{-1}} = 0.4\ \ell$$

2. Change the original equation to the form for finding moles:

$$M = nV^{-1}$$
(multiply the equation by V)
$$VM = n\cancel{V}^{-1}(\cancel{V})$$

or

$$n = M \times V$$
$$\text{moles} = \text{molarity} \times \text{liters}$$

3. Substitute in the equation and solve for moles:

$$n = MV$$
$$n = (0.5 \text{ mol } \ell^{-1})(0.4\ \ell)$$
$$n = 0.2 \text{ mole}$$

EXAMPLE 9.10

How many molecules of glucose ($C_6H_{12}O_6$) are there per milliliter in a 2.5 M solution?

1. Moles are converted to a number of particles by multiplying by Avogadro's number:

$$\text{number of particles} = n \times \mathcal{N}_A$$

2. Organize the data:

$$M = 2.5 \text{ moles } \ell^{-1}$$
$$V = 1.0 \text{ ml} = 10^{-3}\ \ell$$
$$n = ?$$
$$\mathcal{N}_A = 6.02 \times 10^{23} \text{ mole}^{-1}$$
$$\text{number of glucose molecules} = ?$$

3. Write the appropriate equation, substitute, and solve for n:

$$n = MV$$
$$n = (2.5 \text{ mols } \ell^{-1})(10^{-3}\ \ell)$$
$$n = 2.5 \times 10^{-3} \text{ mole}$$

4. Finally calculate the number of molecules:

$$\text{number of molecules} = n \times \mathcal{N}_A$$
$$= (2.5 \times 10^{-3} \text{ mol})(6.02 \times 10^{23} \text{ mol}^{-1})$$
$$\text{number of molecules per ml} = 1.5 \times 10^{21}$$

EXAMPLE 9.11

A bottle of $H_2SO_4(aq)$ is labeled 92.0% by weight with a specific gravity of 1.83. What is the molarity?

1. Convert sp. gr. to density:

$$D = 1.83 \text{ g ml}^{-1}$$

2. Change the M/V ratio from g/ml to g/ℓ.

$$1.83 \text{ g ml}^{-1}\ (10^3 \text{ ml } \ell^{-1}) = 1830 \text{ g } \ell^{-1}$$

3. Find 92.0% of 1830 g ℓ^{-1}, since this is the percentage of pure H_2SO_4 in solution:

$$92.0\% \text{ of } 1830 \text{ g } \ell^{-1} = 0.92 \times 1830 \text{ g } \ell^{-1}$$
$$= 1684 \text{ g } \ell^{-1} \text{ pure } H_2SO_4$$

4. Write the equation that will solve for molarity:

$$g = MW \times M \times \ell$$

$$\frac{g}{MW \times \ell} = \frac{MW \times M \times \ell}{MW \times \ell} \text{ (divide the equation by MW } \times \ell)$$

or

$$M = \frac{g}{MW \times \ell}$$

5. Organize the data:

$$g = 1684 \text{ grams}$$

$$MW = 98.0 \text{ g mole}^{-1}$$

(from the formula weight of H_2SO_4)

$$\ell = 1 \text{ liter}$$

$$M = ?$$

6. Substitute in the equation and solve for M.

$$M = \frac{1684 \text{ g}}{(98 \text{ g mol}^{-1})(1\ \ell)} = 17.2 \text{ moles } \ell^{-1}$$

Answer is written as 17.2 M.

Exercise 9.2

1. What mass (grams) of solute is needed to prepare each of the following solutions from solid reagents?
   a. 35.0 ml of 2.0 M $AgNO_3$
   b. 160.0 ml of 0.5 M $Na_2SO_4 \cdot 10H_2O$

   Note: The weight of the hydrated water is included in calculating the mole weight because it makes up part of the actual weight of salt used.
2. What molar concentration results when 10.0 g of KOH(*s*) is dissolved in water to a final volume of 200.0 ml?
3. Find the molarity of $HNO_3(aq)$ which is 50.0% by weight and has a sp. gr. of 1.3.
4. If 30.0 g of acetone (C_3H_6O) is dissolved in water to a final volume of 0.5 liters, approximately how many molecules of acetone are there in 10.0 ml?

Molar Dilutions

The molarity problems thus far have dealt with solid solutes which can be weighed in grams. However, it is often necessary or convenient to prepare large volumes of dilute solutions from small volumes of concentrated stock solutions. For example, a large volume of 0.1 M HCl(*aq*) could easily be prepared by measuring out a few milliliters of 12.0 M (concentrated) HCl and adding it to enough water so that the desired molarity and volume are obtained.

It is important to recognize first that a dilution process changes only the volume of solvent, and that the original number of moles of solute remains the same. In Figure 9–11, the constancy of the number of solute particles is illustrated by a symbolic 8 molecules of solute being diluted.

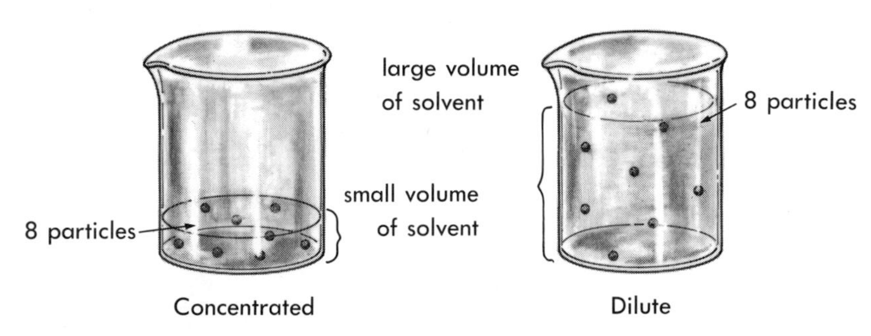

Figure 9–11 Dilution does not alter the number of moles of solute.

Starting with the observation that in a dilution process

$$\underset{\text{(before dilution)}}{\text{number of moles}_1} = \underset{\text{(after dilution)}}{\text{number of moles}_2}$$

and bearing in mind that

$$\text{moles} = \text{molarity} \times \text{volume in liters}$$

or

$$n = MV,$$

it is possible to substitute "equals for equals" and say:

$$M_1V_1 = M_2V_2$$

M_1V_1 = the number of moles before dilution

M_2V_2 = the number of moles after dilution

EXAMPLE 9.12

What volume of 1.0 M stock solution of KCl is needed to prepare 500.0 ml of 0.2 M solution?

1. Write the dilution equation:

$$M_1V_1 = M_2V_2$$

2. Organize the data:

$$M_1 = 1.0 \text{ M}$$
$$V_1 = ?$$
$$M_2 = 0.2 \text{ M}$$
$$V_2 = 500 \text{ ml}$$

3. Notice that the 500 ml does not have to be changed to liters. The answer for V_1 will also be milliliters. This is possible because multiplying both sides of the equation by 10^3 converts liters to milliliters. Substitute in the equation and solve for V_1:

$$(1.0 \text{ M})(V_1) = (0.2 \text{ M})(500 \text{ ml})$$
$$V_1 = \frac{(0.2 \cancel{\text{M}})(500 \text{ ml})}{1.0 \cancel{\text{M}}}$$
$$V_1 = 100.0 \text{ ml}$$

4. The final step is to pour 100.0 ml of the 1.0 M stock solution of KCl into a volumetric flask and add about 400.0 ml of water to yield 500.0 ml of 0.2 M KCl(*aq*).

EXAMPLE 9.13

What volume of 0.2 M KCN(*aq*) is needed to prepare 300.0 ml of a 0.25 *mM* (millimolar) solution?

1. The concentration units must be compatible. Both units must be moles or millimoles. Since the conversion factor must be 10^3 millimoles mole^{-1}, the change for 0.2 M to mM is

$$(0.2 \text{ } \cancel{\text{mole}} \text{ liter}^{-1})(10^3 \text{ millimoles } \cancel{\text{mole}}^{-1}) = 200.0 \text{ mM}$$

2. Write the equation:

$$M_1V_1 = M_2V_2$$

3. Organize the data:

$$M_1 = 200.0 \text{ mM} = 2.00 \times 10^2 \text{ mM}$$
$$V_1 = ?$$
$$M_2 = 0.25 \text{ mM}$$
$$V_2 = 300.0 \text{ ml} = 3.00 \times 10^2 \text{ ml}$$

4. Substitute the data and solve for V_1:

$$(2 \times 10^2 \text{ mM})(V_1) = (0.25 \text{ mM})(3.00 \times 10^2 \text{ ml})$$

$$V_1 = \frac{(0.25 \text{ m}\cancel{\text{M}})(3.00 \times 1\cancel{0}^2 \text{ ml})}{2 \times 1\cancel{0}^2 \text{ m}\cancel{\text{M}}}$$

$$V_1 = 0.38 \text{ ml}$$

5. When 0.38 ml of 0.2 M stock is diluted up to 300.0 ml, the proper concentration is prepared.

EXAMPLE 9.14

What volume of 96.0%, sp. gr. 1.84, bottle of commercial $H_2SO_4(aq)$ is needed to prepare 500.0 ml of 0.2 M dilution?

1. The first step is to calculate the molar concentration of the commercial acid as explained in Example 9.11. The answer should be 18.0 M.
2. Organize the data and solve the dilution equation:

$$M_1 = 18.0 \text{ M}$$

$$V_1 = ?$$

$$M_2 = 0.2 \text{ M}$$

$$V_2 = 500.0 \text{ ml}$$

$$M_1V_1 = M_2V_2$$

$$(18.0 \text{ M})(V_1) = (0.2 \text{ M})(500 \text{ ml})$$

$$V_1 = 5.6 \text{ ml}$$

3. **Caution:** The 5.6 ml of concentrated H_2SO_4 should be added to about 100 ml of water, and then diluted up to 500.0 ml. The great heat of hydration of H_2SO_4 could cause dangerous splattering if the less dense water is poured into the acid.

Exercise 9.3

How should the following dilutions be carried out?

1. 250.0 ml of 0.35 M KNO_3 from 3.2 M stock.
2. 0.65 liters of 70.0 mM NaOH from 5.2 M stock.
3. 3.0 liters of 0.15 M HCl from 6.0 M stock.
4. 500.0 ml of 0.35 M KCl from 6.0 M stock.
5. 350.0 ml of 20.0 mM KCN from 1.2 M stock.

9.8 NORMAL SOLUTIONS*

The term **normal** solutions, symbolized by the capital letter N, indicates another method for expressing the concentration of solutions. This method, which is very useful for solving analytical problems, is based on the *equivalent weight* of the solute. It was pointed out previously that the equivalent weight of a substance is that weight which can combine with 8.0 g of oxygen. Another description is that the equivalent weight of a substance is that weight which may transfer (gain or lose) one mole of electrons. For example, if magnesium is oxidized, the stoichiometry of the reaction indicates that 1 mole of magnesium combines with a 1/2 mole of O_2:

$$Mg + \frac{1}{2} O_2 \rightarrow MgO$$

One mole of magnesium loses 2 moles of electrons in the oxidation. By definition, the equivalent weight of magnesium is that number of grams that loses just 1 mole of electrons. Therefore, the equivalent weight of Mg must be 1/2 of the mole weight:

$$Mg = 24.3 \text{ g mole}^{-1}$$

or

$$Mg = \frac{24.3}{2} \text{ g per equivalent weight}$$

Simplified

$$Mg = 12.15 \text{ g eq}^{-1}$$

In effect, the equivalent weight of Mg was obtained by dividing the mole weight by its common oxidation number or ionic charge. If other similar examples are investigated, it will be observed that this method of calculating equivalent weights is useful in a number of cases.

In order to avoid a misleading oversimplification, it must be emphasized that the equivalent weight of a substance is ultimately related to a specific reaction. The number of moles of electrons transferred by an atom or ion may vary in different reactions. Therefore, each reaction must be considered independently in order to calculate the equivalent weight of a particular species. The general equation used to calculate equivalent weight will then be

$$\text{eq wt} = \frac{\text{mole weight}}{\text{moles of electrons transferred}}$$

This question will be discussed in more detail in Chapter 12. For the time being, it is advantageous to define the equivalent weight of a metal as the mole weight (gram atomic weight) divided by its total positive ionic charge:

$$\text{eq wt} = \frac{\text{mole weight}}{\text{total positive ionic charge}}$$

EXAMPLE 9.15

Calculate the equivalent weights of the following: (1) NaOH, (2) H_2SO_4, (3) $Zn(NO_3)_2$, (4) Al_2S_3.

1. a. NaOH = 40.0 g mole^{-1}

   b. total positive ion charge of $Na_1^+ = 1$

   c. $\text{eq wt} = \frac{40.0 \text{ g}}{1} = 40.0 \text{ g}$

2. a. H_2SO_4 = 98 g mole^{-1}

   b. total positive ion charge of $H_2^+ = 2$

   c. $\text{eq wt} = \frac{98.0 \text{ g}}{2} = 49.0 \text{ g}$

3. a. $Zn(NO_3)_2$ = 189.4 g mole^{-1}

   b. total positive ion charge of $Zn_1^{2+} = 2$

   c. $\text{eq wt} = \frac{189.4 \text{ g}}{2} = 94.7 \text{ g}$

4. a. Al_2S_3 = 150.2 g mole^{-1}

   b. total positive ion charge of $Al_2^{3+} = 6$

   c. $\text{eq wt} = \frac{150.2 \text{ g}}{6} = 25.0 \text{ g}$

A normal solution is defined as one containing an equivalent weight of solute per liter of solution. Expressed in the form of an equation where the normality is equal to the number of equivalent weights per liter of solution:

$$N = \#\text{eq liter}^{-1}$$

normality — number of equivalents — volume

Just as moles could be written as g/g mole^{-1}, so, too, equivalents can be written

$$\boxed{\#\text{eq} = \frac{\text{g}}{\text{g eq}^{-1}}}$$

#eq = number of equivalents

g = grams of solute

g eq^{-1} = grams per one equivalent (i.e., the equivalent weight)

The basic equation can be put in a more useful form if g/g eq^{-1} is substituted for #eq:

$$N = \frac{g}{g\ eq^{-1} \times \text{liters}}$$

Rearranging the equation, it becomes a rather simple job to calculate the number of grams of solute needed to prepare various normal concentrations. The equation becomes

$$g = (g\ eq^{-1})(N)(V)$$

g = grams of solute needed

$g\ eq^{-1}$ = equivalent weight

N = normality

V = volume of solution in liters

EXAMPLE 9.16

How many grams of $Ca(OH)_2(s)$ are needed to prepare 150 ml of 0.25 N solution?

1. Calculate the equivalent weight of $Ca(OH)_2$ from the formula weight:

$$Ca(OH)_2 = 74\ g\ mole^{-1}$$

$$g\ eq^{-1} = \frac{\text{mole weight}}{\text{total positive ionic charge}}$$

total positive ionic charge of $Ca_1^{2+} = 2$

$$g\ eq^{-1} = \frac{74\ g}{2} = 37\ g$$

2. Change the volume to liters:

$$\frac{150.0\ \cancel{ml}}{10^3\ \cancel{ml}\ \ell^{-1}} = 0.15\ \text{liter}$$

3. Substitute in the equation to find the number of grams of $Ca(OH)_2(s)$ needed:

$$g = (g\ eq^{-1})(N)(\text{liters})$$

$$g = (37\ g\ \cancel{eq^{-1}})(0.25\ \cancel{eq}\ \cancel{\ell^{-1}})(0.15\ \cancel{\text{liter}})$$

$$g = 1.4$$

4. Place 1.4 g of $Ca(OH)_2(s)$ in a 150.0 ml volumetric flask and add water up to the volume marker.

While Example 9.16 illustrates a typical method for preparing a solution from a solid solute in which the concentration is expressed in normality units, most normal solutions are prepared by diluting aqueous stock solutions. Furthermore, it is often convenient to interconvert molar and normal concentrations. Dilution problems involving normal concentrations are solved the same way as those involving molar concentrations. The only difference is that in normal concentration dilutions, the number of equivalents of solute remains constant, while in molar solution dilutions the moles of solute constancy are the focal point. Since

$$\#eq_1 = \#eq_2$$

$$\text{(initial number of equivalents)} = \text{(final number of equivalents)}$$

the equation is modified to the more useful form:

$$N_1V_1 = N_2V_2$$

Remembering, that $\#eq = NV$, so it is simply a matter of substituting "equals for equals." These two equations point out the great advantage of the normal system: one ml of a 1 N solution of anything will react exactly with 1 ml of a 1 N solution of anything else, and solutions of equal normality will react volume for volume. This greatly simplifies operations and calculations.

9.9 INTERCONVERTING MOLARITY AND NORMALITY*

Since the only difference between molar and normal concentrations is that molarity uses *moles* of solute per liter of solution while normality uses *#eq* of solute per liter of solution, the ratio of equivalents to moles provides the means of interconverting between normality and molarity. For example, H_2SO_4 has a mole weight of 98 and an equivalent weight of $98/2 = 49$. This means that 1 mole of H_2SO_4 is equal to 2 equivalents. In effect, a 1.0 M solution is equal to a 2.0 N concentration. The total positive ionic charge emerges as the conversion factor. The relationship between M and N concentrations may be summarized by the simple equation:

$$N = \text{(total positive ionic charge)} \times M$$

or

$$M = \frac{N}{\text{(total positive ionic charge)}}$$

EXAMPLE 9.17

What volume of 18.0 M stock $H_2SO_4(aq)$ is needed to prepare 200.0 ml of 0.33 N $H_2SO_4(aq)$?

1. Convert the 18.0 M concentration of stock to N concentration:

$$N = (\text{total positive ionic charge}) \times M$$
$$\text{total positive ionic charge of } H_2SO_4 = H_2^{+} = 2$$
$$N = 2 \times 18.0\ M$$
$$N = 36.0$$

2. Organize the data and substitute in the dilution equation:

$$N_1 = 36.0$$
$$V_1 = ?$$
$$N_2 = 0.33$$
$$V_2 = 200.0\ \text{ml}$$
$$N_1V_1 = N_2V_2$$
$$(36.0\ N)(V_1) = (0.33\ N)(200.0\ \text{ml})$$
$$V_1 = \frac{(0.33\ \cancel{N})(200.0\ \text{ml})}{36.0\ \cancel{N}}$$
$$V_1 = 18.3\ \text{ml}$$

3. Pour the 18.3 ml of concentrated stock solution into a large volume of water and then bring the final volume up to 200.0 ml.

The special usefulness of normality as an expression of concentration will be illustrated in the next chapter under the heading of acid-base neutralization reactions.

Exercise 9.4

1. How many grams of solid solute are needed to prepare each of the following solutions?
   a. 120.0 ml of 0.4 N $CaCl_2$
   b. 2.5 liters of 0.2 N $MgSO_4 \cdot 7H_2O$
2. What volume of 1.5 N stock solution of H_2SO_4 is needed to prepare 50.0 ml of 0.2 N concentration?
3. Calculate the normality of a stock solution of $H_3PO_4(aq)$ which is 84.0% by weight pure and has a sp. gr. of 1.74.
4. Find the molar concentration of the solution in the preceding exercise.

9.10 OTHER METHODS OF EXPRESSING SOLUTION CONCENTRATIONS

Per Cent by Weight Solutions

Although solutions are sometimes prepared on a ratio of weight of solute to volume of solvent (W/V), or a volume-to-volume ratio (V/V), the most common method is to write the ratio of weight of solute to weight of solution (W/W), and express this ratio as a percentage. For example, a 3.0% (W/W) solution of iodine in alcohol is one in which there are 3.0 g of iodine for every 97.0 g of alcohol. In other words, the 3.0 g of iodine make up 3.0% of the total weight of solute and solvent, which is 100.0 g.

EXAMPLE 9.18

If a solution is prepared by dissolving 4.0 g of $KMnO_4(s)$ in 640.0 g of water, how is this expressed as a percent solution (W/W)?

1. The total weight of the solution is solute weight plus solvent weight:

$$4.0 \text{ g } KMnO_4(s) + 640.0 \text{ g water} = 644.0 \text{ g solution}$$

2. $\%$ weight of $KMnO_4 = \dfrac{4.0 \text{ g}}{644.0 \text{ g}} = 0.0062 = 0.62\%$W/W solution.

EXAMPLE 9.19

How many grams of table salt are needed to prepare 200.0 g of a 5.0% NaCl solution (W/W)?

1. Find 5.0% of the 200.0 g total:

$$5.0\% \text{ of } 200.0 \text{ g} = (0.05)(200.0 \text{ g}) = 10.0 \text{ g}$$

2. The solution is prepared by dissolving 10.0 g of $NaCl(s)$ in 190.0 g (effectively the same as 190.0 ml) of water.

Parts Per Million Solutions (ppm)

When solutes are present in solution in very small concentrations, it is often useful to express such concentrations in **ppm**. Since the density of water is 1.0 g ml^{-1}, it is convenient to calculate ppm of solute particles on the basis of milligrams of solute per liter of solution instead of milligrams per kilogram. For example, if 1.4×10^{-3} g of cyanide ion was found in 700.0 ml of solution, the ppm calculation would require the conversion of 1.4×10^{-3} g to milligrams

$$1.4 \times 10^{-3} \cancel{\text{g}} (10^3 \text{ mg } \cancel{\text{g}^{-1}}) = 1.4 \text{ mg}$$

and 700.0 ml of solution to liters (700.0 ml = 0.7 liter). The equation and solution would be

$$\text{ppm} = \frac{\text{mg CN}^-}{\text{liters of solution}} = \frac{1.4}{0.7} = 2.0 \text{ ppm}$$

Formal Concentrations

Some chemists object to expressing the concentrations of ionic solutions in molar concentration terms. They prefer to reserve the mole as a representation of gram *molecular* weights (GMW), and in this sense, the application of GMW to a nonmolecular substance would be misleading. Instead of GMW, the alternative would be GFW where the letter F stands for **formula**. However, this difficulty can be resolved by defining the GMW as the *mole weight* instead of *molecular weight*. For the most part, the concept of **formality** is an attempt to resolve a semantic rather than chemical problem. If a label on a bottle of solution reads 1.0 F NaCl(*aq*) this may be interpreted as 1.0 M concentration, where it means 1.0 mole of NaCl *units* (mass = 58.5 g) per liter of solution.

Two more methods of expressing concentrations of solutions (mole fraction and molality) will be discussed as they relate to the colligative properties of solutions.

9.11 COLLIGATIVE PROPERTIES OF SOLUTIONS*

Colligative is meant to suggest the special properties of solutions, as distinct from the properties of pure solvents. The introduction of solute particles into a solvent involves interionic attractions among solute particles, ion-water molecule interaction, and, in the special case of nonelectrolytes, dipole-water interaction.

The colligative effect is the sum of the ways in which solute particles affect the vapor pressure of the solvent by virtue of the ion-water and dipole-water interactions. In other words, particles of solute bind water molecules, to some extent, so that fewer water molecules will shift from the liquid to the vapor phase at a given temperature. This effective lowering of vapor pressure is illustrated in Figure 9–12, in which it can be seen how fewer molecules of water vapor exert less pressure. Remember, from the equation of state, that pressure is directly proportional to the number of gas molecules.

The lowering of the vapor pressure of a solvent has the effect of shifting its normal boiling and freezing points. However, if the shift of boiling point (ΔT_b) and the shift of freezing point (ΔT_f) are to be mathematically predictable, the solute must be a **nonvolatile** *nonelectrolyte*. A nonvolatile substance is one which has a relatively high boiling point and therefore will not easily move into the vapor phase.

One reason for making a distinction between electrolytes and nonelectrolytes is that the dissolution of nonelectrolytes into nonionizing mole-

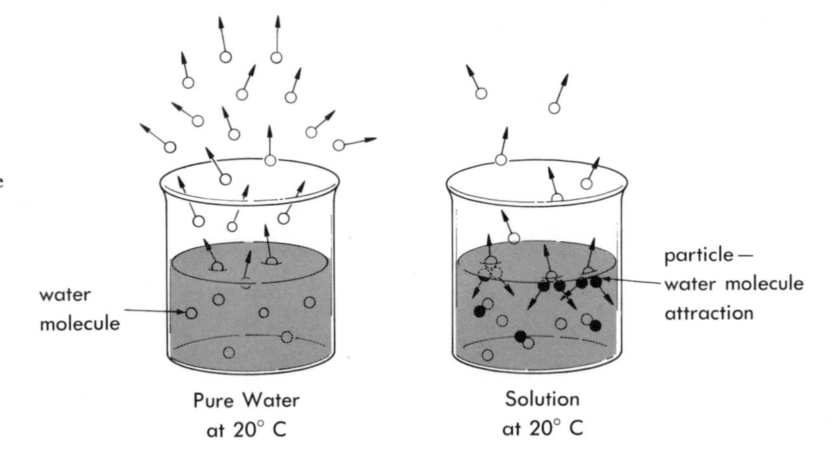

Figure 9–12 The interaction between water molecules and solute particles (ions or molecules) prevents many water molecules from moving into the vapor phase. This is a colligative effect.

cules results in a predictable number of particles that exhibit very weak intermolecular attractions, while ions tend to clump together because of a much stronger interionic attraction. Nonelectrolytes approach what may be described as an **ideal solution** because of their negligible molecular interactions. This is somewhat analogous to the concept of the ideal gas. When one mole of nonelectrolyte is dissolved, the dissociated molecules approach Avogadro's number of individual particles. However, this model is limited to relatively dilute solutions, since the molecular interactions do become significant at high concentrations.

The abnormal colligative properties of electrolytes were explained by the theory of Peter Debye and Erich Hückel in 1923. The **Debye-Hückel theory** says, in effect, that the interionic attractions among dissociated electrolytes create clumps of ions, so that one mole of an ionic compound will not produce as many moles of individual, free-moving particles in solution as expected. This does not mean that NaCl, for example, will not produce 2 moles of ions:

$$\underset{\text{1 mole}}{Na^+Cl^-(s)} \xrightarrow{\text{dissociate}} \underbrace{\underset{\text{1 mole}}{Na^+} + \underset{\text{1 mole}}{Cl^-}}_{\text{2 moles of ions}}$$

But, the Na^+ and Cl^- ions will interact so that the *effective* number of moles exhibits a colligative effect which is characteristic of fewer particles. In Figure 9–13, the clumping of ions is illustrated and shows the interionic effect on the number of solute particles.

It should be emphasized that the colligative effect in dilute solutions is related to the *number* of individual particles, and not to the size, weight, or shape of the particles. One large particle composed of many clumped ions will have the same effect on the vapor pressure as one separate molecule.

For the time being, the following discussions of vapor pressure lowering, the effect of vapor pressure lowering on the ΔT_b and ΔT_f of solvents, and the expressions of concentrations of solutions will be restricted to dilute solutions of nonelectrolytes.

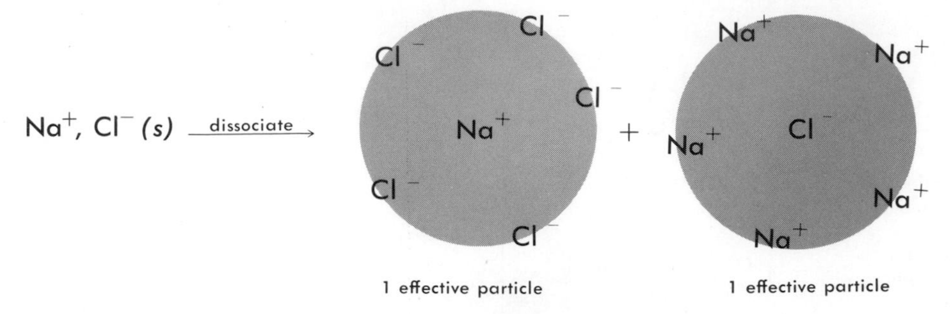

Figure 9–13 Interionic attractions reduce the number of effective particles.

9.12 THE MOLE FRACTION AND RAOULT'S LAW*

In 1886, François Raoult observed that the vapor pressure lowering of a solvent was directly proportional to the **mole fraction** of solute. In an ideal solution, an increase in the number of grams of nonvolatile, nonelectrolyte solutes per gram of solution will lower the pressure according to the equation

$$\boxed{P = P^\circ X}$$

P = vapor pressure lowering of solution

P° = vapor pressure of pure solvent

X = mole fraction

In case there should be more than one solute, the total effect on the vapor pressure of the solution is the *sum* of the separate effects due to the mole fraction of each component. The equation $P = P^\circ X$, which expresses **Raoult's law,** may be modified as follows:

$$P_{total} = (P^\circ X)_A + (P^\circ X)_B \ldots$$

P_{total} = total solution vapor pressure lowering

P° = vapor pressure of pure solvent

X_A = mole fraction of solute A

X_B = mole fraction of solute B

EXAMPLE 9.20

Calculate the vapor pressure of a solution of glucose ($C_6H_{12}O_6$) in which 30.0 g of glucose is dissolved in 200.0 g of water at 23.0°C. (Vapor pressure of water at 23.0°C = 21.0 torr.)

1. Calculate the moles of glucose and water:
   a. glucose ($C_6H_{12}O_6$) = 180.0 g mole^{-1}

$$n = \frac{g}{g\ mole^{-1}} = \frac{30.0\ \cancel{g}}{180.0\ \cancel{g}\ mol^{-1}} = 0.167\ \text{mole glucose}$$

b. water (H_2O) = 18.0 g mole⁻¹

$$n = \frac{g}{g\ mole^{-1}} = \frac{200.0\ \cancel{g}}{18.0\ \cancel{g}\ mol^{-1}} = 11.1\ mole\ H_2O$$

2. The mole fraction, X, of glucose is

$$X = \frac{\text{moles of solute}}{\text{moles of solute + solvent}}$$

$$X = \frac{0.167}{0.167 + 11.1} = \frac{0.167}{11.267} = 0.0148$$

3. The vapor pressure lowering is calculated from the equation:

$$P = P°X$$

$$P = (21.0\ torr)(0.0148)$$

$$P = 0.31\ torr$$

4. The vapor pressure of the solution at 23.0°C is:

$$P = 21.0\ torr - 0.31\ torr$$

$$P = 20.7\ torr$$

The ΔT_b and ΔT_f of solvent are a direct result of the lowering of vapor pressure. Figure 9–14 graphically represents the relationship between the vapor pressure of water and the shifting of the normal boiling and freezing points of water when it functions as a solvent.

When vapor pressure lowering data are collected from other solvents, the colligative effect is observed to be similar. However, the extent to which the freezing point is lowered and the boiling point is raised may vary considerably. The quantitative investigation of the colligative effect requires a standardization of the way in which concentrations are expressed. Since the relationship between the lowering of vapor pressure and the mole fraction

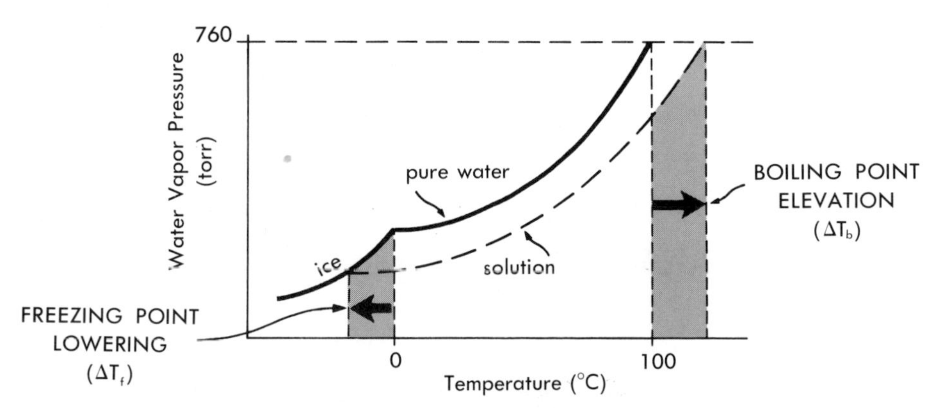

Figure 9–14 The effect of vapor pressure lowering on the freezing and boiling points of water.

of solute has been demonstrated, the best way to express concentrations of solutions for uniform comparisons is to adopt a system in which all solutes of the same concentration can have the same mole fraction. The molar solution measure cannot serve this purpose. While molar solutions of sugar and urea will have equal numbers of moles, they will not necessarily be dissolved in equal numbers of moles of water when their volumes are equal. The objective of comparing equal mole fractions is achieved by the use of **molal** solutions.

Molality

A molal solution, symbolized by *m*, is one in which there is a concentration of *one mole of solute per kilogram of solvent*. Expressed in the form of an equation

$$\boxed{m = n\ kg^{-1}}$$

$$m = \text{molality}$$

$$n = \text{moles of solute}$$

$$kg = \text{kilograms of solvent}$$

EXAMPLE 9.21

What is the molality of a solution in which 6.2 g of NaCl(*s*) is dissolved in 86.0 g of water?

1. Calculate the moles of NaCl:

$$NaCl = 58.5\ g\ mole^{-1}$$

$$n = \frac{g}{g\ mole^{-1}} = \frac{6.2\ \cancel{g}}{58.5\ \cancel{g}\ mol^{-1}} = 0.106\ mole$$

2. Convert 86.0 g to kg:

$$86.0\ g = 0.086\ kg$$

3. Substitute in the equation for molality and solve:

$$m = n\ kg^{-1} = \frac{n}{kg}$$

$$m = \frac{(0.106\ mols)}{(0.086\ kg)}$$

$$m = 1.23\ molal$$

EXAMPLE 9.22

What mass of urea, $CO(NH_2)_2$, is needed to prepare a 0.1 m solution using 500.0 g of benzene, C_6H_6?

1. Calculate the mole weight of urea and convert the weight of benzene to kg:

$$\text{urea} = CO(NH_2)_2 = 60 \text{ g mole}^{-1}$$

$$\text{benzene} = C_6H_6 = 500.0 \text{ g} = 0.5 \text{ kg}$$

2. Solve the equation for moles of urea:

$$m = n \text{ kg}^{-1}$$

$$\text{kg} \times m = n \cancel{\text{kg}}^{-1} \times \cancel{\text{kg}} \text{ (multiply the equation by kg)}$$

$$n = m \times \text{kg}$$

3. Substitute in the equation:

$$n = (0.1 \text{ molal})(0.5 \text{ kg})$$

$$n = (0.1 \text{ mols } \cancel{\text{kg}}^{-1})(0.5 \cancel{\text{kg}})$$

$$n = 0.05 \text{ mole}$$

4. Calculate the weight of urea:

$$g = \text{moles} \times \text{g mole}^{-1}$$

$$g = (0.05 \cancel{\text{mol}})(60 \text{ g } \cancel{\text{mol}}^{-1})$$

$$g = 3.0 \text{ grams of urea}$$

Exercise 9.5

1. What mass of $CuSO_4(s)$ is needed to prepare 150.0 g of a 7.0% solution (W/W)?
2. If 0.020 mg of $Hg^{2+}(aq)$ was found in 1.4×10^2 liters of solution, calculate the concentration of mercury in ppm.
3. To what extent will the water vapor pressure be lowered at 29.0°C if 10.0 g of urea, $CO(NH_2)_2$, is dissolved in 250.0 g of water?
4. What is the molality of a solution in which 20.0 g of CCl_4 is dissolved in 400.0 ml of benzene (density = 0.88 g ml^{-1})?
5. How many grams of $CS_2(\ell)$ are required to prepare a 0.25 molal solution in 0.30 kg of chloroform, $CHCl_3$?

Relating ΔT_f, ΔT_b, and Molecular Weight

Figure 9–14 graphically represents the colligative effect of solute particles on the vapor pressure of the solvent. The vapor pressure curve of the solution clearly indicates that the freezing point of the solvent will be lowered and the boiling point will be elevated. On the basis of these observations, chemists have found it extremely useful to standardize the mole fraction to the extent that the ΔT_f and the ΔT_b of a variety of solvents are measured for 1.0 molal concentrations. For any solvent, the ΔT_f in a 1.0 m solution is called the **molal freezing point depression,** symbolized K_f, and the 1.0 m ΔT_b is called the **molal boiling point elevation,** K_b. If the molality of the nonelectrolyte solution is more than 1.0 m, the ΔT_f will be more than the K_f for that solution and the ΔT_b will be higher than the boiling point for the pure solvent. A concentration less than 1.0 m will have the same effect but to a lesser degree, which is to say that the ΔT_f and the ΔT_b will be less than the K_f and K_b values. The equation that relates these terms is

$$\Delta T_f = mK_f$$

and

$$\Delta T_b = mK_b$$

ΔT_f and ΔT_b = number of degrees deviation from boiling and freezing points of pure solvent

m = molality

K_f and K_b = molal T_f and T_b constants

The great usefulness of these equations comes about because of the possibility of calculating molecular weights of unknown solutes by the ΔT_f or ΔT_b measurement. An example of a very practical application is the use of water-soluble nonelectrolytes for antifreeze solutions, in which low molecular weight, nonvolatile substances are used to lower the freezing point of automobile radiator water to a safe level. Table 9–3 lists for selected solvents the freezing and boiling point data that may be helpful in solving related problems.

TABLE 9–3 BOILING AND FREEZING POINT DATA FOR SELECTED SOLVENTS

Solvent	Formula	T_f(°C)	K_f(°C)	T_b(°C)	K_b(°C)
benzene	C_6H_6	5.48	5.12	80.15	2.53
carbon disulfide	CS_2	—	—	46.25	2.34
carbon tetrachloride	CCl_4	−22.8	29.8	76.8	5.02
acetic acid	HOAc	16.6	3.90	118.1	3.07
chloroform	$CHCl_3$	−63.5	4.68	61.2	3.63
water	H_2O	0	1.86	100	0.52
naphthalene	$C_{10}H_8$	80.2	6.80	—	—
camphor	$C_{10}H_{16}O$	178.4	37.7	208.3	5.95
cyclohexane	C_6H_{12}	6.5	20.0	80.9	2.79

In the process of defining molality, the equation that emerged was $m = n\ kg^{-1}$ (molality = moles of solute per kilogram of solvent). A slight modification of the equation can help shorten the method of calculating molecular weights of solutes. Remember that $n = g/g\ mole^{-1}$. If the expression $g/g\ mole^{-1}$ is substituted for n in the equations $\Delta T_f = (n\ kg^{-1}) \times K_f$ and $\Delta T_b = (n\ kg^{-1}) \times K_b$, the new equations are

$$\Delta T_f = \frac{g\ kg^{-1}}{g\ mole^{-1}} \times K_f$$

and

$$\Delta T_b = \frac{g\ kg^{-1}}{g\ mole^{-1}} \times K_b$$

$g\ kg^{-1}$ = grams of solute per kilogram of solvent (C)

$g\ mole^{-1}$ = mole weight (MW)

Simplifying the equation, since C/MW = molality:

$$\Delta T_f = \frac{C}{MW} \times K_f$$

and

$$\Delta T_b = \frac{C}{MW} \times K_b$$

EXAMPLE 9.23

What is the boiling point of a solution of 2.4 g of urea in 50.0 g of water?

1. Write the appropriate equation:

$$\Delta T_b = \frac{C}{MW} \times K_b$$

2. Organize the data:

$$\Delta T_b = ?$$

$$C = \frac{2.4\ g}{0.050\ kg} = 48.0\ g\ kg^{-1}$$

$$\text{urea MW} = 60.0\ g\ mole^{-1}$$

$$\text{water } K_b = 0.52^\circ$$

3. Substitute in the equation and solve for T_b:

$$\Delta T_b = \frac{48.0\ \cancel{g}\ \cancel{kg}^{-1}}{60.0\ \cancel{g}\ mol^{-1}} \times 0.52^\circ$$

$$\Delta T_b = 0.42^\circ$$

4. The boiling point of the solution is elevated by 0.42°:

$$100.0^\circ + 0.42^\circ = 100.42^\circ C$$

EXAMPLE 9.24

What is the molecular weight of heptene if 0.49 g dissolved in 10.0 g of benzene lowers the freezing point of the solution to 2.93°C?

1. Write the equation that solves the problem for molecular weight (MW):

$$\Delta T_f = \frac{C}{MW} \times K_f$$

becomes

$$MW = \frac{C}{\Delta T_f} \times K_f$$

2. Organize the data:

$$MW = ?$$

$$C = \frac{0.49 \text{ g heptene}}{0.01 \text{ kg benzene}} = 49.0 \text{ g kg}^{-1}$$

$$T_f \text{ benzene} = \text{(from Table 9–3) } 5.48°$$

$$\Delta T_f = 5.48° - 2.93° = 2.55°$$

$$K_f \text{ of benzene} = \text{(from Table 9–3) } 5.12°$$

3. Substitute in the equation and solve for MW:

$$MW = \frac{49.0 \text{ g kg}^{-1}}{2.55°} \times 5.12°$$

$$\text{MW of heptene} = 98.0 \text{ g mole}^{-1}$$

EXAMPLE 9.25

When 0.65 g of phenylenediamine is dissolved in 20.0 g of water, the T_f of the solution is $-0.56°C$. If the percentage composition of phenylenediamine is 66.64% carbon, 7.46% hydrogen, and 25.90% nitrogen, what is the true molecular formula?

1. The percentage composition is used to find the empirical formula. The freezing point depression data will yield the actual molecular weight. The actual molecular weight will be a small, whole-number multiple of the empirical formula weight. The true molecular formula will reflect this multiplication.
2. Organize the freezing point data and solve for the MW of phenylenediamine:

$$MW = ?$$

$$\Delta T_f = 0° - (-0.56°) = 0.56°$$

$$K_f = 1.86°$$

$$C = \frac{0.65 \text{ g}}{0.02 \text{ kg}} = 32.5 \text{ g kg}^{-1}$$

$$MW = \frac{C}{\Delta T_f} = K_f$$

$$MW = \frac{32.5}{0.56°} \times 1.86° = 108 \text{ g mole}^{-1}$$

3. Find the empirical formula from the percentage composition:

$$C = \frac{66.64\,\not{g}}{12.0\,\not{g}\,\text{mol}^{-1}} = 5.55 \text{ moles}, \frac{5.55}{1.85} = 3.00$$

$$H = \frac{7.46\,\not{g}}{1.0\,\not{g}\,\text{mol}^{-1}} = 7.46 \text{ moles}, \frac{7.46}{1.85} = 4.03$$

$$N = \frac{25.90\,\not{g}}{14.0\,\not{g}\,\text{mol}^{-1}} = 1.85 \text{ moles}, \frac{1.85}{1.85} = 1.00$$

The empirical formula is C_3H_4N.
C_3H_4N = 54 g per formula weight.

4. Compare the empirical formula weight to the actual molecular weight to find the small, whole-number multiple for the empirical formula:

$$\frac{108 \text{ g}}{54 \text{ g}} = 2$$

The empirical formula is doubled to produce the actual molecular formula of phenylenediamine: $C_6H_8N_2$.

EXAMPLE 9.26

Approximately what mass of ethanol, CH_3CH_2OH, is needed to lower the freezing point of 200 g of $CCl_4(\ell)$ by 25°C? From Table 9–3, the K_f of CCl_4 is 29.8°C.

1. Develop the equation that will solve for concentration of ethanol:

$$\Delta T_f = \frac{C}{MW} \times K_f$$

$$C = \frac{(MW)(\Delta T_f)}{K_f}$$ (multiply the equation by MW and divide by K_f)

2. Organize the data, substitute and solve:

$MW = CH_3CH_2OH = 46 \text{ g mole}^{-1}$

$\Delta T_f = 25°$

$K_f = \sim 30°$

$C = ?$

$$C = \frac{(46)(25°)}{30°} = \sim 38 \text{ g kg}^{-1}$$

---

3. Since the concentration of ethanol is $\sim$38 g kg^{-1}, the weight per 200 g (i.e., 0.2 kg) is found by multiplying:

$$38 \text{ g } \cancel{kg}^{-1}(0.2 \cancel{kg}) = 7.6 \text{ g},$$

4. The answer is that approximately 8 grams of ethanol in 200 g of CCl_4 will lower the freezing point by 25°C.

Exercise 9.6

1. What is the molecular weight of an unknown nonelectrolyte if 68.4 g dissolved in 600.0 g of benzene lowers the freezing point from +5.53° C to −2.15°C?
2. How many grams of naphthalene ($C_{10}H_8$) have to be dissolved in 5.0 g of $CS_2(\ell)$ in order to raise the boiling point of the solution to 48.0°C?
3. If 1 mole of $CaCl_2$ were assumed to have no interionic attractions upon dissociation in 1.0 kg of water, what ΔT_f would be expected? If the actual ΔT_f is 4.28°, how many effective moles of particles are in solution? (Hint: divide the ΔT_f by the K_f. This is sometimes known as the **mole number** or **van't Hoff factor,** and is symbolized by the letter, *i*.)
4. A carbohydrate was assayed to be 40.0% carbon, 6.7% hydrogen, and 53.3% oxygen. If dissolving 2.5 g of the carbohydrate in 50.0 g of chloroform raises the boiling point of the solution to 62.4°C, find the true molecular formula of the carbohydrate.

QUESTIONS AND PROBLEMS

9.1 What is the difference between a true solution and a colloidal suspension?

9.2 Define the following terms:

a. solute
b. solvent
c. miscible
d. electrolyte
e. saturated
f. ionization
g. solubility
h. dissociation

9.3 Distinguish between the meanings of the terms enhanced dipole and induced dipole.

9.4 Write equations which predict the species in solution when the following compounds are dissolved in water. Roughly estimate the number of moles of particles in solution. Assume no interionic attraction.

$$\text{Example: } Ba(NO_3)_2 \xrightarrow{H_2O} \underbrace{Ba^{2+} + 2NO_3^-}_{\text{3 moles of particles}}$$

a. $SrCl_2 \rightarrow$
b. $HCN \rightarrow$
c. $C_{12}H_{22}O_{11} \rightarrow$
d. $CO(NH_2)_2 \rightarrow$
e. $Fe_2(SO_4)_3 \rightarrow$

9.5 Write net ionic equations for the following reactions:
a. $(NH_4)_2S + Pb(NO_3)_2 \rightarrow$
b. $AlCl_3 + LiOH \rightarrow$
c. $H_2CrO_4 + Al \rightarrow$
d. $Hg_2(NO_3)_2 + CaCl_2 \rightarrow$
e. $NiSO_4 + Ba(OAc)_2 \rightarrow$

9.6 Complete and balance the following ionic equations:
a. $C_2O_4^{2-} + Fe^{3+} \rightarrow$
b. $Zn(s) + H^+(aq) \rightarrow$
c. $PO_4^{3-} + Ag^+ \rightarrow$
d. $Al + Cu^{2+} \rightarrow$
e. $S^{2-} + Co^{3+} \rightarrow$

9.7 What mass of $KBr(s)$ is needed to prepare 250.0 ml of a 0.05 molar solution?

9.8 What mass of $KOH(s)$ is required to prepare 0.20 liter of 40.0 mM solution?

9.9 What is the molarity of a solution when 3.0 g of $CaSO_4 \cdot 5H_2O$ is dissolved to make 75.0 ml of solution?

9.10 What is the molarity of an $HCl(aq)$ solution that is 60% by weight HCl and has a sp. gr. of 1.10?

9.11 How many milliliters of 6.0 M stock $HCl(aq)$ is needed to prepare 250.0 ml of 0.02 M solution?

9.12 What volume (ml) of 80.% $HNO_3(aq)$, sp. gr. 1.24, is needed to prepare 0.5 litres of 0.1 M solution?

9.13 How many grams of $Ca(NO_3)_2$ are needed to prepare 50.0 ml of 3.0 N solution?★

9.14 What is the normal concentration of 210.0 ml of $Ba(NO_3)_2(aq)$ which contains 16.0 g of solute?★

9.15 What volume of 3.0 N $H_2SO_4(aq)$ is needed to prepare 0.75 liter of 4.0×10^{-3} N solution?★

9.16 Convert the following concentrations as indicated:★
a. 4.0 N $NaCl(aq)$ = ? M
b. 0.22 M $FeCl_3(aq)$ = ? N
c. 0.32 N $CaSO_4(aq)$ = ? M
d. 0.08 M $SrBr_2(aq)$ = ? N
e. 2.5 M $NaOAc(aq)$ = ? F

9.17 How many grams of NaOH (s) are needed to prepare a 6.0% (W/W) solution in 62.0 g of water?

9.18 What is the mole fraction of ethanol, CH_3CH_2OH, when 20.0 g is dissolved in 100.0 ml of water?★

9.19 Calculate the vapor pressure of the solution in Problem 9.18 at 23.0°C.★

9.20 How many grams of glucose, $C_6H_{12}O_6$, are required to prepare a 0.2 molal solution in 80.0 g water?★

9.21 What is the boiling point elevation of an acetic acid solution if 6.2 g of urea, $CO(NH_2)_2$, is dissolved in 31.0 g of acetic acid?★

9.22 When 5.30 g of a nonelectrolyte is dissolved in 200.0 g of benzene, the freezing point of the solution is 4.63°C. If the percentage composition of the nonelectrolyte is 9.4% hydrogen and 90.6% carbon, find the true molecular formula.★

CHAPTER 10

It is generally said that science, which is the search for truth, is neither moral nor immoral, but only that those who put its applications into practices are faced with ethical decisions.

W. Heitler, *Man & Science**

Behavioral Objectives:

At the completion of this chapter, the student should be able to:

1. List four prominent characteristics of acids and bases.
2. Summarize the Arrhenius concept of acids and bases.
3. Define acids and bases in terms of the Brønsted-Lowry theory.
4. Distinguish between strong and weak acids.
5. Write conjugate acid-base pairs of common acids and bases.
6. Use a table of conjugate acid-base pairs to predict products of acid-base reactions.
7. Complete and balance ionic acid-base equations.
8. Define amphiprotism (amphoterism) and identify amphiprotic substances.★
9. State the difference between neutralization and hydrolysis.★
10. Name two classes of hydrolysis reactions.★
11. Predict the nature of the solution resulting from a hydrolysis reaction.★
12. Complete and balance the ionic equations for neutralization and hydrolysis reactions.★
13. State the Lewis concept of acids and bases.★
14. Identify Lewis acids and bases.★
15. Write complete and balanced equations for Lewis acid-base reactions.★
16. Calculate hydrogen ion and hydroxide ion concentrations in solution.
17. Calculate pH and pOH values of solutions.
18. Explain the significance of pH and pOH in terms of ion concentrations and of relative acidity or alkalinity.
19. Determine hydrogen ion or hydroxide ion concentrations from pH or pOH values.
20. Use a slide rule to find logarithms and antilogarithms.
21. Solve stoichiometric problems involving solutions of acids or bases.★
22. State the meaning and uses of acid-base titrations.★
23. Calculate unknown molar or normal concentrations of acids or bases from titration data.★
24. Find the concentrations, volumes, and masses of acids or bases from titration data when the balanced equation is not available.★
25. Sketch and label the setup of the titration apparatus.★

---

* New York, Basic Books, Inc., Publishers, 1963

ACIDS AND BASES

10.1 INTRODUCTION

In the past, acids and bases had been defined operationally. An acid was described as a compound that produced hydrogen ions in water solution; the relative strength of a particular acid was said to be related to the vigor and extent to which molecular ionization resulted in the production of hydrogen ions. Acids also were characterized as compounds in which the symbol for hydrogen usually appeared first in the formula. Acidic anhydride was the label applied to water-soluble, nonmetallic oxides which proceeded to form compounds having classical acid formulas in aqueous solution.

Simple common bases were described as compounds that furnished hydroxide ions. The strong bases of this type were said to be those ionic compounds that dissociated extensively into metal ions and hydroxide ions in water solution. Weak bases of this type dissociated only slightly.

Acids and bases of the types discussed above were often characterized by the following sets of classical properties, which expanded the list of operational definitions:

Acids

1. Acids have a sour taste. (However, indiscriminate tasting is not recommended as a safe test for acids.) In Latin, the word *acid* means "sour."
2. Blue litmus paper turns red in acids.
3. Acids react with metals above hydrogen in the activity series to produce hydrogen gas. For example

$$Zn(s) + 2H^+(aq) \longrightarrow Zn^{2+}(aq) + H_2(g)$$

4. Basic solutions are *neutralized* by acids. Neutralization may be defined as the reaction between $H^+(aq)$ and $OH^-(aq)$ to form water:

$$H^+(aq) + OH^-(aq) \longrightarrow H_2O(\ell)$$

Bases

1. Bases have a bitter taste. (Once again, tasting is not advised.) Unsweetened cooking chocolate gives a fair example of a bitter taste.
2. Red litmus paper turns blue in basic solutions. Another common dye, phenolphthalein, turns from colorless to red in the presence of sufficient $OH^-(aq)$ concentration.

3. Bases react with heavy metal ions to form insoluble hydroxides or oxides. For example

$$Fe^{3+}(aq) + 3OH^-(aq) \longrightarrow Fe(OH)_3(s)$$

4. Basic solutions feel slippery to the touch.
5. Acid solutions are neutralized by bases.

A variety of operational definitions and a familiarity with the classical properties of simple, common acids and bases are very useful. These two major classes of solutions have been known and used for centuries. While their various definitions are not wrong (in the hard, inflexible sense of that word), they have become inadequate. The inadequacy of these classical definitions becomes apparent when laboratory tests indicate that some metal ions in aqueous solution test as acids. And, curiously, a number of anions, such as $CN^-(aq)$, $CO_3^{2-}(aq)$, $PO_4^{3-}(aq)$, and $S^{2-}(aq)$, among others, behave like moderately strong bases. Obviously, the classical theory of acids and bases, proposed by Svante Arrhenius toward the latter part of the 19th century had to be revised.

10.2 THE ARRHENIUS CONCEPT

Arrhenius, on the basis of his theory of ionization, said that acids were compounds that ionize to yield hydrogen ion in water. A strong acid ionizes greatly while a weak acid ionizes slightly:

$$\underset{\text{general strong acid formula}}{HX} \xrightarrow{100\%} \underset{\text{in water}}{H^+ + X^-}$$

$$\underset{\text{general weak acid formula}}{HA} \xrightarrow{\text{slight degree}} \underset{\text{in water}}{H^+ + A^-}$$

$$\underset{\text{general strong base formula}}{MOH(aq)} \xrightarrow{\backsim 100\%} \underset{\text{in water}}{M^+ + OH^-}$$

$$\underset{\text{general weak base formula}}{MOH(s)} \xrightarrow{\text{slight degree}} \underset{\text{in water}}{M^+ + OH^-}$$

The remnants of the Arrhenius concept are still with us today. The big difference, however, is that the classical terminology is a matter of convenience (since water solutions are used most of the time) and it is no longer mistaken as ultimate truth. It is well established at this time that hydrogen ions do not exist as such in water solution.

Svante Arrhenius

Remembering that a hydrogen ion is a proton, a single unit of positive charge, it would be difficult to imagine it as not interacting with the molecular dipoles of water.

A major advance occurred in 1923 when Johannes Brønsted and Thomas Lowry independently proposed a broader theory of acids and bases that was not restricted to water solutions.

10.3 THE BRØNSTED-LOWRY THEORY

By definition, *a Brønsted-Lowry acid is a proton donor, and a base is a proton acceptor*. For example, when HCl gas is placed in contact with water, hydrochloric acid is formed. The proton transfer is illustrated in Figure 10–1. The equation that summarizes the proton transfer shown in Figure 10–1 is

$$\underset{\text{acid}}{\overset{\text{proton transfer}}{HCl(g)}} + \underset{\text{base}}{H_2O(\ell)} \longrightarrow \underset{\text{hydronium ion}}{H_3O^+(aq)} + Cl^-(aq)$$

The hydronium ion is usually written as the empirical formula H_3O^+, but it should be pointed out that experimental evidence indicates that the transferred proton bonds to more than one water molecule in the

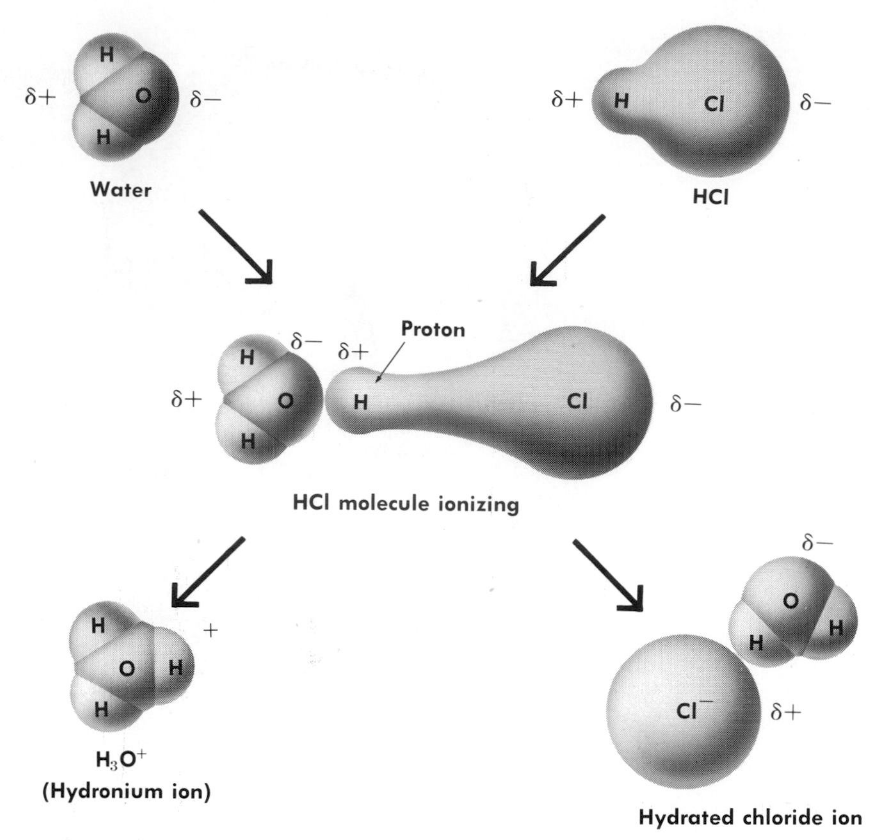

Figure 10–1 Proton transfer between water and HCl.

process of hydration. The actual formula is more likely to be $(H_2O)_4H^+$ (or, $H_9O_4^+$) or an even larger aggregate. Because of the variability in the hydronium ion structure, the empirical H_3O^+ is commonly used.*

In the proton transfer between HCl and H_2O, notice that the water is a proton acceptor. This only means that in this particular type of acid-base interaction, water is a base. This is not to suggest that water is a strong base. In fact, the lack of vigor demonstrated by water in proton donating and accepting classifies it as both a weak acid and a weak base.

Observe the same equation again:

$$\overset{\frown}{HCl(g) + H_2O}(\ell) \longrightarrow \overset{\frown}{H_3O^+(aq) + Cl^-}(aq)$$

There is a question that presents itself: Can the $H_3O^+(aq)$ act as a proton donor to the $Cl^-(aq)$, so that the reaction moves in reverse?

$$HCl(g) + H_2O(\ell) \longleftarrow \overset{\frown}{H_3O^+(aq) + Cl^-}(aq)$$

To a very slight extent this is possible. The basis for such a conclusion

* Many chemists prefer to indicate the hydrated proton as $H^+(aq)$ rather than use the empirical formula for the hydronium ion. Good arguments can be made for either formula, both of which are used in this text.

comes from experimental evidence. Chemists have collected and organized data of this type, so that a list of acids and bases according to strength is available for everyone's convenience. That is to say, there is a listing of proton donors and acceptors that indicates, by the position of the acids and bases on the list, the tendency to lose or to gain protons. On this list of **conjugate acid-base pairs,** as they are commonly called, HCl is observed to be a stronger acid than H_3O^+, and $Cl^-(aq)$ is a weaker base than water. This means, in effect, that the proton transfer between HCl and H_2O is markedly dominant as compared to the H_3O^+ and Cl^- interaction. At equilibrium, the species in solution would be almost 100% H_3O^+ and Cl^-, even though an equilibrium reaction could be indicated:

$$HCl + H_2O \rightleftharpoons H_3O^+ + Cl^-$$

The term *conjugate acid-base pair* was just used, and its meaning will now be investigated. Notice that the transfer of a proton by the HCl changed the structure from a strong acid molecule to a weak ion base (a proton acceptor):

$$\underset{\text{acid}}{HCl} + \underset{\text{base}}{H_2O} \rightleftharpoons \underset{\text{acid}}{H_3O^+} + \underset{\text{base}}{Cl^-(aq)}$$

At the same time, the weak molecular base H_2O became a strong ionic acid, H_3O^+. If this observation is generalized, the essence of the Brønsted-Lowry concept may be stated as follows:

1. When an acid transfers a proton, it becomes a base.
2. When a base accepts a proton, it becomes an acid.

The two related species, in each proton-transfer operation, are known as a conjugate acid-base pair. Furthermore, another generalization can be made:

1. A strong acid will have a weak conjugate base.
2. A weak acid will have a strong conjugate base. For example

$$\underset{\text{strong acid}}{HCl} \longrightarrow \underset{\text{weak base}}{Cl^-}$$

$$\underset{\text{weak base}}{H_2O} \longrightarrow \underset{\text{strong acid}}{H_3O^+}$$

In an acid-base reaction, the conjugate acid-base pairs are easy to recognize. *The only difference between an acid and its conjugate base is the transferred proton.* For example, in the reaction between $HCl(aq)$ and $CO_3^{-2}(aq)$, the reaction is

$$\underset{\text{acid}_1}{HCl(aq)} + \underset{\text{base}_2}{CO_3^{2-}(aq)} \rightleftharpoons \underset{\text{base}_1}{Cl^-(aq)} + \underset{\text{acid}_2}{HCO_3^-(aq)}$$

The only difference between $acid_1$ and $base_1$ (a conjugate pair) is the proton in HCl, which is absent in Cl^-. The difference between the carbonate ion and the hydrogen carbonate ion is also a proton, CO_3^{2-} and HCO_3^-. Once again, the student must be reminded to pay special attention to the conservation of charge in any ionic equation. In the equation

$$HCl + CO_3^{2-} \rightleftharpoons Cl^- + HCO_3^-$$

the net charge on both sides of the equation is observed to be minus 2.

Table 10–1 is a list of Brønsted-Lowry acid-base conjugate pairs, indicating the direction of the increasing tendency for an ion to donate or to accept a proton. This list applies to dilute aqueous solutions. In concentrated aqueous solutions and in nonaqueous solutions, the order of the compounds is different.

TABLE 10–1 BRØNSTED-LOWRY ACID-BASE CONJUGATE PAIRS

Acid	Base
$HClO_4$	ClO_4^-
HI	I^-
H_2SO_4	HSO_4^-
HNO_3	NO_3^-
HCl	Cl^-
H_3O^+	H_2O
H_2SO_3	HSO_3^-
HSO_4^-	SO_4^{2-}
H_3PO_4	$H_2PO_4^-$
HF	F^-
HNO_2	NO_2^-
HOAc	OAc^-
$Al(H_2O)_6^{3+}$	$Al(H_2O)_5(OH)^{2+}$
H_2S	HS^-
HSO_3^-	SO_3^{2-}
NH_4^+	NH_3
HCN	CN^-
HCO_3^-	CO_3^{2-}
HPO_4^{2-}	PO_4^{3-}
HS^-	S^{2-}
H_2O	OH^-
CH_3OH	CH_3O^-
NH_3	NH_2^-
OH^-	O^{2-}
H_2	H^-

INCREASING STRENGTH (acids, upward) — INCREASING STRENGTH (bases, downward)

EXAMPLE 10.1

Write the Brønsted-Lowry equation for the reaction between $HNO_3(aq)$ and $OH^-(aq)$. Label the conjugate pairs.

1. Write the reactants:

$$HNO_3(aq) + OH^-(aq) \rightleftharpoons$$

2. Obviously, the strong acid HNO_3 will be the proton donor:

$$\underset{\text{acid}_1}{HNO_3(aq)} + \underset{\text{base}_2}{OH^-(aq)} \rightleftharpoons \underset{\text{base}_1}{NO_3^-(aq)} + \underset{\text{acid}_2}{H_2O(\ell)}$$

3. The distinction made between pair number 1 and pair number 2 is completely arbitrary. The equation could just as well be written

$$\underset{\text{acid}_2}{HNO_3} + \underset{\text{base}_1}{OH^-} \rightleftharpoons \underset{\text{base}_2}{NO_3^-} + \underset{\text{acid}_1}{H_2O}$$

4. Notice that the only difference between HNO_3 and NO_3^- is a proton. The same observation is true when comparing OH^- and H_2O.
5. The net charge of minus 1 on both sides of the equation satisfies the conservation law.

EXAMPLE 10.2

Write the equation for the first proton transfer in the acid-base reaction between $H_3PO_4(aq)$ and CH_3O^- (methoxide ion).

1. Find the reacting species in Table 10–1 in order to determine which one will be the proton donor. H_3PO_4 definitely appears to be the acid.
2. Write the equation:

$$H_3PO_4(aq) + CH_3O^-(aq) \rightleftharpoons \underset{\substack{\text{dihydrogen}\\\text{phosphate ion}}}{H_2PO_4^-(aq)} + \underset{\text{methyl alcohol}}{CH_3OH(aq)}$$

3. Notice the conservation of charge.

EXAMPLE 10.3

Complete and balance the following acid-base equations.

1. a. $HClO_4(aq) + NO_2^-(aq) \rightleftharpoons$
   b. $HOAc(aq) + CN^-(aq) \rightleftharpoons$
   c. $OH^-(aq) + HCO_3^-(aq) \rightleftharpoons$
   d. $Al(H_2O)_6^{3+}(aq) + NH_3^-(aq) \rightleftharpoons$
   e. $H_2SO_4(\ell) + HF(\ell) \rightleftharpoons$
2. Note the stronger proton donor in each reaction:

$$HClO_4,\ HOAc,\ HCO_3^-,\ Al(H_2O)_6^{3+},\ \text{and}\ H_2SO_4$$

3. Complete the equations, paying careful attention to the conservation of charge:

a. $HClO_4(aq) + NO_2^-(aq) \rightleftharpoons ClO_4^-(aq) + HNO_2(aq)$
acid$_1$ base$_2$ base$_1$ acid$_2$

b. $HOAc(aq) + CN^-(aq) \rightleftharpoons HCN(g) + OAc^-(aq)$
acid$_1$ base$_2$ acid$_2$ base$_1$

c. $OH^-(aq) + HCO_3^-(aq) \rightleftharpoons H_2O(\ell) + CO_3^{2-}(aq)$
base$_1$ acid$_2$ acid$_1$ base$_2$

d. $Al(H_2O)_6^{3+}(aq) + NH_2^-(aq) \rightleftharpoons Al(H_2O)_5(OH)^{2+}(aq) + NH_3(g)$

Note: In the hydrated aluminum ion, a proton is transferred from one of the hydrated water molecules to the amide ion (base). This leaves the aluminum ion with 5 water molecules and the remaining hydroxide ion from the water molecule that lost the proton. Also, the proton transfer reduces the net charge from 3+ to 2+, since the proton is a unit of positive charge.

e. $H_2SO_4(\ell) + HF(g) \rightleftharpoons HSO_4^- + H_2F^+$

The strange-looking dihydrogen fluoride ion is a hypothetical product in a nonaqueous system.

Exercise 10.1

Complete and balance the following Brønsted-Lowry equations. Use an arrow to show the direction of proton transfer in the forward reaction. Label the conjugate acid-base pairs. The stoichiometry involved is equimolar.

1. $HI + NH_3 \rightleftharpoons$
2. $HCN + NH_2^- \rightleftharpoons$
3. $O^{2-} + HF \rightleftharpoons$
4. $HS^- + H_3O^+ \rightleftharpoons$
5. $S^{2-} + HSO_4^- \rightleftharpoons$
6. $HCO_3^- + H^- \rightleftharpoons$
7. $Fe(H_2O)_6^{3+} + OH^- \rightleftharpoons$
8. $CH_3O^- + Zn(H_2O)_4^{2+} \rightleftharpoons$
9. $HClO_4 + HSO_4^- \rightleftharpoons$
10. $HNO_2 + CO_3^{2-} \rightleftharpoons$

10.4 AMPHOTERISM *

Amphoterism, or **amphiprotism** as a more descriptive label, refers to the remarkable property of some substances to behave either as acids or as bases; that is, to be either proton donors or proton acceptors. Amphiprotic substances, generally speaking, are of two classes:

1. *Hydrated metal hydroxides,* where the metal ion is small in relation to its charge. In other words, metal ions having a large charge-to-radius ratio.
2. *Conjugate bases of some oxyacids,* which because of the nature of their structure can accept or donate protons.

Class 1. Hydrated Metal Hydroxides

When aluminum ion is added to NaOH(*aq*), hydrated aluminum hydroxide is produced. The small (ionic radius is ~0.5 Å) aluminum ion has a charge of 3+. The Al^{3+}(*aq*), having a large charge-to-radius ratio, becomes vigorously hydrated, forming an octahedral structure with 6 water molecules per aluminum ion (Fig. 10–2).

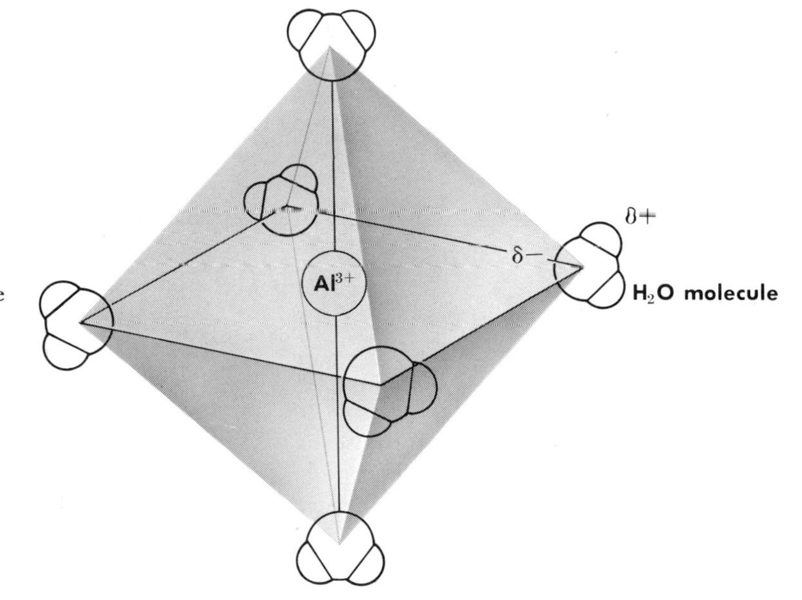

Figure 10–2 The octahedral structure of the hydrated aluminum ion.

The formula for the aluminum (III) ion hexahydrate is $Al(H_2O)_6^{3+}$. Other metal ions follow a similar pattern. Some examples are $Fe(H_2O)_6^{3+}$, $Zn(H_2O)_4^{2+}$, and $Co(H_2O)_6^{3+}$.

When the hydrated aluminum ion reacts with the strong base, NaOH, the Brønsted-Lowry reaction is between the hydrated metal ion and the $OH^-(aq)$, the sodium ion having a spectator-ion status:

$$Al(H_2O)_6^{3+}(aq) + 3OH^-(aq) \rightleftharpoons Al(H_2O)_3(OH)_3(s) + 3H_2O(\ell)$$

transfer of 3 protons

The formula $Al(H_2O)_3(OH)_3(s)$ is clearly the formula for hydrated aluminum hydroxide. While the formula is commonly written $Al(OH)_3(s)$, the gelatinous consistency of the compound tends to indicate its hydrated structure.

The amphiprotic nature, or amphoterism, of the $Al(H_2O)_3(OH)_3$ is demonstrated by its ability to react with *both strong acids and strong bases*. For example

$$Al(H_2O)_3(OH)_3(s) + 3H_3O^+(aq) \rightleftharpoons Al(H_2O)_6^{3+}(aq) + 3H_2O(\ell)$$

strong acid
3 proton transfer .

$$Al(H_2O)_3(OH)_3(s) + OH^-(aq) \rightleftharpoons Al(H_2O)_2(OH)_4^-(aq) + H_2O(\ell)$$

strong base complex ion*

Either way, the solid aluminum hydroxide dissolves.

Class 2. Conjugate Bases of Some Oxyacids

When H_2SO_4 ionizes in water, it does so in two steps:

$$1.\ H_2SO_4 + H_2O \rightarrow HSO_4^- + H_3O^+$$

$$2.\ HSO_4^- + H_2O \rightarrow SO_4^{2-} + H_3O^+$$

Analysis of the species in solution shows that the first step is markedly dominant and the HSO_4^- ion concentration is much greater than the SO_4^{2-} ion concentration. However, in acid-base reactions, the proton-accepting properties of bases push the ionization process to completion, so that the product, $2H^+ + SO_4^{2-}$, which usually appears in such equations, is understandable. If salts, such as $NaHSO_4$ or $KHSO_4$, are dissolved, there is no question about the abundance of $HSO_4^-(aq)$ ions. Similarly structured ions, such as $H_2PO_4^-$, HPO_4^{2-}, and HCO_3^-, behave much alike insofar as they have the capacity to regain the lost proton (acting as conjugate bases), or to donate still another proton. Clearly, these ions are amphiprotic, and their

* The special nomenclature of complex ions is beyond the scope of this text. However, just for the record, the name of $Al(H_2O)_2(OH)_4^-$, usually written as $Al(OH)_4^-$ is tetrahydroxoaluminate (III) ion.

behavior as acids or bases depends on the nature of the species with which they react. Some examples are

$$SO_4^{2-} + H_3O^+ \rightleftharpoons HSO_4^- + H_2O$$
acid

$$HSO_4^- + CN^- \rightleftharpoons HCN(g) + SO_4^{2-}(aq)$$
base

$$H_2PO_4^- + CO_3^{2-} \rightleftharpoons HPO_4^{2-} + HCO_3^-$$
base

$$HPO_4^{2-} + HNO_2 \rightleftharpoons H_2PO_4^- + NO_2^-$$
acid

$$HCO_3^- + HOAc \rightleftharpoons OAc^- + H_2CO_3^*$$
acid

$$HCO_3^- + OH^- \rightleftharpoons CO_3^{2-} + H_2O$$
base

Exercise 10.2

Complete and balance the following equations to illustrate the amphiprotic nature of the amphoteric species:

1. $Cr(H_2O)_3(OH)_3 + H_3O^+ \rightleftharpoons$
2. $Pb(H_2O)_2(OH)_2 + HClO_4 \rightleftharpoons$
3. $Zn(H_2O)_2(OH)_2 + OH^- \rightleftharpoons$
4. $Co(H_2O)_3(OH)_3 + CO_3^{2-} \rightleftharpoons$
5. $HSO_4^- + NH_3 \rightleftharpoons$
6. $H_2PO_4^- + HS^- \rightleftharpoons$
7. $HCO_3^- + HBr \rightleftharpoons$
8. $HPO_4^{2-} + H_3O^+ \rightleftharpoons$

10.5 HYDROLYSIS ⋆

When a few drops of phenolphthalein are added to a solution of Na_2CO_3, the red color indicates a distinctly basic solution. If a sufficient volume of an $AlCl_3$ solution is added, the red color disappears. This indicates that the base has been largely neutralized. If enough $AlCl_3$ is added to make the solution acidic, then two examples of **hydrolysis** reactions have been observed. In short, *hydrolysis is the opposite of neutralization.* The word hydrolysis means "water splitting." However, the splitting of water (hydrolysis) as opposed to the formation of water (neutralization) only comes about if the products contain at least one weak electrolyte, which may be

* There is no experimental evidence supporting the existence of an H_2CO_3 molecule, or an H_2SO_3 molecule, or $NH_4OH(aq)$. These species are often, and perhaps preferably, written as $[CO_2(aq)]$, $[SO_2(aq)]$, and $[NH_3(aq)]$.

either a weak acid or a weak base. If hydrolysis could lead to the formation of both H_3O^+ and OH^- simultaneously, it would be impossible to explain how a strong proton donor and a strong proton acceptor could coexist without neutralization. Hydrolysis must result in the formation of at least one weak electrolyte, or weak proton donor or acceptor, or hydrolysis will not occur. It is useful to generalize two classes of hydrolysis reactions:

Class 1. Strong proton acceptors will cause hydrolysis.

Class 2. Hydrated metal ions that form weak bases will cause hydrolysis.

Class 1

Some strong proton acceptors are NH_3, CN^-, CO_3^{2-}, PO_4^{3-}, S^{2-}, CH_3O^-, NH_2^-, O^{2-}, and H^-. Examples of hydrolysis reactions are

1. $H_2O + CN^- \rightleftharpoons HCN + OH^-$ (HCN: weak acid)

Note: The solution tests decidedly basic. While there is a considerable buildup of $OH^-(aq)$, the $H^+(aq)$ is bound to CN^- in the formation of the weak electrolyte HCN. The mechanism is shown in Figure 10–3.

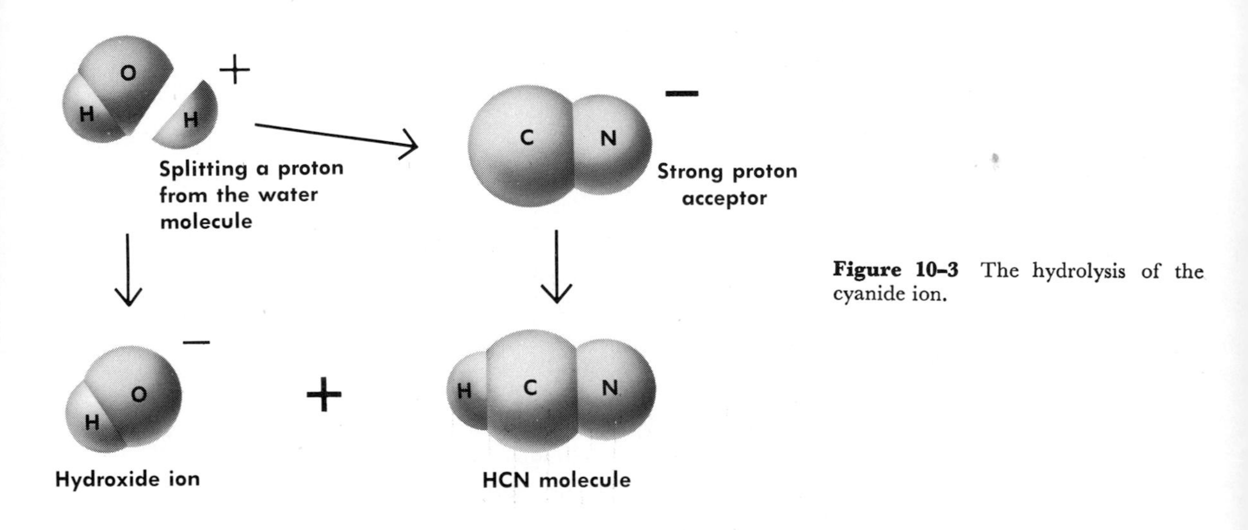

Figure 10–3 The hydrolysis of the cyanide ion.

2. $H_2O + CO_3^{2-} \rightleftharpoons HCO_3^- + OH^-$ (HCO_3^-: weak proton donor)

3. $H_2O + CH_3O^- \rightleftharpoons CH_3OH + OH^-$ (CH_3OH: weak electrolyte)

4. $H_2O + S^{2-} \rightleftharpoons HS^- + OH^-$

weak proton donor

Class 2

Since the hydrated aluminum ion is a stronger acid than water, hydrolysis will occur, as the splitting off of a proton from one of the hydrated water molecules leads to the formation of a hydronium ion, water acting in the role of a proton acceptor:

$$Al(H_2O)_6{}^{3+} + H_2O \rightleftharpoons Al(H_2O)_5(OH)^{2+} + H_3O^+$$

weak proton acceptor

Note: The formation of H_3O^+ in solution characterizes the acidity of the solution of aluminum ion. Figure 10–4 illustrates the hydrolysis process.

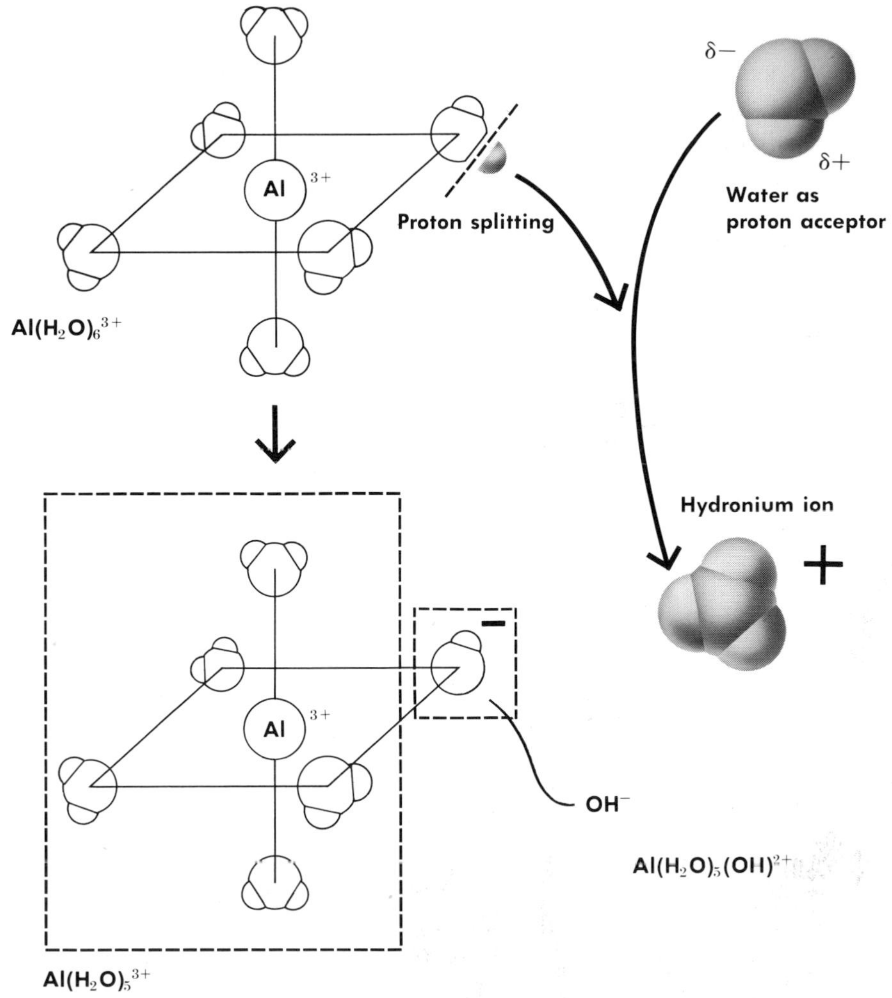

Figure 10–4 The hydrolysis of the hydrated aluminum ion.

Some other examples of Class 2 hydrolysis reactions are

1. $Cu(H_2O)_4{}^{2+} + H_2O \rightleftharpoons Cu(H_2O)_3(OH)^+ + H_3O^+$

weak proton acceptor

2. $Zn(H_2O)_4{}^{2+} + H_2O \rightleftharpoons Zn(H_2O)_3(OH)^+ + H_3O^+$

weak proton acceptor

3. $Cr(H_2O)_6{}^{3+} + H_2O \rightleftharpoons Cr(H_2O)_5(OH)^{2+} + H_3O^+$

weak proton acceptor

Exercise 10.3

Complete and balance the following equations for hydrolysis reactions:

1. $PO_4{}^{3-} + H_2O \rightleftharpoons$
2. $H_2O + NH_2{}^- \rightleftharpoons$
3. $C_2O_4{}^{2-} + H_2O \rightleftharpoons$
4. $H_2O + H^- \rightleftharpoons$
5. $Fe(H_2O)_6{}^{3+} + H_2O \rightleftharpoons$
6. $Pb(H_2O)_4{}^{2+} + H_2O \rightleftharpoons$
7. $H_2O + Ni(H_2O)_6{}^{3+} \rightleftharpoons$
8. $Sn(H_2O)_4{}^{2+} + H_2O \rightleftharpoons$

10.6 THE LEWIS CONCEPT *

In 1923, Gilbert N. Lewis proposed a still broader concept of acids and bases involving electron pairs. The Lewis concept is more inclusive than the

Gilbert Lewis

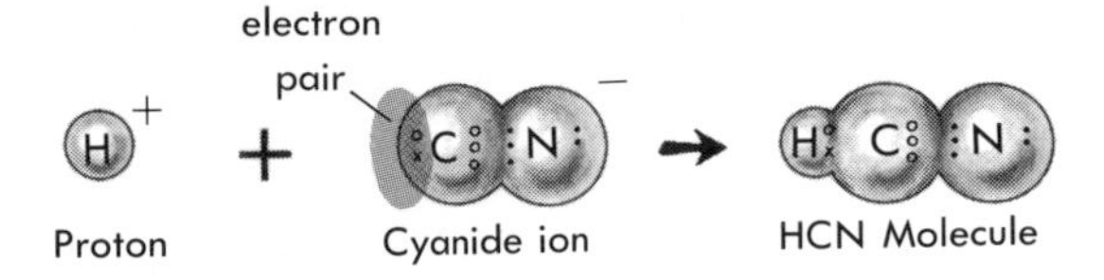

Figure 10–5 The donation of an electron pair by the cyanide ion.

Brønsted-Lowry theory, because the definitions of acids and bases are not restricted to the proton-transfer factor. There is no contradiction, however, because the essential mechanism in a proton transfer is the formation of a covalent bond. When CN^- accepts a proton, it is, in effect, donating a pair of electrons that forms a covalent bond between the hydrogen nucleus and the cyanide ion (Fig. 10–5).

In terms of the Lewis concept, the proton, or hydrogen ion, is the acid and the cyanide ion is the base. Many chemists routinely call the hydrogen ion a *Lewis acid*. Therefore, the *Lewis concept defines an acid as an electron pair acceptor, and a base as an electron pair donor*. The covalent bond formation is the difference between the Lewis and the Brønsted classifications. The Lewis concept is broader, because, in addition to protons, there are many ions, atoms, and molecules that can function as electron pair acceptors in forming covalent bonds.

An example of a Lewis acid-base reaction that does not fit the Brønsted theory is the reaction between BF_3 and NH_3. While NH_3 is consistent with Brønsted's proton acceptor (base), the BF_3 molecule has no protons to donate. However, the BF_3 molecule can be an electron pair acceptor, which classifies it as a Lewis acid:

$$BF_3 + :NH_3 \rightarrow F_3B:NH_3$$

electron pair acceptor (acid) — electron pair donor (base) — covalent bond

The definition of Lewis acids may be generalized to include atoms, ions, and molecules that do not have complete octets of valence electrons. Other examples of Lewis acid-base reactions are

$$Ca^{2+}, \ddot{O}:^{2-} + SO_3 \rightarrow Ca^{2+}, SO_4^{2-}$$

calcium oxide	+	sulfur trioxide	→ calcium sulfate
electron pair donor (base)		electron pair acceptor (acid)	

$$Cu^{2+} + 4\,:NH_3 \rightarrow Cu(NH_3)_4^{2+}$$

copper ion	+	ammonia	→ copper-ammonia complex
electron pair acceptor (acid)		electron pair donor (base)	

After these introductions to broader theories, it is appropriate now to return to the common and practical aspects of acids and bases in water solution.

10.7 HYDROGEN ION CONCENTRATION

Pure water is the standard for acid-base neutrality. If an acid is added to water, the hydrogen ion concentration naturally will be higher than the $H^+(aq)$, owing to the ionization of water alone. Water molecules do ionize to a slight degree. In a liter of chemically pure water, the hydrogen ion concentration is found experimentally (by electrical conductivity) to be 1.0×10^{-7} moles per liter at 24°C. (The ionization of water varies slightly with the temperature.) Of course, the concentration of hydroxide ion will be the same as that of the hydrogen ion, since each ionizing molecule produces 1 hydrogen ion and 1 hydroxide ion. The Brønsted equation

$$H_2O + H_2O \rightleftharpoons H_3O^+ + OH^-$$

may be simplified to

$$H_2O \rightleftharpoons H^+ + OH^-$$

The product of the ion concentrations, K_w (the ion product of water), is

$$K_w = [H^+][OH^-]$$

A bracket, [ ], around the ion formula means molar concentration. Hence,

$$K_w = (1.0 \times 10^{-7}\ M)(1.0 \times 10^{-7}\ M)$$
$$K_w = 1.0 \times 10^{-14}$$

When an acid is added to the water, the $[H^+]$ becomes larger as the $[OH^-]$ becomes smaller because it reacts with the excess H^+ to form water. The net effect is that the ion product, K_w, remains constant.

EXAMPLE 10.4

Calculate the hydroxide ion concentration in a 0.1 M solution of HCl.

1. Since HCl is a strong electrolyte, ~100% ionization is assumed. Therefore, the $[H^+]$ is 1.0 M. In one liter

$$\underset{0.1\text{ mole}}{HCl} \xrightarrow[\text{ionization}]{\sim 100\%} \underset{0.1\text{ mole}}{H^+} + \underset{0.1\text{ mole}}{Cl^-}$$

2. The $K_w = [H^+][OH^-] = 1.0 \times 10^{-14}$

$$1.0 \times 10^{-14} = [H^+][OH^-]$$

$$1.0 \times 10^{-14} = (0.1)[OH^-]$$

$$[OH]^- = 1.0 \times 10^{-13}$$

The $OH^-(aq)$ is very small, as expected.

EXAMPLE 10.5

What is the $[H^+]$ in a 0.002 M solution of NaOH?

1. The $[OH^-]$ will be 0.002 M, since the ionic NaOH dissociates completely into $Na^+(aq)$ and $OH^-(aq)$.
2. The $K_w = [H^+][OH^-] = 1.0 \times 10^{-14}$

$$1.0 \times 10^{-14} = [H^+](2.0 \times 10^{-3})$$

$$[H^+] = \frac{1.0 \times 10^{-14}}{2.0 \times 10^{-3}}$$

$$[H^+] = 5.0 \times 10^{-12}$$

The $H^+(aq)$ is small, as expected.

Because of the great importance attached to the control of acidity and alkalinity in chemistry and biology, a system proposed by Peder Sørensen in 1909 is used to express the degree of acidity. This system converts the values for $[H^+]$ from the cumbersome values just observed to small numbers between 0 and 14. This is accomplished by expressing the molar concentrations of hydrogen ion as *exponents of 10*, while eliminating the negative sign. For example, $[H^+] = 10^{-5}$ M is written as pH = 5.

10.8 THE CONCEPT OF pH

The symbol pH means the power (exponent) of the Hydrogen ion concentration and expresses the "strength" of the solution. It should be emphasized that pH is a *logarithmic* expression of the hydrogen ion concentration in a solution. It does not measure directly the actual strength of an acid. A strong acid such as HCl will ionize completely, but the total hydrogen ion concentration will be small if only a drop is added to 20 liters of water. By contrast, the hydrogen ion concentration due to the addition of a large volume of acetic acid (a weak acid) to water could result in a much higher hydrogen ion concentration.

In order to convert the negative exponents of the base 10 to positive numbers, the concept of negative logarithms is applied and the equation is

$$pH = -\log [H^+]$$

pH is the negative logarithm of the hydrogen ion concentration. The same type of approach can be used for the measurement of the hydroxide ion concentration. In this case, the $[OH]^-$ would be expressed as pOH, and the equation would be

$$pOH = -\log [OH^-]$$

pOH is the negative logarithm of the hydroxide ion concentration. The interconversion between pH and pOH is very simply accomplished when the equation for the ion product of water is treated logarithmically. This may be understood more clearly after reading Example 10.6.

EXAMPLE 10.6

Calculate the pH and the pOH of chemically pure water in which the $[H^+]$ and $[OH^-]$ are both 1.0×10^{-7} M.

1. Write the equation:

$$pH = -\log [H^+]$$

2. Substitute the 1.0×10^{-7} M for $[H^+]$ and note the logarithms of 1.0 and 10^{-7}. The log of 1.0 = 0, and the log of 10^{-7} = the exponent, −7. Since the log of a product of two numbers equals the sum of the two separate logs

$$pH = -[\log 1.0 + \log 10^{-7}]$$

$$pH = -[0.0 + (-7)]$$

Simplify the expression:

$$pH = -(0.0 - 7)$$

$$pH = 7.0$$

3. Since the $[OH^-]$ is the same as the $[H^+]$, the pOH will equal 7.0 also.

From Example 10.6 it can be seen that in the case of water solutions

$$pH + pOH = 14.0$$

The value, 14.0, is called the pK_w of water. Since the ion product of water solutions must always be 1.0×10^{-14} M, the pK_w will always be 14.0. In other words, if the pH of a solution is 4.3, the pOH will be:

$$pK_w - pH = pOH$$
$$14.0 - 4.3 = 9.7$$
$$pOH = 9.7$$

As a simple practical operation, the subtraction of pH from 14.0 will yield the pOH. Conversely, subtracting the pOH of a solution from 14.0 will give the pH.

EXAMPLE 10.7

Calculate the pH of a solution having a $[H^+] = 4.6 \times 10^{-3}$ M.

1. Write the equation:

$$pH = -\log [H^+]$$

2. Substitute the numerical value for $[H^+]$:

$$pH = -[\log 4.6 + \log 10^{-3}]$$

3. Set the hairline on the number 4.6 on the D scale of the slide rule and find the logarithm on the L scale, under the hairline.* The log of 4.6 = 0.66.
4. The log of 10^{-3} is -3.
5. Simplify the equation:

$$pH = -[0.66 + (-3)]$$
$$pH = -(0.66 - 3)$$
$$pH = -(-2.34)$$
$$pH = 2.34$$

EXAMPLE 10.8

Find the pH of a solution of sodium hydroxide in which the $[OH^-]$ is 5.1×10^{-2}.

* See Appendix B for an explanation of the use of the slide rule.

1. An efficient method is to calculate the pOH first. Then subtract the pOH from 14.0 to obtain the pH.
2. Write the equation:

$$pOH = -\log [OH^-]$$

3. Substitute the data:

$$pOH = -[\log 5.1 + \log 10^{-2}]$$

4. Find the logarithmic values as previously described:

$$\log 5.1 = 0.71$$
$$\log 10^{-2} = -2$$

5. Find the pOH:

$$pOH = -(0.71 - 2)$$
$$pOH = -(-1.29)$$
$$pOH = 1.29$$

6. The pH is obtained by subtraction:

$$pK_w - pOH = pH$$
$$14.0 - 1.29 = pH$$
$$pH = 12.71$$

EXAMPLE 10.9

Calculate the pH of a 6.0 M solution of HCl. Assume 100% ionization.

1. Write the equation:

$$pH = -\log [H^+]$$

2. Substitute and solve:

$$pH = -(\log 6.0)$$
$$pH = -(0.78)$$
$$pH = -0.78$$

Example 10.9 was presented to demonstrate the proper application of the concept of pH. It is not appropriate for high concentrations of

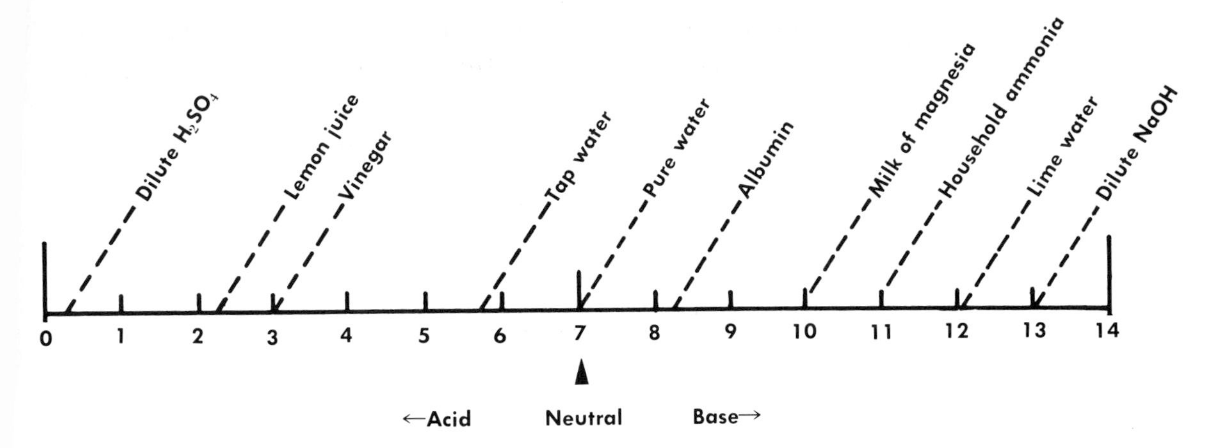

Figure 10–6 pH values of some common solutions.

$H^+(aq)$ in acids nor is pOH appropriate for high $OH^-(aq)$ concentrations in bases. The concept of pH has greatest usefulness when the application is made to the measurement of hydrogen ion concentrations that are relatively low. The useful range of the scale of pH values (and pOH values) is between 0 and 14, with the pH of chemically pure water, 7.00, taken as both the midpoint of the scale and the measure of perfect neutrality at 24°C. Figure 10–6 illustrates the relative pH values of a number of common substances in water solution.

EXAMPLE 10.10

If the pH of solution is found to be 4.5, what is the hydrogen ion concentration?

1. Since pH = −log [H^+], taking the **antilog** of both sides of the equation will produce the [H^+] in moles per liter:

$$[H^+] = \text{antilog}\ (-\text{pH})$$

2. Express the −pH, −4.5, as a whole number and a decimal for convenience in finding the antilog of the decimal on the slide rule:

$$-4.5 = -5 + 0.5$$

3. Substitute in the equation:

$$[H^+] = (\text{antilog}\ -5 \times \text{antilog}\ 0.5)$$

4. Find the antilogs:

$$\text{antilog of}\ -5 = 10^{-5}$$

Antilog of 0.5 is found by setting the hairline on 0.5 on the L scale and finding the answer on the D scale, also under the hairline:

$$\text{antilog of } 0.5 = 3.16$$

5. Complete the equation:

$$[H^+] = 10^{-5} \times 3.16$$

or

$$[H^+] = 3.16 \times 10^{-5}$$

EXAMPLE 10.11

Find the $[H^+]$ of a solution in which the pOH = 6.3.

1. Convert the pOH to pH by subtracting from 14.0:

$$pH = 14.0 - 6.3 = 7.7$$

Then

$$-pH = -7.7$$

2. Find the $[H^+]$ by separating the value, −7.7, into the two convenient parts:

$$-7.7 = -8.0 + 0.3$$

3. Substitute in the equation and solve for $[H^+]$:

$$[H^+] = \text{antilog } -pH$$

$$[H^+] = \text{antilog of } -7.7 = \text{antilog } (-8.0 \times 0.3)$$

$$\text{the antilog of } -8.0 = 10^{-8}$$

and

$$\text{the antilog of } 0.3 = 2.0$$

$$[H^+] = 2.0 \times 10^{-8}$$

Exercise 10.4

1. If a water solution has a hydrogen ion concentration of 0.0032 M, what is the hydroxide ion concentration?

2. Calculate the pH of the following solutions:
   a. 0.2 M HCl (assume 100% ionization)
   b. 0.04 M H_2SO_4 (assume complete ionization, $H_2SO_4 \rightarrow 2H^+ + SO_4^{2-}$)
   c. 4.4×10^{-3} M $[H^+]$
   d. 1.7×10^{-9} M $[H^+]$
   e. 8.2×10^{-4} M $[OH^-]$
3. Calculate the pOH of the following solutions:
   a. 0.0002 M NaOH
   b. 0.05 M $Sr(OH)_2$ (assume 100% dissociation, $Sr(OH)_2 \rightarrow Sr^{2+} + 2\,OH^-$)
   c. 3.7×10^{-8} M $[OH^-]$
   d. 5.3×10^{-2} M $[OH^-]$
   e. 7.2×10^{-11} M $[H^+]$
4. Given pH and pOH values, calculate the hydrogen ion concentrations of the following solutions:
   a. pH = 9.3
   b. pOH = 2.4
   c. pH = 0.4
   d. pH = 12.2
   e. pOH = 7.8

10.9 SOLUTION STOICHIOMETRY

The fact that a correctly balanced equation for a stoichiometric reaction indicates the quantitative relationships among reactants and products has already been discussed. A number of problems, classified as mole-to-mole, mole-to-weight, weight-to-weight, weight-to-volume, and volume-to-volume, have been presented. However, many reactants, especially acid solutions, are not conveniently expressed in grams. Of course, the mass of acid solute in grams can be calculated if the volume and molarity or normality are known:

$$g = \text{mole weight} \times \text{molarity} \times \text{liters of solution}$$

Nevertheless, it is simpler to convert the data on concentration and volume to moles or to equivalents for calculation purposes, since

$$n = MV$$

$$\text{moles} = \text{molarity} \times \text{liters of solution}$$

and

$$\#\,eq = NV$$

$$\text{number of equivalents} = \text{normality} \times \text{liters of solution}$$

Such calculations are often placed under the heading of *solution stoichiometry*.

EXAMPLE 10.12

Calculate the mass of zinc needed to react completely with 50.0 ml of 0.5 M HCl.

1. Write a balanced equation for the reaction in order to determine the stoichiometry:

$$Zn(s) + 2HCl(aq) \longrightarrow ZnCl_2(aq) + H_2(g)$$

2. The equation indicates that 1 mole of zinc will react with 2 moles of HCl.
3. Since the molarity and volume of the HCl are given, the number of moles of dissolved HCl can be found from the equation:

$$n = MV$$
$$n = (0.5 \text{ mol} \not{\ell}^{-1})(0.05 \not{\ell})$$
$$n = 0.025 \text{ mole HCl}$$

4. Zinc is 65.4 g mole^{-1}.
5. Using the factor-label method

$$(0.025 \text{ mol } \cancel{HCl}) \times \left(\frac{1 \text{ mol Zn}}{2 \text{ mols } \cancel{HCl}}\right) \times (65.4 \text{ g mol}^{-1}) = \chi$$

moles × mole ratio × mole weight of zinc

$$\chi = 0.82 \text{ g of zinc}$$

6. The calculation could be modified slightly by incorporating the values for molarity and solution volume directly, instead of changing to moles in a preliminary step:

$$[(0.5 \text{ mol } \not{\ell}^{-1})(0.05 \not{\ell})\cancel{HCl}] \times \left(\frac{1 \text{ mol Zn}}{2 \text{ mols } \cancel{HCl}}\right) \times (65.4 \text{ g mol}^{-1}) = \chi$$

moles of HCl × mole ratio × mole weight of zinc

$$\chi = 0.82 \text{ g of zinc}$$

EXAMPLE 10.13

What volume of 0.02 M HCl is necessary to react with an excess of zinc in order to produce 600.0 ml of hydrogen gas at STP?

1. Write the balanced equation for the reaction stoichiometry:

$$Zn(s) + 2HCl(aq) \longrightarrow ZnCl_2(aq) + H_2(g)$$

2. The presence of excess zinc makes the HCl the reaction-limiting factor; i.e., the volume of gas depends directly on the number of moles of HCl reacting.
3. The stoichiometry indicates that 2 moles of HCl will produce 1 mole of $H_2(g)$:

$$\begin{array}{ccc} 2HCl(aq) & \longrightarrow & H_2(g) \\ 2 \text{ moles} & & 1 \text{ mole} \end{array}$$

4. Using the factor-label method

$$\chi = \left(\frac{0.6\ \ell\ H_2}{22.4\ \ell\ mol^{-1}}\right) \times \left(\frac{2\ mols\ HCl}{1\ mol\ H_2}\right)$$

moles of H_2 × mole ratio

$$\chi = 0.0536 \text{ mole of HCl}$$

5. The volume of HCl can then be calculated, since n = MV, or $V = \frac{n}{M}$.

$$V = \frac{0.0536\ mol}{0.02\ mol\ \ell^{-1}} = 2.68 \text{ liters of HCl}$$

EXAMPLE 10.14

Suppose 1.68 g of zinc were added to 2.68 liters of 0.02 M HCl. Would 600.0 ml of H_2 gas at STP be obtained, as it was in Example 10.13?

1. The data in this problem do not indicate whether or not the zinc is in excess. Before the calculation can be made, it must be determined whether the HCl(*aq*) or the Zn(*s*) is the limiting factor in the reaction.
2. The mole ratio, indicated by the balanced equation, states that 1 mole of zinc is needed to react completely with 2 moles of HCl. Calculate the number of moles of the two reactants in order to see whether their ratios are 1:2.

$$Zn(s) + 2HCl(aq)$$

$$65.4 \text{ g mole}^{-1} + 2 \text{ moles}$$

3. Calculate the number of moles of Zn and HCl for zinc:

$$n = \frac{g}{g\ mole^{-1}} = \frac{1.68\ g}{65.4\ g\ mol^{-1}} = 0.026\ mole$$

for HCl,

$$n = MV = (0.02\ M)(2.68\ \ell) = 0.054\ mole$$

4. Compare the whole-number mole ratios by dividing both values by the smaller number:

$$Zn = \frac{0.026}{0.026} = 1\ mole$$

$$HCl = \frac{0.054}{0.026} = 2.1\ moles$$

5. The ratio is not 1 mole of zinc to 2 moles of HCl. The calculations indicate an excess of HCl. Since there is not enough zinc to react completely with the acid, the final volume of gas would be less than 600 ml at STP. If 600 ml were required, it would be necessary to add 0.027 mole of zinc:

$$(65.4\ g\ \cancel{mol}^{-1})(0.027\ \cancel{mol}) = 1.77\ g\ of\ zinc$$

In reality, chemical reactions hardly ever take place at STP. A laboratory maintained at 0°C is not a reasonable environment. In order to adjust calculated results to actual conditions, the gas laws are commonly employed.

EXAMPLE 10.15

What volume of 0.2 M H_2SO_4 would have to react with an excess of calcium in order to produce 0.300 liter of hydrogen gas measured at 23°C and 750 torr?

1. Write the balanced equation for the solution stoichiometry:

$$Ca(s) + H_2SO_4(aq) \longrightarrow CaSO_4(aq) + H_2(g)$$

2. The stoichiometry indicates that 1 mole of H_2SO_4 will produce 1 mole of hydrogen:

$$\underset{1\ mole}{H_2SO_4(aq)} \longrightarrow \underset{1\ mole}{H_2(g)}$$

3. The simplest method of expressing the number of moles ' $H_2(g)$ at 23°C and 750 torr is to make use of the equation oı state:

$$PV = nRT$$

where

$$n = \frac{PV}{RT}$$

4. Organize the data:

$$P = 750 \text{ torr} = \frac{750 \text{ torr}}{760 \text{ torr atm}^{-1}} = 0.99 \text{ atm}$$

$$V = 0.300 \text{ liter}$$

$$R = 0.082 \text{ liter-atm mole}^{-1} \text{ K}^{-1}$$

$$T = 23°C = 23 + 273 = 296 \text{ K}$$

5. Substitute in the equation and solve for moles of $H_2(g)$.

$$n = \frac{(0.99 \text{ atm})(0.300\ \ell)}{(0.082\ \ell\text{-atm mol}^{-1} \text{ K}^{-1})(296 \text{ K})}$$

$$n = 0.0012 \text{ mole of } H_2(g)$$

6. Since the stoichiometry is

$$1 \text{ mole } H_2SO_4 \longrightarrow 1 \text{ mole } H_2,$$

the moles of H_2SO_4 must also be 0.0012, or 1.2×10^{-3} mole.

7. The volume of 0.2 M H_2SO_4 can be calculated from the basic equation:

$$n = MV$$

where

$$V = \frac{n}{M}$$

8. Substitute and solve:

$$V = \frac{1.2 \times 10^{-3} \text{ mol}}{0.2 \text{ mol } \ell^{-1}}$$

$$V = 6.0 \times 10^{-3} \text{ liter of } H_2SO_4$$

9. In other words, 6.0 ml of 0.2 M H_2SO_4 would be sufficient to

produce 0.300 ℓ of hydrogen gas when measured at 23°C and 750 torr.

A variation of Example 10.15 would be the calculation of the volume of gas produced under nonstandard conditions. If, for example, an excess of zinc metal were added to 50.0 ml of 0.05 M H_2SO_4, then the number of moles of hydrogen produced could be determined from the balanced equation. The equation of state, once again, would provide the answer for the volume of gas under the stated conditions of temperature and pressure.

EXAMPLE 10.16

What volume of hydrogen gas, measured at 36.0°C and 1.02 atm pressure, would be obtained by the reaction of an excess of zinc with 50.0 ml of 0.05 M H_2SO_4?

1. Determine the mole ratios from the balanced equation:

$$Zn(s) + H_2SO_4(aq) \longrightarrow ZnSO_4(aq) + H_2(g)$$

2. The stoichiometry indicates that the acid and gas concentrations are equimolar:

$$H_2SO_4 \longrightarrow H_2$$

1 mole   1 mole

3. Find the moles of acid used from the equation:

$$50.0 \text{ ml} = 0.05 \text{ liter}$$
$$n = MV$$
$$n = (0.05 \text{ M})(0.05\ \ell)$$
$$n = 0.0025 \text{ mole of acid}$$

4. Since the acid and gas are equimolar, 0.0025 mole of hydrogen gas is produced.
5. Use the equation of state to find the volume occupied by 0.0025 mole of gas at 36.0°C and 1.02 atm:

$$PV = nRT$$

becomes

$$V = \frac{nRT}{P}$$

6. Organize the data and substitute in the equation, solving for volume:

$$P = 1.02 \text{ atm}$$

$$V = ?$$

$$n = 0.0025 = 2.5 \times 10^{-3} \text{ mole}$$

$$R = 0.082 \ \ell\text{-atm mole}^{-1} \text{ K}^{-1}$$

$$T = 36.0°C = 309.0 \text{ K}$$

$$V = \frac{(2.5 \times 10^{-3} \text{ mol})(0.082 \ \ell\text{-atm mol}^{-1} \text{ K}^{-1})(309.0 \text{ K})}{1.02 \text{ atm}}$$

$$V = 62.1 \times 10^{-3} \text{ liter} = 62.1 \text{ ml}$$

7. The answer is that 62.1 ml of hydrogen would be measured under the stated conditions.

Exercise 10.5

1. How many grams of magnesium are required to react completely with 75.0 ml of 1.4 M HCl?
2. What volume of 2.3 M hydrochloric acid is needed to react completely with 50.0 mg of iron in the formation of iron (II) chloride?
3. If 2.0 g of CaO is added to 25.0 ml of 0.1 M H_2SO_4, what weight of $CaSO_4(s)$ could be obtained? (Hint: Check the mole ratios of the reactants in order to find the reaction-limiting component.)
4. What volume of gas (STP) can be obtained if an excess of magnesium is added to 50.0 ml of 2.5 M HCl?
5. How many milliliters of 0.075 M H_2SO_4 must have reacted with an excess of aluminum if 450.0 ml of gas is collected at 27.0°C and a pressure of 735 torr?

10.10 ACID-BASE TITRATION ★

An acid-base titration is a laboratory method of controlling the combination of solutions of acids and bases. The purposes of the controlled combination may be to determine the previously unknown concentration of either the acid or the base, or the method of titration might be used to adjust the pH of a solution to a desired value.

The solutions, called **titrants**, are usually added drop by drop into a constantly stirred flask or beaker until the **end point** is reached. The end point may be determined by a color change, if dye indicators are used, or by the indicated pH if a pH meter is preferred. When the end point is supposed to signal an acid-base neutralization, the type of indicator used should be one that undergoes a color change at the **equivalence point.** The equivalence point is reached when the numbers of moles of $H^+(aq)$ and $OH^-(aq)$ in the solution is equal. If a titration is designed to adjust the pH of solution to some value other than 7.00, the end point and equivalence point are quite different.

The apparatus used to perform titration is illustrated in Figure 10–7. The burets are actually precision-graduated pipets equipped with **stopcocks** to control the flow of the solutions.

While the control of the pH of a solution is primarily a manual and visual process, the determination of the unknown concentration of an acid or a base requires some calculation from the observed data.

Remember that neutralization is the reaction between equivalent numbers of $H^+(aq)$ and $OH^-(aq)$ to form water. For this reason, many

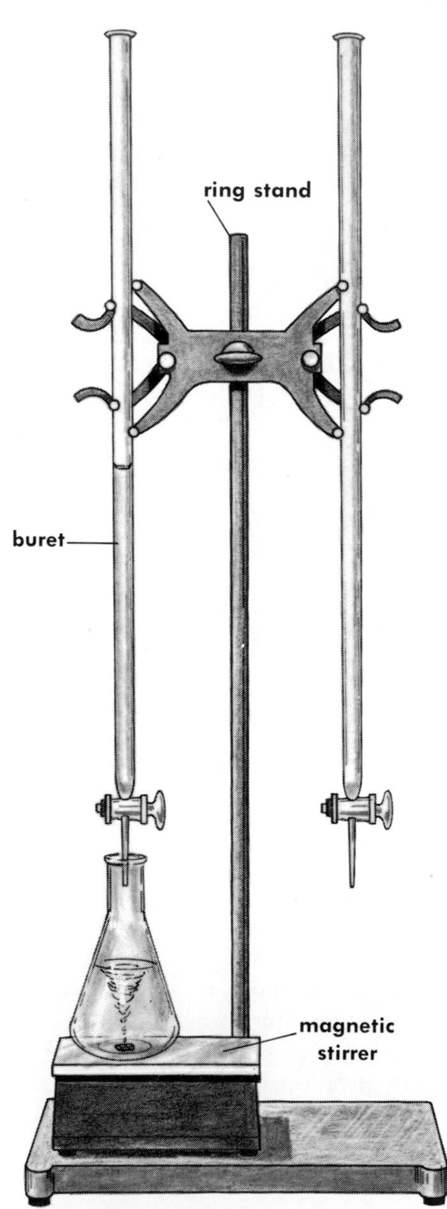

Figure 10–7 A common arrangement of burets, flask, and magnetic stirrer used to perform titration.

chemists find it simpler to handle the calculations in terms of equivalents of acid and base; i.e., to use normal concentrations rather than molar concentrations. While a liter of 1.0 M NaOH(*aq*) *will not* neutralize a liter of 1.0 M H_2SO_4(*aq*), the same volumes of 1.0 N solutions *will* result in neutralization. Since one mole of H_2SO_4 can produce two moles of H^+(*aq*), a half mole, which is the equivalent weight, produces 1 mole of H^+(*aq*):

$$\underset{\text{1 mole}}{H_2SO_4} \xrightarrow[\text{excess } OH^-(aq)]{\text{[In the presence of}]} \underset{\text{2 moles}}{2H^+} + \underset{\text{1 mole}}{SO_4^{2-}}$$

In effect, neutralization occurs when the number of equivalents of acid equals the number of equivalents of base:

$$\#\ \text{eq acid} = \#\ \text{eq base}$$

The number of equivalents of acids or bases may further be expressed by equations previously developed in Chapter 9.

$$\#\ \text{eq} = NV$$
$$\text{equivalents} = \text{normality} \times \text{volume in liters}$$

and,

$$\#\ \text{eq} = \frac{g}{g\ eq^{-1}}$$

(g: grams of solute; g eq^{-1}: equivalent weight)

The practical application of these equations will now be demonstrated in several examples.

EXAMPLE 10.17

What is the normal concentration of an unknown acid if 30.2 ml is required to neutralize 61.5 ml of 0.5 N NaOH(*aq*)?

1. Since both titrants are in solution, the equation

$$\#\ \text{eq acid} = \#\ \text{eq NaOH}(aq)$$

may be expressed as

$$N_A V_A = N_B V_B$$

$$(\text{normality acid}) \times (\text{vol acid}) = (\text{normality base}) \times (\text{vol base})$$

2. Organize the data and substitute in the equation:

$$N_A = ?$$
$$V_A = 30.2 \text{ ml}$$
$$N_B = 0.5 \text{ N}$$
$$V_B = 61.5 \text{ ml}$$

(Note that the volumes do not have to be changed to liters, since both sides of the equation have effectively been multiplied by 10^3.)

$$N_A(30.2 \text{ ml}) = (0.5 \text{ N})(61.5 \text{ ml})$$

3. Solve for normality of the acid.

$$N_A = \frac{(0.5 \text{ N})(61.5 \text{ ml})}{(30.2 \text{ ml})}$$

$$N_A = 1.02 \text{ normal acid}$$

EXAMPLE 10.18

Calculate the molar concentration of $Ca(OH)_2(aq)$ if 17.4 ml is titrated with 83.2 ml of 0.25 N HCl to the $CaCl_2$ neutralization end point.

1. Use the same basic equation, $N_AV_A = N_BV_B$, to determine the normality of the basic solution. Organize the data first:

$$N_A = 0.25 \text{ N}$$

$$V_A = 83.2 \text{ ml}$$

$$N_B = ?$$

$$V_B = 17.4 \text{ ml}$$

2. Solve the equation for N_B:

$$(0.25 \text{ N})(83.2 \text{ ml}) = (N_B)(17.4 \text{ ml})$$

$$N_B = \frac{(0.25 \text{ N})(83.2 \text{ ml})}{(17.4 \text{ ml})}$$

$$N_B = 1.2 \text{ normal}$$

3. Recall from Chapter 9 that the equation for interconverting normality and molarity is

$$N = M \times \text{total positive ionic charge}$$

or

$$M = \frac{N}{\text{total positive ionic charge}}$$

In the case of acids and bases, the total positive ionic charge is sometimes equal to the number of replaceable hydrogen ions (acids) or of replaceable hydroxide ions (bases). Therefore

$$M_{Ca(OH)_2} = \frac{N}{2}$$

$$M_{Ca(OH)_2} = \frac{1.2}{2} = 0.6 \text{ molar}$$

EXAMPLE 10.19

What volume of 0.3 N acid is needed to neutralize 1.4 g of CaO dissolved in an unmeasured volume of water?

1. Notice that the nature of the acid is unknown. This problem cannot be solved by the usual stoichiometric calculations. However, the method of equivalents will work in any case:

$$\# \text{ eq acid} = \# \text{ eq base}$$

$$\underset{\text{(acid)}}{N_A V_A} = \underset{\text{(base)}}{\frac{g}{g\ eq^{-1}}}$$

2. Organize the data:

$$N_A = 0.3 \text{ N}$$
$$V_A = ?$$
$$g = 1.4 \text{ grams}$$

$$\text{Equivalent weight of CaO} = \frac{56 \text{ g mol}^{-1}}{2(\text{total positive ionic charge})} = 28 \text{ g eq}^{-1}$$

3. Substitute in the equation and solve for the volume of acid. Note: The volume will be in liters.

$$(0.3 \text{ N})(V_A) = \frac{1.4 \not{g}}{28 \not{g} \text{ eq}^{-1}}$$

$$V_A = \frac{1.4}{(0.3 \not{eq}\ \ell^{-1})(28 \not{eq}^{-1})}$$

$$V_A = 0.17 \text{ liter}$$

4. The answer is that 0.17 liter, or 170 ml, of 0.3 N acid is required.

EXAMPLE 10.20

What is the molar concentration of $Ca(OH)_2(aq)$ if 43.4 ml titrates 3.60 g of oxalic acid (in water solution) to the equivalence end point for CaC_2O_4?

1. The appropriate expression to represent the number of equivalents of $Ca(OH)_2(aq)$ is $N_B V_B$, while the equivalents of the oxalic acid are g/g eq^{-1}.

2. Write the equation and organize the data:

$$N_BV_B = \frac{g}{g\ eq^{-1}}$$

g ← grams oxalic acid; $g\ eq^{-1}$ ← equivalent weight

$N_B = ?$

$V_B = 43.4\ ml = 0.0434\ liter$

$g = 3.60\ grams$

$$\text{Equivalent weight of } H_2C_2O_4 = \frac{90.0\ g\ mol^{-1}}{2} = 45.0\ g\ eq^{-1}$$

(2 ← replaceable hydrogens for this reaction)

3. Substitute in the equation and solve for the normality of the base:

$$(N_B)(0.0434\ liter) = \frac{3.60\ \not{g}}{45.0\ \not{g}\ eq^{-1}}$$

$$N_B = \frac{3.60}{(0.0434\ liter)(45.0\ eq^{-1})}$$

$$N_B = 1.84\ eq\ \ell^{-1} = 1.84\ N$$

4. Convert 1.86 N $Ca(OH)_2$ to molar concentration:

$$M = \frac{N}{\text{replaceable hydroxides for this reaction}} = \frac{1.84}{2}$$

$$M = 0.92\ \text{molar}$$

Exercise 10.6

1. Find the normal concentration of an unknown acid if 126.2 ml is required to titrate 43.8 ml of 1.4 M KOH(*aq*) to the equivalence end point.
2. If 84.0 mg of oxalic acid crystals is dissolved in water, what is the molar concentration of NaOH(*aq*) when 43.7 ml is required to adjust the pH of the resulting solution to 7.00? Assume that, at this pH, the NaOH was added in an amount equivalent to the titration of the oxalic acid to oxalate.
3. If, on three successive titrations, 20.0 ml samples of 0.22 N KOH(*aq*) are apparently neutralized by 31.3 ml, 30.9 ml, and 31.4 ml of H_2SO_4(*aq*), calculate the average normal and molar concentrations of the acid. Assume that H_2SO_4 was neutralized to K_2SO_4.

QUESTIONS AND PROBLEMS

10.1 Define acids and bases in terms of the Arrhenius concept.

10.2 What are Brønsted-Lowry acids and bases?

10.3 Complete and balance the following acid-base equations. Label the conjugate acid-base pairs:
a. $HSO_4^- + OH^- \rightleftharpoons$
b. $H_2S + CN^- \rightleftharpoons$
c. $O^{2-} + Zn(H_2O)_4^{2+} \rightleftharpoons$
d. $HCO_3^- + CH_3O^- \rightleftharpoons$
e. $OAc^- + HI \rightleftharpoons$

10.4 What is an amphoteric or amphiprotic substance?★

10.5 Complete and balance the following equations to illustrate amphoterism.★
a. $Fe(H_2O)_3(OH)_3 + HCl \rightleftharpoons$
b. $CH_3O^- + Ni(H_2O)_3(OH)_3 \rightleftharpoons$
c. $NH_2^- + HSO_4^- \rightleftharpoons$
d. $HClO_4 + HCO_3^- \rightleftharpoons$
e. $HNO_3 + HPO_4^{2-} \rightleftharpoons$

10.6 Define hydrolysis. What are two classes of hydrolysis reactions?★

10.7 Complete and balance the following equations for hydrolysis reactions:★
a. $H_2O + CO_3^{2-} \rightleftharpoons$
b. $CH_3O^- + H_2O \rightleftharpoons$
c. $Al(H_2O)_6^{3+} + H_2O \rightleftharpoons$
d. $H_2O + OAc^- \rightleftharpoons$
e. $S^{2-} + H_2O \rightleftharpoons$

10.8 Define acids and bases in terms of the Lewis concept.★

10.9 Select Lewis acids from the following list: S^{2-}, $H^+(aq)$, H_2O, NH_3, BF_3, O^{2-}, $Al^{3+}(aq)$.★

10.10 Define pH and pOH. Express the relationship between pH and hydrogen ion concentration in the form of an equation.

10.11 If a water solution has a hydrogen ion concentration of 0.068 M, what are the pH and the pOH?

10.12 Calculate the pH of the following solutions:
a. 0.04 M HBr
b. 1.3×10^{-4} M $[H^+]$
c. 0.0032 M HCl
d. 6.2×10^{-8} M $[H^+]$
e. 4.5×10^{-5} M $[OH^-]$

10.13 Given the following pH and pOH values, calculate the hydrogen ion concentrations of the following solutions:
a. pH = 4.2
b. pOH = 1.7
c. pOH = 11.6
d. pH = 8.3
e. pH = 0.8

10.14 How many grams of barium are needed to react completely with 60.0 ml of 0.75 M HBr?★

10.15 What volume of 0.65 M HCl is required to react completely with 80.2 mg of calcium?★

10.16 If 35.0 ml of 0.012 M HNO_3 reacts with an excess of zinc, what volume of hydrogen gas will be produced when measured at 40.0°C and 1.20 atm?★

10.17 Calculate the normal concentration of an acid if 23.2 ml neutralizes 81.5 ml of 0.83 N $NaOH(aq)$ in a titration.★

10.18 Find the molar and normal concentration of a solution of $H_2SO_4(aq)$ if 216.0 ml is required to neutralize 2.7 g of CaO that has dissolved in water, the product being $CaSO_4$.★

CHAPTER 11

In science there are unending horizons to be explored and no limit to the discoveries to be made. It is difficult to explain to one who has never experienced it, the incomparable thrill, excitement, and satisfaction of original discovery. In science this satisfaction is frequent.

Geo. W. Beadle, *Listen to Leaders in Science**

Behavioral Objectives

At the completion of this chapter, the student should be able to:

1. Define equilibrium in a reversible reaction.
2. List five factors that may affect the rates of chemical reactions.
3. Predict changes in equilibrium systems related to alteration of concentration, temperature, and pressure.
4. State Le Chatelier's principle.
5. Apply Le Chatelier's principle to the control of reversible reactions.
6. State the law of chemical equilibrium.
7. Write an equilibrium expression for a reversible reaction.
8. Explain the origin and meaning of an equilibrium constant.
9. Identify the variety of symbols used as equilibrium constants.
10. Describe the relative strengths of acids in terms of their dissociation constants.
11. Explain the meaning of the term solubility product constant (K_{sp}).*

---

* Atlanta, Georgia, Tupper and Love, Inc., 1965.

CHEMICAL EQUILIBRIUM

11.1 INTRODUCTION

The topic *chemical equilibrium* deals with the large number of chemical reactions that are described as **reversible** Reversible reactions, usually designated by a double arrow ($\rightleftharpoons$), are those in which the products formed by the original reactants will also interact to restore the original reactants to some degree. Finally, after a period of time that varies with the particular reactants, a dynamic balance between the two reactions is achieved. If, for example, A and B react to form C and D, and then C and D react to re-form A and B, the equilibrium reaction would be

$$A + B \rightleftharpoons C + D$$

Conventionally, the initial reaction between A and B is called the **forward** reaction and is indicated by the small letter f over the arrow pointing toward the right:

$$A + B \xrightarrow{f} C + D$$

The secondary reaction between C and D is called the **reverse** reaction, and is symbolized by the small letter r:

$$A + B \xleftarrow[r]{} C + D$$

The reversibility may be clearly indicated by combining the forward and reverse notations:

$$A + B \underset{r}{\overset{f}{\rightleftharpoons}} C + D$$

When the **rates** of the forward and reverse reactions become stabilized and equal, the number of moles of each species present in the system also becomes constant because A and B continue to react to form C and D just as fast as C and D reconstitute A and B. This balance in a state of constant change is known as **dynamic equilibrium.** Although the *rates* of the two reactions are equal, the *concentrations* of reactants and products will seldom be equal. They vary widely.

The rate of a chemical reaction refers to the number of moles of reactants that undergoes chemical change per unit of time. The progress of the reversible reaction, $A + B \rightleftharpoons C + D$, toward equilibrium is illustrated in Figure 11–1, where it can be seen that the rate of the reverse reaction increases until equilibrium is reached. Notice how the rates of the forward and reverse reactions become equal and constant

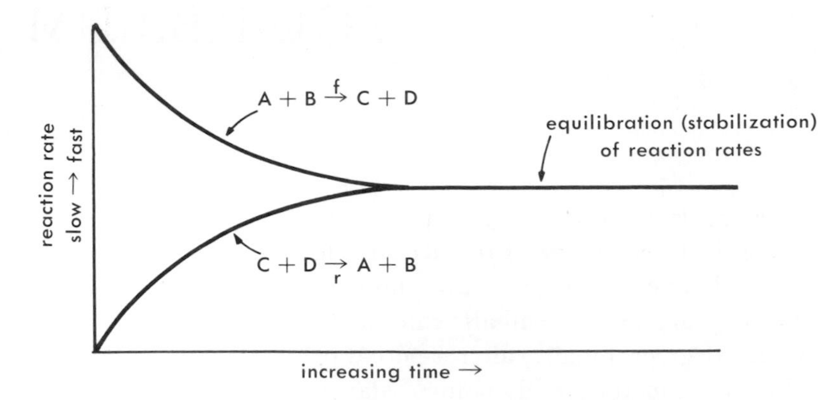

Figure 11–1 The stabilization of the reaction rates in an equilibrium system.

at equilibrium. Since the quantitative considerations of equilibrium reactions deal with comparisons between the stabilized rates of the forward and reverse reactions of particular systems, it would be advantageous to review the primary factors that affect reaction rates.

11.2 FACTORS AFFECTING REACTION RATES (KINETICS)

The study of those factors that affect the rates of chemical reactions is often called **kinetics.** The collision theory, which is central to the kinetic molecular theory of gases (Chapter 4), is a properly related concept. However, the elastic collisions between noninteracting gas molecules are not the same as the collisions between chemically reacting molecules, atoms, and ions. The factors that merely affected the volume, temperature, and pressure characteristics of gases now affect the rates of chemical change. Five principal factors will be considered: **concentration, temperature, pressure, catalysis,** and the **nature of the reactants.**

Concentration

The rate of the reaction between A and B depends on the number of moles of A and B present per unit volume. The initial concentrations of A and B are always expressed in terms of *moles per liter* or, conveniently, as *molar* if they are in solution. The symbolism that indicates moles per liter is a bracket, [ ], as was described in the previous chapter. The more particles of A and B present per unit volume, the more collisions there are. More collisions per unit of time naturally enhance the possibility that more particles will collide with sufficient energy and favorable geometry to result in a chemical change. The need for sufficient energy and favorable geometry when reactive species collide may be observed in the forward reaction between $H_2(g)$ and $I_2(g)$ in the formation of $HI(g)$:

$$H_2(g) + I_2(g) \rightleftharpoons 2HI(g)$$

Figure 11–2 illustrates the possible interactions between the hydrogen

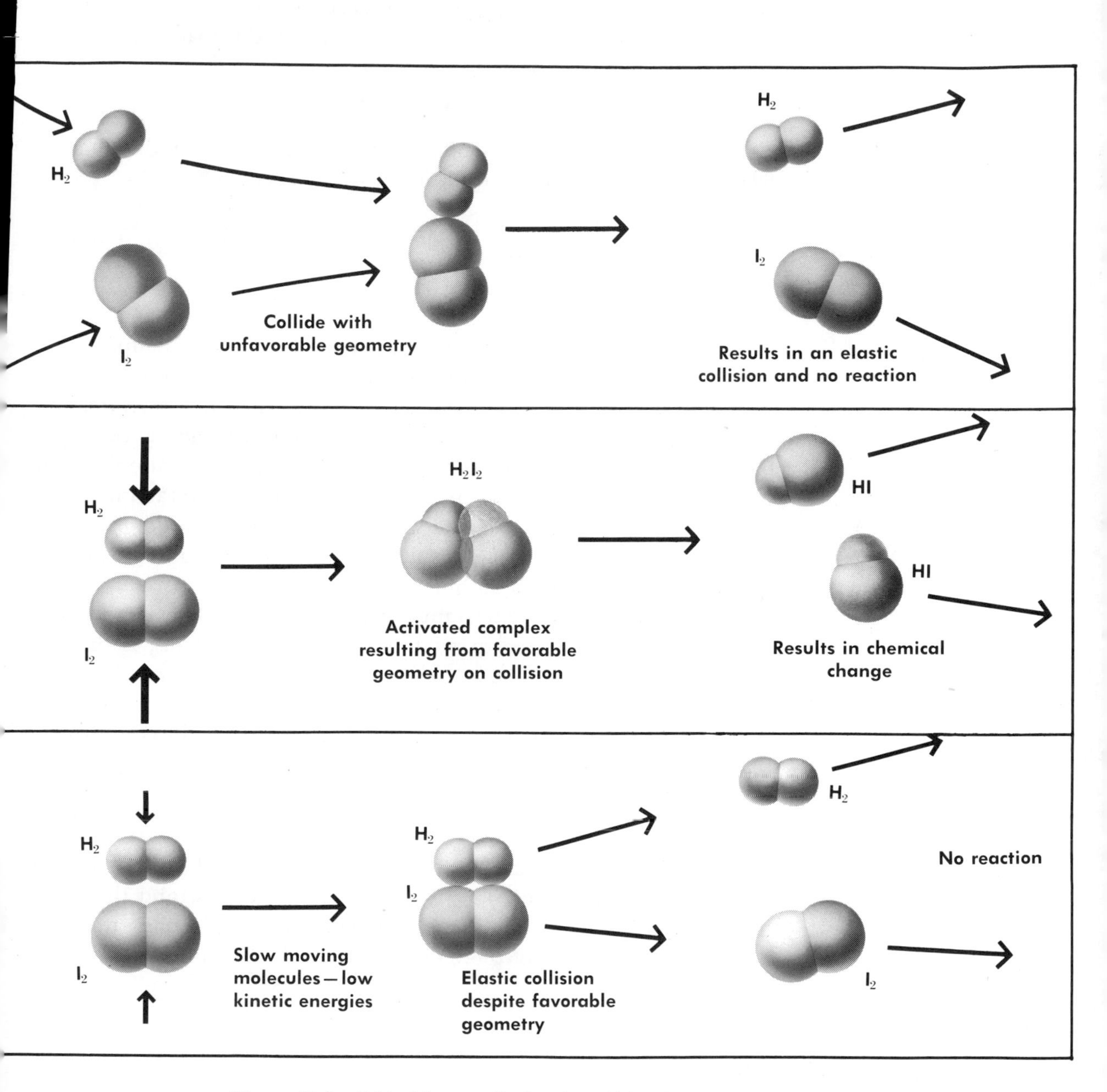

Figure 11–2 Critical factors affecting the collisions between reactants.

and iodine molecules. The actual interaction between H_2 and I_2 is more complex than the models indicate, however.

The rates of many common chemical reactions are sometimes empirically found to be directly proportional to the *product* of the molar concentrations. The rate expression for the reaction between A and B in such cases may be represented as

$$\text{rate} \propto [A][B]$$

This relationship will be expanded mathematically later in this chapter when the law of chemical equilibrium is discussed.

Temperature

Another critical factor affecting reaction rate is temperature. Many reactions that apparently do not occur or that happen slowly at room temperature will proceed vigorously when the reactants are heated. This does not mean that most reactions are endothermic. In fact, most reactions that "go" because the external application of heat has provided the activation energy required by the interacting molecules, atoms, or ions, are decidedly exothermic reactions. For example, a mixture of powdered iron and sulfur will glow with an intense heat of reaction once activated by a burner flame.

However, the most important observation to be retained, especially as it will be demonstrated to relate to equilibrium systems, is that *the application of heat favors endothermic reactions, while the removal of heat (cooling) favors exothermic reactions*, once they are started.

Temperature is therefore a very important factor in equilibrium reactions. The symbolic reaction

$$A + B \underset{r}{\overset{f}{\rightleftharpoons}} C + D + \text{heat}$$

indicates that the forward reaction is exothermic while the reverse reaction is endothermic. The rates of these reactions, and the equilibrium concentrations of all the components decided by the rates, will depend on the control of the temperature of the system.

Pressure

In nonvapor systems, the effect of pressure is small and often negligible. However, if some of the interacting species in a reaction are in the gas state, pressure may be an important factor. An increase in external pressure on the system clearly implies that the gas molecules will interact at an increased rate because of an increase in the number of collisions per unit time.

In an equilibrium system, the changes in the rates of the forward or reverse reactions depend on the number of moles of gas per unit volume among the initial reactants and products. If there are more moles of gas per liter present as initial reactants than as products, the rate of the forward reaction will be increased when pressure is applied. When the initial products consist of more moles of gas per liter than do the reactants, the reverse reaction is favored. Some examples are as follows:

1. $A(g) + B(g) \rightleftharpoons C(s) + D(s)$

   1 mole of gas 1 mole of gas no gas
   The rate of forward reaction *increases* with pressure.

2. $A(s) + B(g) \rightleftharpoons C(g) + D(g)$

   1 mole of gas 2 moles of gas

   The *reverse* reaction is favored by increase of pressure.

3. $A(s) + B(g) \rightleftharpoons C(g) + D(\ell)$

   1 mole of gas 1 mole of gas
   The increase of pressure has no significant effect on either the forward or the reverse reaction rate, so neither reaction is favored.

Catalysis

The rates of many reactions may be affected markedly by the use of a catalyst. While the mechanisms of many catalyzed reactions are not fully understood, a catalyst seems to increase the rate of a reaction by lowering the activation energy required for the first step, or by providing a faster alternate pathway for the reacting species as they undergo chemical change. Another factor, which partially defines a catalyst, is that the catalyst is recoverable at the end of the reaction, often in its original condition.

Some substances are described as *contact* catalysts while others are classified as *carrier* catalysts. A contact catalyst provides a focal point for interacting gas molecules. For example, when nitrogen and hydrogen react under the most favorable conditions, ammonia is formed, but *very* slowly. When the same reaction occurs in the presence of platinum as the contact catalyst, the change is rapid:

$$N_2(g) + 3H_2(g) \xrightarrow{Pt} 2NH_3(g)$$

The carrier catalysts apparently enter into the reaction as they form temporary unstable structures that rapidly decompose as the final products are produced. Although the carrier catalyst does take part in the reaction as it provides the alternate and more rapid pathway, it is not apparently changed in the final analysis. The addition of MnO_2 to $KClO_3$ forms an unstable intermediate, $MnO_2 \cdot O_2$, which rapidly decomposes into $MnO_2 + O_2$:

$$(1)\ KClO_3 + MnO_2 \longrightarrow KCl + MnO_2 \cdot O_2$$

$$(2)\ MnO_2 \cdot O_2 \longrightarrow MnO_2 + O_2$$

The overall reaction, then, is

$$KClO_3 \xrightarrow{MnO_2} KCl + \frac{3}{2} O_2$$

It must be emphasized that catalysts cannot change the molar concentrations at equilibrium. They alter the rate of the forward and reverse reactions in equal proportions, so that the time required for the reversible reaction to reach the equilibrium state is shortened.

The Nature of the Reactants

The nature of the reactants is probably the most obvious condition affecting reaction rates. The fact that some compounds and elements are extremely reactive and are observed to undergo chemical change very rapidly is a matter affecting laboratory safety. The addition of copper to concentrated sulfuric acid results in a very slow reaction until the mixture is heated, while a mixture of $KClO_3$ and sulfuric acid can produce a dangerously explosive reaction because of the rapidity of the change.

11.3 LE CHATELIER'S PRINCIPLE

In a reversible reaction, the molar concentrations of the initial reactants and products reach certain fixed values when a state of dynamic equilibrium finally exists. However, since the factors of concentration, temperature, and possibly pressure (when gases are involved) may affect forward and reverse reaction rates unequally, these factors must be taken into account as agents of change in equilibrium systems.

The French chemist Henri Le Chatelier, in 1884, proposed an explanation of the altering effects of changes in concentration, temperature, and pressure on equilibrium systems. What **Le Chatelier's principle** says is that *when a stress is applied to a system in equilibrium, the concentrations of the interacting species will change in such a way as to relieve or cancel the effect of the stress.* The special significance of Le Chatelier's principle to the chemist is that by thoughtfully selecting a particular type of stress, such as changing the temperature, selectively increasing or decreasing concentrations of some of the reacting species, or altering the pressure on gases, the forward or the reverse reaction can be favored decisively as the reaction moves toward a restoration of the equilibrium.

EXAMPLE 11.1

In the reaction $N_2(g) + 3H_2(g) \overset{Pt}{\rightleftharpoons} 2NH_3(g)$ + heat, what kinds of changes in concentrations, temperature, and pressure could be used to favor the *forward* reaction? What effect would the catalyst, platinum, have?

1. List the pertinent observations on this equilibrium system.
   a. All of the reactants and products are in the gas state.
   b. The forward reaction is exothermic while the reverse reaction is endothermic.
   c. The addition of more N_2 or H_2 to the same reaction volume would increase the collision rate in the forward reaction.
2. The Pressure Change.
   Since $3H_2(g)$ and $N_2(g)$ represent 4 moles of gas that react to form 2 moles of ammonia, the forward reaction could be favored by increasing the pressure. More collisions between H_2 and N_2 molecules would be expected. In terms of Le Chatelier's principle, the response to the stress of increased pressure would be a favoring of the reaction that yields products that occupy less volume. Two moles of ammonia occupy half the volume that would be occupied by a total of 4 moles of H_2 and N_2.
3. The Temperature Change.
   Since the forward reaction is exothermic, any heat that exceeds the minimal activation energy would aid the reverse, or endothermic, reaction. Therefore, the forward reaction would be favored by keeping the temperature at the threshold activation level while removing *heat* as it is formed.
4. The Concentration Changes.
   The constant addition of both $H_2(g)$ and $N_2(g)$ would favor the forward reaction by increasing the number of collisions of reactants while simultaneously increasing the pressure on the system. If ammonia could be removed (perhaps by cooling and condensation to the liquid state), the forward reaction would be favored even more, as Le Chatelier's principle would relate to increased production of ammonia in response to the stresses in equilibrium restoration.
5. The only effect attributed to the use of the platinum catalyst would be a proportional speeding up of both reactions—favoring neither the forward nor the reverse directions. There

is no such thing as a one-way catalyst, for that would allow us to get something for nothing.

EXAMPLE 11.2

In the reaction

$$CO(g) + Cl_2(g) \rightleftharpoons COCl_2(g) + \text{heat}$$

what effects would an increase in temperature, a decrease in pressure, and a removal of $COCl_2(g)$ have on the equilibrium system?

1. Temperature Increase.
   Since the forward reaction is exothermic, an increase in temperature would favor the reverse, or endothermic, reaction. In other words, the equilibrium would be shifted to the left, so that the concentrations of CO and Cl_2 would be higher at equilibrium.
2. Pressure Decrease.
   Since the CO and Cl_2 represent 2 moles of gas, while the $COCl_2$ is only 1 mole of gas, a decrease in pressure would favor the formation of those products that occupy more volume. Therefore, the equilibrium would be shifted to the left (reverse reaction) as a reaction to the pressure stress.
3. Removal of $COCl_2(g)$.
   The reduction of the concentration of the $COCl_2$ places a stress on the system such that the forward reaction would be favored as a move to restore the original equilibrium concentrations.

Exercise 11.1

1. In the following reactions, indicate whether an increase or a decrease of pressure will favor the forward reaction:
   a. $H_2(g) + I_2(g) \rightleftharpoons 2HI$
   b. $H_2(g) + 1/2\ O_2(g) \rightleftharpoons H_2O(\ell)$
   c. $2NO_2(g) \rightleftharpoons N_2O_4(g)$
   d. $CuO(s) + H_2(g) \rightleftharpoons Cu(s) + H_2O(\ell)$
   e. $COCl_2(g) \rightleftharpoons Cl_2(g) + CO(g)$
2. What effect would heating have on the following reactions?
   a. $CO_2(g) + C(s) \rightleftharpoons 2CO(g) - \text{heat}$
   b. $1/2\ N_2(g) + 1/2\ O_2(g) \rightleftharpoons NO(g) - \text{heat}$
   c. $(NH_4)_2CO_3(s) \rightleftharpoons 2NH_3(g) + CO_2(g) + H_2O(\ell) - \text{heat}$
   d. $PCl_3(g) + Cl_2(g) \rightleftharpoons PCl_5(g) + \text{heat}$
3. Indicate an equilibrium shift to the right (forward) or left

(reverse) in the following reactions for each of the specified conditions:

a. $SO_2(g) + 1/2\ O_2(g) \rightleftharpoons SO_3(g) + \text{heat}$
   1. temperature is raised
   2. pressure is reduced
   3. catalyst is added

b. $CO(g) + 3H_2(g) \rightleftharpoons CH_4(g) + H_2O(g) - \text{heat}$
   1. temperature is lowered
   2. pressure is increased
   3. CH_4 is removed from the system

c. $NOCl(g) \rightleftharpoons NO(g) + 1/2\ Cl_2(g) - \text{heat}$
   1. temperature is raised
   2. pressure is increased
   3. extra chlorine gas is added

A dynamic equilibrium clearly implies that in a simple reversible reaction, the rates of the forward and the reverse reactions are equal, since the initially faster reaction slows down while the slower reaction speeds up. When the rates of the reactions are equal, the concentrations of all the interacting species are constant. It must be remembered, however, that a change in temperature, or in pressure, if gases are involved, will change the equilibrium concentration. When equilibria data are treated mathematically, the temperature and pressure factors must be taken into account. The quantitative treatment of equilibrium concentrations is governed by the **law of chemical equilibrium,** sometimes called the law of mass action, and mathematically described as the **equilibrium expression.**

11.4 THE LAW OF CHEMICAL EQUILIBRIUM

In the symbolic reaction $A + B \rightleftharpoons C + D$, it was pointed out previously that the rate of the forward reaction is proportional to the **product** of the molar concentrations:

$$\text{rate}_f \propto [A][B]$$

Conversely, the rate of the reverse reaction is proportional to the product of the molar concentrations of the reverse reactants:

$$\text{rate}_r \propto [C][D]$$

If the generalized reaction had a different form, such as

$$\underset{\text{1 mole}}{A_2} + \underset{\text{2 moles}}{2B_2} \rightleftharpoons \underset{\text{2 moles}}{2AB_2}$$

then the expressions of the rate proportionalities would be

$$\text{rate}_f \propto [A_2][B_2][B_2]$$

or,

$$\text{rate}_f \propto [A_2][B_2]^2$$

and,

$$\text{rate}_r \propto [AB_2][AB_2]$$

or,

$$\text{rate}_r \propto [AB_2]^2$$

Notice that the *coefficients in the equation become exponential values* in the expressions of rate proportionality. The reasoning used to explain these empirical observations is that the presence of more than one mole of a reacting species creates a much greater probability for favorable collisions to occur per unit of time.

The proportionality of rates and the products of molar concentrations can be easily expressed mathematically for a simple one-step process. For the equation

$$A + B \rightleftharpoons C + D,$$

$$\text{rate}_f = k_f[A][B]$$

where k_f is the proportionality constant for the forward reaction, and

$$\text{rate}_r = k_r[C][D]$$

where k_r is the proportionality constant for the reverse reaction. At equilibrium, the equalization of the rates

$$\text{rate}_f = \text{rate}_r$$

permits an equating of the expressions for the rates:

$$k_f[A][B] = k_r[C][D]$$

Conventionally, the two rate constants are combined into a single constant called the **equilibrium constant,** and symbolized by K_c. The right-hand side of the equation is written as a *numerator*, while the left-hand side becomes the *denominator:*

$$\frac{k_f}{k_r} = K_c = \frac{[C][D]\ (\text{products})}{[A][B]\ (\text{reactants})}$$

In the second generalized equation, $A_2 + 2B_2 \underset{r}{\overset{f}{\rightleftharpoons}} 2AB_2$, the equilibrium expression is

$$K_c = \frac{[AB_2]^2}{[A_2][B_2]^2}$$

From the above examples, the law of chemical equilibrium can be extracted! *In an equilibrium system (at constant temperature), the ratio of the product of the molar concentrations of the products to the product of the molar concentrations of the reactants is equal to a constant value called the equilibrium constant.*

11.5 THE SIGNIFICANCE OF THE EQUILIBRIUM CONSTANT

The most important question at this point is what is the *significance* of an equilibrium constant. The ability to calculate or use an equilibrium constant in a more involved calculation is really quite meaningless if an equilibrium constant is interpreted as nothing more than a number. For example, in the reaction $H_2 + I_2 \rightleftharpoons 2HI$, K_c values for the equilibrium expression

$$K_c = \frac{[HI]^2}{[H_2][I_2]}$$

are found to be different at various temperatures:

$$K_c \text{ at } 445°C = 64$$
$$K_c \text{ at } 490°C = 46$$
$$K_c \text{ at } 25°C = 808$$

The significance of the K_c values is that they clearly indicate the temperature range most appropriate if a large yield of HI is desired. Remember, the K_c is the ratio of the molar concentrations of products to reactants at equilibrium. The value 808 for the K_c at 25°C means that a low temperature provides the greatest yield of product. In effect, the ratio

$$\frac{[HI]^2}{[H_2][I_2]} = \frac{808}{1}$$

clearly demonstrates a better yield than does the K_c at 445°C, in which case

$$K_c = \frac{64}{1} = \frac{[HI]^2}{[H_2][I_2]}$$

What is the significance of the symbol K_w and its equivalence to 1.0×10^{-14} as it was introduced in the previous chapter in the discussion of pH? The K_w is related to the equilibrium constant for the reaction

$$H_2O(\ell) \rightleftharpoons H^+(aq) + OH^-(aq)$$

The equilibrium expression is

$$K_c = \frac{[H^+][OH^-]}{[H_2O]}$$

and since so little $H_2O(\ell)$ actually ionizes, the molar concentration is essentially constant (55.6 moles ℓ^{-1}):

$$[K_c][H_2O] = [K_c][55.6] = K_w$$
$$K_w = [H^+][OH^-] = 1.0 \times 10^{-14} \text{ at } 24°C.$$

The value 1.0×10^{-14} clearly illustrates how small the ion concentration is in pure water:

$$K_w = 1.0 \times 10^{-14} = [H^+][OH^-]$$

A weak acid, such as acetic acid, may be found to have an equilibrium dissociation constant at 25°C of 1.8×10^{-5} in a reference table where equilibrium constants, symbolized as K_a, K_i, or K_{diss}, are listed. What does this mean? In the first place, the subscripts, *a*, *i*, and *diss* are practically synonymous for weak acids. The fact that *a* stands for *acid*, *i* means *ionization*, and *diss* represents *dissociation* only means that chemists are likely to use the notation systems they have learned. Probably the most popular symbol for the equilibrium constant of a weak acid is K_a. The value for the K_a, experimentally determined, indicates the ratio of the product of the molar concentration of the ions compared to the molar concentration of the un-ionized molecules. For example, the equation for the dissociation (or, ionization) of a mole of HOAc per liter of solution is

$$HOAc(aq) \rightleftharpoons H^+(aq) + OAc^-(aq)$$

The equilibrium expression is

$$K_a = \frac{[H^+][OAc^-]}{[HOAc]} \quad \begin{matrix} \text{ions} \\ \leftarrow \text{un-ionized molecules} \end{matrix}$$

The value at 25°C for the K_a, being 1.8×10^{-5}, means that the ratio of ions to molecules is very small:

$$K_a = \frac{1.8 \times 10^{-5}}{1} = \frac{[H^+][OAc^-]}{[HOAc]}$$

Therefore, by any definition of an acid, HOAc(*aq*) must be described as weak. It should be mentioned that water is not included in the equilibrium expression regardless of the fact that the ionization of acetic acid takes place in water solution. This is due to the fact that the concentration of water molecules remains relatively constant. The only outstanding exception is in the case of certain hydrolysis reactions.

Table 11–1 gives a comparison of some acids with regard to a quantitative description of their strengths in terms of K_a values.

TABLE 11–1 EQUILIBRIUM CONSTANTS OF WEAK ACIDS AT 25°C

	Name	First Ionization Reaction	K_a
INCREASINGLY WEAK ACIDS ↓	oxalic	$H_2C_2O_4 \rightleftharpoons H^+ + HC_2O_4^-$	5.6×10^{-2}
	sulfurous	$H_2SO_3 \rightleftharpoons H^+ + HSO_3^-$	1.7×10^{-2}
	phosphoric	$H_3PO_4 \rightleftharpoons H^+ + H_2PO_4^-$	5.9×10^{-3}
	hydrofluoric	$HF \rightleftharpoons H^+ + F^-$	6.7×10^{-4}
	nitrous	$HNO_2 \rightleftharpoons H^+ + NO_2^-$	5.1×10^{-4}
	acetic	$HOAc \rightleftharpoons H^+ + OAc^-$	1.8×10^{-5}
	hydrocyanic	$HCN \rightleftharpoons H^+ + CN^-$	4.8×10^{-10}

The equilibrium constants (also referred to as dissociation constants) of weak bases communicate the same type of information as the K_a values for acids. However, a distinction is usually made in the symbol, which is most often K_b. For example, the value for the K_b of $NH_3(aq)$ is 1.8×10^{-5}. The equation

$$NH_3 + H_2O \rightleftharpoons NH_4^+ + OH^-$$

appears in the equilibrium expression as

$$K_b = \frac{[NH_4^+][OH^-]}{[NH_3]}$$

The K_b value means that the ratio of the ion product concentration to the ammonia concentration is very small. In other words, the aqueous solution of ammonia yields relatively few hydroxide ions in solution:

$$K_b = \frac{1.8 \times 10^{-5}}{1} = \frac{[NH_4^+][OH^-]}{[NH_3]} \begin{matrix} \leftarrow \text{ions} \\ \leftarrow \text{molecules} \end{matrix}$$

The majority of metal hydroxides, which are characterized as weak bases, tend to be only slightly soluble in water. Examples are $Mg(OH)_2$, $Fe(OH)_3$, $Fe(OH)_2$, $Ca(OH)_2$, $Cd(OH)_2$, and $Zn(OH)_2$. The equilibrium constants that reflect the tendency of these compounds to dissociate to a slight degree will be introduced shortly, under the heading **solubility products.**

When the dissociation of diprotic or triprotic acids is considered, the meaning of the K_a values becomes somewhat more complex. For example, the ionization of H_2SO_3 can be shown to occur in two steps:

$$(1)\ H_2SO_3 \rightleftharpoons H^+ + HSO_3^-$$

$$(2)\ HSO_3^- \rightleftharpoons H^+ + SO_3^{2-}$$

Each step in the dissociation will naturally have its own K_a value, since there is no reason to expect both H_2SO_3 and HSO_3^- to have the same dissociation tendencies. A reference table will indicate the following at 25°C:

$$H_2SO_3 \quad K_{a_1} = 1.2 \times 10^{-2}$$

$$HSO_3^- \quad K_{a_2} = 1.0 \times 10^{-7}$$

If these K_a values are viewed within the framework of the equilibrium expression, the species of ions and molecules in greatest abundance at equilibrium can be determined:

$$K_{a_1} = \frac{1.2 \times 10^{-2}}{1} = \frac{[H^+][HSO_3^-]}{[H_2SO_3]}$$

$$K_{a_2} = \frac{1.0 \times 10^{-7}}{1} = \frac{[H^+][SO_3^{2-}]}{[HSO_3^-]}$$

Clearly, the K_{a_1} value is much larger than the K_{a_2}. Therefore, it is reasonable to say that the first ionization step is far more significant than the second. The conclusions are as follows:

1. Most of the H_2SO_3 does not dissociate. This is consistent with the labeling of H_2SO_3 as a weak acid.
2. $H^+(aq)$ and $HSO_3^-(aq)$ will be vastly more abundant than $SO_3^{2-}(aq)$ at equilibrium.
3. The $HSO_3^-(aq)$ could be characterized as an extremely weak acid.

11.6 SOLUBILITY PRODUCTS *

When an excess of a slightly soluble compound is added to water, an equilibrium develops between the undissolved solid compound (which may be molecular or ionic) and the relatively small molar concentration of ions. This is true of many salts and hydroxide compounds that normally appear as precipitates in reactions. For example, the addition of sufficient chloride ion to a solution containing silver ion produces the precipitate AgCl(*s*).

Silver chloride is obviously not very soluble in water. In fact, a concentration in the millimolar range represents a saturated solution. It would be quite impossible to prepare a molar, a tenth molar, or even a thousandth molar solution of AgCl in water.

The equilibrium state for the dissociation of AgCl(*s*) is represented by the equation

$$AgCl(s) \rightleftharpoons Ag^+(aq) + Cl^-(aq)$$

The equilibrium expression would appear to be

$$K_{diss} = \frac{[Ag^+][Cl^-]}{[AgCl]}$$

However, since so little AgCl(*s*) actually dissociates, the original number of moles of AgCl may be considered to be constant. This is analogous to the conditions which lead to the development of a K_w value for water ionization. Combining the two constants, K_{diss} and [AgCl], the equilibrium expression can be simplified:

$$K_{diss} \times [AgCl] = [Ag^+][Cl^-]$$

The combined constants are provided with a new symbol, K_{sp}, called the **solubility product constant.** The term **solubility product** is quite appropriate, since the value for K_{sp} is indeed the product of the ion concentrations in a saturated solution at equilibrium.

The Significance of K_{sp} Values

A glance at Table 11–2, showing the solubility product constants, quickly points out markedly varying degrees of ion dissociation. A compound

TABLE 11-2 SOLUBILITY PRODUCT CONSTANTS OF SOME SLIGHTLY SOLUBLE COMPOUNDS AT 25°C

	Name	Dissociation Reaction	Equilibrium Expression	K_{sp}
DECREASING SOLUBILITY ↓	silver acetate	$AgOAc \rightleftharpoons Ag^+ + OAc^-$	$K_{sp} = [Ag^+][OAc^-]$	2×10^{-3}
	lead(II) chloride	$PbCl_2 \rightleftharpoons Pb^{2+} + 2Cl^-$	$K_{sp} = [Pb^{2+}][Cl^-]^2$	1.7×10^{-5}
	barium fluoride	$BaF_2 \rightleftharpoons Ba^{2+} + 2F^-$	$K_{sp} = [Ba^{2+}][F^-]^2$	2×10^{-6}
	magnesium carbonate	$MgCO_3 \rightleftharpoons Mg^{2+} + CO_3^{2-}$	$K_{sp} = [Mg^{2+}][CO_3^{2-}]$	2×10^{-8}
	silver chloride	$AgCl \rightleftharpoons Ag^+ + Cl^-$	$K_{sp} = [Ag^+][Cl^-]$	1.6×10^{-10}
	silver iodide	$AgI \rightleftharpoons Ag^+ + I^-$	$K_{sp} = [Ag^+][I^-]$	1×10^{-16}
	cadmium sulfide	$CdS \rightleftharpoons Cd^{2+} + S^{2-}$	$K_{sp} = [Cd^{2+}][S^{2-}]$	1×10^{-26}
	iron(III) hydroxide	$Fe(OH)_3 \rightleftharpoons Fe^{3+} + 3OH^-$	$K_{sp} = [Fe^{3+}][OH^-]^3$	5×10^{-38}
	mercury (II) sulfide	$HgS \rightleftharpoons Hg^{2+} + S^{2-}$	$K_{sp} = [Hg^{2+}][S^{2-}]$	1×10^{-52}
	bismuth (III) sulfide	$Bi_2S_3 \rightleftharpoons 2Bi^{3+} + 3S^{2-}$	$K_{sp} = [Bi^{3+}]^2[S^{2-}]^3$	1.6×10^{-72}

in which the product of ion concentrations (K_{sp}) is of the order 10^{-5}, for example, is much more soluble than another compound which has a K_{sp} of the order 10^{-20}.

With the availability of K_{sp} values for slightly soluble compounds, it is possible to calculate exactly what molar concentration of a selected ion would be necessary to cause precipitation. That is to say, if the ion product were adjusted to exceed the K_{sp}, a precipitate would form. In the same manner, ion additions may be calculated so that precipitation can be avoided. Such calculations are common in analytical chemistry when it is advantageous to separate ions by precipitation techniques.

QUESTIONS AND PROBLEMS

11.1 Define the term *dynamic equilibrium* in reversible chemical reactions.

11.2 List five factors that may affect the rate of a reaction.

11.3 In what type of equilbrium reaction does pressure become a necessary factor for calculation purposes?

11.4 If platinum acts as a catalyst in the formation of ammonia from nitrogen and hydrogen gas, what effect will the removal of the platinum have on the concentration of NH_3 at equilbrium? Explain your answer.

11.5 In the following reactions, explain the effects of pressure changes on the forward and reverse reactions:
a. $H_2(g) + Cl_2(g) \rightleftharpoons 2HCl(g)$
b. $CO(g) + CuO(s) \rightleftharpoons Cu(s) + CO_2(g)$
c. $C_6H_6(\ell) + 15/2\ O_2(g) \rightleftharpoons 6CO_2(g) + 3H_2O(g)$

11.6 What effect would an increase in temperature be expected to have on the following reactions?
a. $N_2O_4 \rightleftharpoons 2NO_2 - \text{heat}$
b. $H_2O \rightleftharpoons H_2 + 1/2\ O_2 - \text{heat}$
c. $C_2H_4 + H_2 \rightleftharpoons C_2H_6 + \text{heat}$

11.7 Given the reaction

$$CO_2(g) \rightleftharpoons CO(g) + \frac{1}{2} O_2(g) - \text{heat}$$

indicate the direction of shift (forward or reverse) when
a. the temperature is raised
b. the pressure is increased
c. oxygen leaks out of the system

11.8 Explain the meanings of the following symbols:
a. K_c
b. K_i
c. K_a
d. K_h
e. K_{diss}
f. K_{sp}
g. K_w
h. K_b

11.9 Write a sample equation and equilibrium expression for each of the symbols listed in Question 11.8.

11.10 Write equilibrium expressions for the following reactions:

a. $2NOCl(g) \rightleftharpoons 2NO(g) + Cl_2(g)$
b. $HCO_3^-(aq) \rightleftharpoons H^+(aq) + CO_3^{2-}(aq)$
c. $CO_3^{2-} + H_2O \rightleftharpoons HCO_3^- + OH^-$
d. $HNO_2(aq) \rightleftharpoons H^+(aq) + NO_2^-(aq)$
e. $Fe(OH)_2(s) \rightleftharpoons Fe^{2+}(aq) + 2OH^-(aq)$

11.11 Write equilibrium expressions for the two steps in the ionization of H_2SO_3.

11.12 If the reaction $X + Y \rightleftharpoons XY$ has the values K_c at 20°C = 17 and K_c at 80°C = 42, what should be done about the temperature if the production of XY is undesirable?

CHAPTER 12

Science is man's exploration of his universe, and to exclude himself even in principle is certainly not objective realism. . . . And to say that we cannot learn anything materially factual about a situation if we ourselves are in it is utter and nonsensical negation of the very meaning of learning.

Geo. Gaylord Simpson, *This View of Life**

Behavioral Objectives:

At the completion of this chapter, the student should be able to:

1. Define oxidation, reduction, oxidizing agents, and reducing agents.
2. Determine oxidation numbers of selected atoms in binary and ternary compounds.
3. Find the oxidation numbers of selected atoms from ions in solution.
4. Balance redox equations by the oxidation number method.
5. Identify the specific atoms involved in oxidation-reduction, given a skeleton equation.
6. Write and balance half-reactions, indicating the loss, gain, and conservation of charge.
7. List the steps in the systematic balancing of redox equations, both molecular and ionic.
8. Balance ionic redox equations by the ion-electron method.
9. Solve stoichiometric problems in the case of redox reactions.
10. Calculate molar and normal concentrations of compounds in redox solution stoichiometry.★
11. Determine concentrations of unknowns from data supplied by redox titration measurements.★

---

* New York, Harcourt Brace Jovanovich, Inc., 1963.

OXIDATION AND REDUCTION

12.1 INTRODUCTION

The type of chemical reaction described as oxidation-reduction, often contracted to **redox**, is a topic of major importance in chemistry. The tendency of elements to lose and gain electrons in the drive toward stability and lower energy content was introduced in Chapter 5 under the heading *Ionization Energy* (p. 152).

An understanding of the principles of oxidation and reduction is especially critical because of the physical hazards involved. The chronicles of scientific activity are filled with stories of explosions, damage to skin and eyes, fires, and billions of dollars lost by corrosion, and gruesome accounts of what surgical metals have done to teeth and bones. On the other hand, knowledge has permitted great strides forward in the development of electrochemical cells, safety procedures in laboratories, and modern dental and surgical methods, and a growing understanding of the use and production of energy for life.

12.2 REDOX PRINCIPLES REVIEWED

Oxidation of a substance is an *increase* in its **oxidation number.** This is accomplished by a *loss* of electrons. **Reduction** is a *decrease* in the oxidation number, which is achieved by a *gain* of electrons. Any substance that promotes the loss of electrons because of its own tendency to gain them is called an **oxidizing agent.** The substance which readily gives up its electrons is called the **reducing agent.** Figure 12–1 illustrates the relationship between the change in oxidation number and the processes of oxidation and reduction.

$$\xrightarrow{\text{Reduction}}$$

7+ 6+ 5+ 4+ 3+ 2+ 1+ 0 1− 2− 3− 4− 5− 6− 7−

oxidation numbers

$$\xleftarrow{\text{Oxidation}}$$

Figure 12–1 How an increase in oxidation number is defined as oxidation (loss of electrons) and how a decrease in oxidation number (gain of electrons) is defined as reduction.

12.3 DETERMINATION OF OXIDATION NUMBERS

The oxidation numbers assigned to free (uncombined) elements and to elements in binary compounds present no problem. Since uncombined elements have neither lost nor gained electrons, their oxidation numbers are always zero, regardless of their state of complexity (S_2 or S_6 or S_8, for example). The elements may also be said to be in

TABLE 12-1

Binary Compound Formula	Nomenclature	Oxidation Numbers
NaCl	sodium chloride	Na is 1+, Cl is 1−
CaF_2	calcium fluoride	Ca is 2+, F is 1−
FeS	iron (II) sulfide	Fe is 2+, S is 2−
Fe_2S_3	iron (III) sulfide	Fe is 3+, S is 2−
PbO	lead (II) oxide	Pb is 2+, O is 2−
PbO_2	lead (IV) oxide	Pb is 4+, O is 2−
PCl_3	phosphorus trichloride	P is 3+, Cl is 1−
V_2O_5	divanadium pentoxide	V is 5+, O is 2−
H_2O_2	hydrogen peroxide	H is 1+, O_2 is 2−
NaH	sodium hydride	Na is 1+, H is 1−

the *zero oxidation state.* The oxidation numbers of elements in binary compounds are the usual values that have been discussed and illustrated in Chapter 7. Table 12-1 presents a brief review. Notice that the oxidation number of oxygen as a peroxide and of hydrogen as a hydride are unusual. However, while the exceptions must be remembered, they do not detract from the usefulness of the generalizations that combined hydrogen is usually in the 1+ oxidation state, and that combined oxygen is usually in the 2− oxidation state.

Exercise 12.1

Name the following compounds and indicate the oxidation number of the underlined element.

a. $K\underline{Br}$
b. $\underline{Cd}Cl_2$
c. $\underline{Ni}Cl_2$
d. $\underline{S}O_2$
e. $Ca\underline{H}_2$
f. $\underline{P}_2O_5$
g. $\underline{C}O$
h. $\underline{Si}Br_4$
i. $\underline{S}O_3$
j. $Na_2\underline{O}_2$

The oxidation numbers of the individual elements in ternary compounds require a more careful investigation. For example, the ionic charges of iron and sulfate in the compound, $FeSO_4$ (iron (II) sulfate), are obvious. Iron is 2+ and sulfate is 2−. These values are clearly listed in the tables of oxidation numbers and ionic charges in Chapter 7. The important question, however, is the oxidation state of the *sulfur.* The varying oxidation number of combined sulfur, often called the "apparent valence," may be determined from the common oxidation numbers of the iron and oxygen. In other words, the oxidation number of an element in the *middle* of a ternary compound must be such that the sum of the oxidation numbers of all the combined elements is zero. This is a matter of the conservation of charge. In the compound

$$Fe\ S\ O_4$$

the oxidation number of iron (II) is clearly 2+, that of the oxygen is 2−, *but* the presence of 4 units of oxygen yields a total charge value of 8−. (The individual oxidation number of a single unit is always multiplied by the number of units indicated in the formula.)

Total charge	2+	?	8−
Individual oxidation number	2+	?	2−
	Fe	S	O
Number of units of each element	1	1	4

Since the sum of the charges in a compound must equal zero, the oxidation number of sulfur must be such that when added to the 2+ of the iron and the 8− of the oxygen, the sum is zero. Obviously, the oxidation number of sulfur in the compound $FeSO_4$ must be 6+:

Total charge	2+	**6+**	8−	= 0
Individual oxidation number	2+	**6+**	2−	
	Fe	S	O	
Number of units of each element	1	1	4	

EXAMPLE 12.1

Find the oxidation number of chromium in $K_2Cr_2O_7$.

1. Establish the individual oxidation numbers of potassium and oxygen:

1+		2−
K_2	Cr_2	O_7

2. Calculate the total charge on the K and O:

Total	2+		14−
	1+	?	2−
	K_2	Cr_2	O_7

3. Determine the total charge on Cr necessary to make the K, Cr, and O charges add up to zero:

$2+$	**$12+$**	$14-$	$= 0$
$1+$	?	$2-$	
K_2	Cr_2	O_7	

4. The oxidation number of the chromium must be $6+$, since 2 units of Cr^{6+} will equal the necessary charge total of $12+$:

$2+$	$12+$	$14-$	$= 0$
$1+$	($6+$)	$2-$	
K_2	Cr_2	O_7	

Exercise 12.2

Calculate the oxidation number of the underlined element in the following compounds:

a. $H\underline{Cl}O_4$
b. $H\underline{Cl}O_3$
c. $K\underline{Mn}O_4$
d. $Al_2(\underline{S}O_4)_3$
e. $Na_2\underline{S}_2O_3$
f. $K_2H\underline{P}O_4$
g. $Na_3\underline{As}O_4$
h. $K\underline{I}O_3$
i $Na_2\underline{Cr}_2O_7$
j. $Na_2\underline{Cr}O_4$

Ions in solution present a slight twist to the previous examples. The oxidation number of sulfur in Fe_2SO_4 has been demonstrated to be $6+$. Regardless of the particular example, the oxidation number of sulfur in any sulfate compound will always be $6+$. In ionic equations, where spectator ions are routinely eliminated, the sulfate compound may simply be written as $SO_4^{2-}(aq)$, assuming the metal ion to be the spectator. The problem is to determine the charges (oxidation numbers) of the elements as they are combined in the form of a polyatomic ion. The sum of the charges obviously does *not* add up to zero. The net charge on the ion is now the sum of the individual oxidation numbers. In the case of $SO_4^{2-}(aq)$, what are the oxidation numbers? The calculation may be handled in the same manner as compound formulas:

Total charge	?	$8-$	$=$	$2-$
Oxidation number	?	$2-$		
	S	O		
Number of units	1	4		

Since 4 oxygens, each having an oxidation of 2−, produce a total charge of 8−, the single sulfur must have a charge of 6+, since the indicated net charge on the sulfate radical is 2−:

Total charge	6+	8−	=	2−
Oxidation number	(6+)	2−		
	S	O		
Number of units	1	4		

Note how the oxidation number of sulfur in the sulfate structure is 6+, and therefore identical with the previously calculated value. It remains the same, necessarily, whether the sulfate ion is in a crystal or in water solution.

EXAMPLE 12.2

Find the oxidation number of arsenic in the arsenite ion.

1. Write the formula for arsenite:

$$AsO_3^{3-}(aq)$$

2. Determine the total charge contributed by the oxygen:

?	6−	= 3−
	2−	
As	O_3	

3. Since the net charge on the ion (the sum of the positive and negative totals) is 3−, the oxidation number of the As can be determined:

(3+)	6−	= 3−
As_1	O_3	

The sum of 3+ and 6− is 3−.

EXAMPLE 12.3

What is the oxidation number of carbon in the oxalate ion?

1. Write the formula for the oxalate ion:

$$C_2O_4^{2-}$$

2. Calculate the total charge due to 4 units of oxygen:

?	8−	= 2−
?	2−	
C_2	O_4	

3. The total charge on the carbon must therefore be 6+, since the net charge on the ion is 2−:

6+	8−	= 2−
?	2−	
C_2	O_4	

4. The fact that there are 2 units of carbon means that the oxidation number of carbon in oxalate compounds must be 3+.

Exercise 12.3

Find the oxidation number of the underlined element in the following ions:

a. $\underline{N}H_4^{+}$
b. $\underline{As}O_4^{3-}$
c. $\underline{S}_2O_3^{2-}$
d. $\underline{N}O_3^{-}$
e. $\underline{N}O_2^{-}$
f. $\underline{P}O_3^{3-}$
g. $\underline{Cr}_2O_7^{2-}$
h. $\underline{Mn}O_4^{-}$
i. $H_2\underline{P}O_4^{-}$
j. $\underline{C}O_3^{2-}$

12.4 OXIDATION-REDUCTION REACTIONS

Consider the reaction between zinc metal and a copper (II) sulfate solution. When the zinc is placed in a tube of the blue solution, it darkens immediately as reddish clumps of solid form. The zinc gradually disappears and the solution turns from blue to colorless (Fig. 12–2). This is an oxidation-reduction reaction. The zinc is a good reducing agent, which is another way of saying that zinc becomes oxidized quite easily. The copper (II) ion has changed to plain copper metal, which was accomplished by gaining electrons (i.e., undergoing reduction). The equation for the entire redox reaction is the sum of the oxidation part and the reduction part. The number of electrons lost must equal the number of electrons gained, since electric charge is also subject to the law of conservation:

oxidation number ↙ ↘ oxidation number

$$\underset{\text{metallic zinc}}{Zn^{0}(s)} \longrightarrow \underset{\text{zinc ion}}{Zn^{2+}(aq)} + \underset{\text{two electrons lost}}{2e^{-}}$$

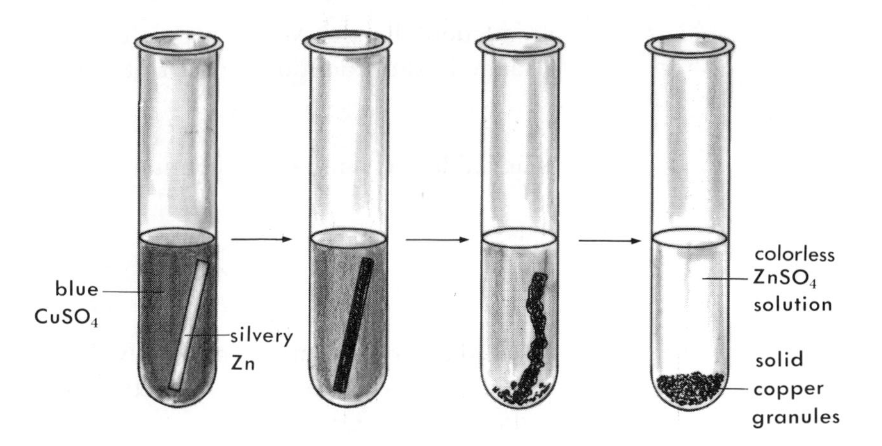

Figure 12-2 The oxidation-reduction reaction between zinc and copper (II) sulfate.

The rise in oxidation number from zero to positive two fits the definition of oxidation. The equation is said to represent a *half-reaction*, since only the oxidation half is given. The equation

$$\underset{\text{(from CuSO}_4\text{ solution)}}{\underset{\text{copper ion}}{Cu^{2+}(aq)}} + \underset{\text{gained}}{\underset{\text{two electrons}}{2e^-}} \longrightarrow \underset{\text{metallic copper}}{Cu^0(s)}$$

represents a *reduction half-reaction.* The copper gains two electrons, so the lowering of the oxidation number from 2+ to zero is reduction. One generally accepted convention is to transpose the $+2e^-$ from the left side of the equation to the right side for purposes of uniformity and simplicity:

$$Cu^{2+}(aq) \longrightarrow Cu^0 - 2e^-$$

When the two half-reactions (half oxidation and half reduction) are added, the net redox equation is obtained:

$$\begin{array}{l} Zn^0 \longrightarrow Zn^{2+} + 2e^- \\ Cu^{2+} \longrightarrow Cu^0 - 2e^- \\ \hline \underset{\text{zinc + copper (II) ion}}{Zn^0 + Cu^{2+}} \longrightarrow \underset{\text{zinc ion + copper}}{Zn^{2+} + Cu^0} \end{array}$$

The resulting reaction explains the observations. The formation of reddish clumps is the appearance of copper. The loss of blue color occurs as the copper (II) ion disappears, and the zinc is "eaten away" as it changes to zinc ion.

In any redox equation, the species being oxidized and reduced can be determined by the increase and decrease in the oxidation numbers. The number of electrons lost and gained will always equal the changes in the oxidation number of each species.

For example, a solution of permanganate ion which has been acidified is a good oxidizing agent. The $MnO_4^-(aq)$ becomes reduced to $Mn^{2+}(aq)$ in the process. Observe how the number of electrons gained (reduction) by the $MnO_4^-(aq)$ is equal to the difference between the change in the oxidation number of the manganese:

$$MnO_4^-(aq) \longrightarrow Mn^{2+}$$

The oxidation number of the manganese in MnO_4^- is 7+

7+	8− = 1−
(7+)	2−
Mn_1	O_4

while that in the Mn^{2+} is obviously 2+. The difference between 7+ and 2+ is 5+. Therefore, 5 electrons must be gained when MnO_4^- is reduced to Mn^{2+}:

$$MnO_4^- \longrightarrow Mn^{2+} - 5e^-$$

EXAMPLE 12.4

In an acid environment, $Cr_2O_7^{2-}(aq)$ tends to react with many substances, so that it becomes changed to $Cr^{3+}(aq)$. Label the change oxidation or reduction, and calculate the total number of electrons gained or lost per unit of $Cr_2O_7^{2-}$.

1. Find the oxidation number of chromium before and after:

12+	14− = 2−	
(6+)	2−	
Cr_2	O_7	Cr^{3+} is clearly chromium (III) ion

2. **Note:** There are 2 units of chromium in $Cr_2O_7^{2-}$. Therefore, the conservation laws require that 2 units of Cr^{3+} be produced:

$$Cr_2O_7^{2-} \longrightarrow 2Cr^{3+}$$

3. The difference in oxidation number between *single* units of chromium is 3+:

$$Cr^{6+} \longrightarrow Cr^{3+}$$

However, since 2 units of chromium change from 6+ to 3+, the difference must be doubled:

$$Cr_2^{6+} \longrightarrow 2Cr^{3+}$$

$$12+ \longrightarrow 6+$$

4. The decrease in the oxidation number of chromium indicates that it is a *reduction* change and the number of electrons *gained* must be 6, which is the difference between 12+ and 6+.
5. The change may be summarized by the equation:

$$Cr_2O_7^{2-} \longrightarrow 2Cr^{3+} - 6e^-$$

Exercise 12.4

Label each of the following changes as oxidation or reduction and indicate the number of electrons gained or lost:

a. $Mn^{2+} \rightarrow MnO_2{}^0$
b. $2IO_3{}^- \rightarrow I_2{}^0$
c. $2I^- \rightarrow I_2{}^0$
d. $SO_4{}^{2-} \rightarrow S^{2-}$
e. $S_2O_3{}^{2-} \rightarrow 2S^0$
f. $SO_4{}^{2-} \rightarrow SO_3{}^{2-}$
g. $2Cr^{3+} \rightarrow Cr_2O_7{}^{2-}$
h. $Mn^{2+} \rightarrow MnO_4{}^-$
i. $ClO_4{}^- \rightarrow Cl^-$
j. $SO_4{}^{2-} \rightarrow S^0$

12.5 BALANCING OXIDATION-REDUCTION EQUATIONS

Many redox reactions that are normally encountered by chemists are not as simple as the example of zinc and copper (II) ions. There are two common methods suited to the task of completing and balancing redox reactions: The **oxidation number method** for molecular equations, and the **ion-electron method** for ionic equations. The first method is easier and therefore a good starting point, while the latter method is more useful because it omits "spectator" ions.

12.6 THE OXIDATION NUMBER METHOD

It is experimentally established that potassium permanganate reacts with iron (II) sulfate in an acid environment (sulfuric acid is suitable) to produce manganese (II) sulfate, iron (III) sulfate, potassium sulfate, and water. The equation

$$KMnO_4 + FeSO_4 + H_2SO_4 \longrightarrow MnSO_4 + K_2SO_4 + Fe_2(SO_4)_3 + H_2O$$

may be described as "molecular," since the spectator ions are included.

Redox reactions should *not* be balanced by inspection. The numerical coefficients, which represent the mole ratios, are often so varied that a "trial and error" method often results in acute frustration.

One method is to look for common oxidizing and reducing agents. $KMnO_4$ is quickly recognized as an oxidizing agent, but the iron (II) in $FeSO_4$ is not so easily identified. There is an alternative method based on three clues.

Clues Used to Identify Redox Participants

1. When an element is in the *middle* of a ternary (3 element) compound on one side of the equation but not on the other side.

Example:

$KMnO_4$ — in the middle of a ternary compound

$MnSO_4$ — not in the middle

2. When an element is in a compound on one side of the equation but is in the "free" (uncombined) state on the other side. Example:

$CuSO_4$ — combined with sulfate ion

Cu — free (uncombined)

3. When a metal ion exhibits one oxidation number on one side of the equation and a different oxidation number on the other side. Example:

$FeSO_4$ — iron (II)

$Fe_2(SO_4)_3$ — iron (III)

Using the original equation, the clues may be applied to find the elements undergoing oxidation and reduction:

$$K(Mn)O_4 + (Fe)SO_4 + H_2SO_4 \longrightarrow (Mn)SO_4 + (Fe)_2(SO_4)_3 + K_2SO_4 + H_2O$$

iron (II) ⟶ iron (III)
change in oxidation number (clue #3)
middle of ternary compound ⟶ not in the middle (clue #1)

The next step is to write the oxidation-reduction **half-reactions** apart from the equation, determine the electron loss and gain, balance the half-reactions, and, finally, replace the data in the equation. The electron loss and gain require preliminary calculations.

I. Calculate the oxidation number of manganese in the compound $KMnO_4$.

a. The formula indicates that the oxidation number of potassium is 1+:

$$K^+, \quad MnO_4^-$$

(from the table of common oxidation numbers)

b. The oxidation number of the oxygen is 2−. There are a total of 4 oxygen atoms. Therefore, the *total* oxidation value for the oxygen is minus eight. 4 times (2−) = 8−:

$$\overset{1+}{K_1}, \quad \overset{8-}{MnO_4^{2-}}$$

c. Since the sum of the oxidation numbers of any compound must equal zero, the value for the Mn in the formula has to be 7+:

$$\begin{array}{ccc} 1+ & \textcircled{7+} & 8- \\ K_1^{+}, & Mn & O_4^{2-} \end{array}$$

$$1+ \text{ and } 7+ \text{ and } 8- = \text{zero.}$$

II. The oxidation number of manganese in $MnSO_4$ is 2+, and this is clearly indicated by the name of the compound, manganese (II) sulfate:

$$Mn^{2+}, \quad (SO_4)^{2-}$$

III. Iron (II) sulfate and iron (III) sulfate also indicate the oxidation numbers by their names.

IV. Write the half-reactions.

a. Mn^{7+} is reduced to Mn^{2+}. This requires a *gain* of five electrons (negative charges):

$$Mn^{++\not{+}\not{+}\not{+}\not{+}\not{+}} \longrightarrow Mn^{++}$$

5+ and 5− cancel out

$$Mn^{7+} + 5e^- \longrightarrow Mn^{2+}$$

transpose the $+5e^-$

$$Mn^{7+} \longrightarrow Mn^{2+} - 5e^- \text{ (reduction)}$$

b. Fe^{2+} is oxidized to Fe^{3+}, which means a *loss* of one electron per unit. Since two units of Fe^{3+} are produced, two units of Fe^{2+} must be used. Two units of Fe^{2+} undergoing oxidation will require a loss of two electrons:

$$2Fe^{2+} \longrightarrow Fe_2^{3+} + 2e^- \text{ (oxidation)}$$

V. Write the two half-reactions together for addition and balancing:

$$Mn^{7+} \longrightarrow Mn^{2+} - 5e^-$$

$$2Fe^{2+} \longrightarrow Fe_2^{3+} + 2e^-$$

The law of conservation requires a balanced loss and gain of electrons. To accomplish this, the half-reactions must be multiplied through by the smallest possible number that will equalize the number of electrons gained and lost. This means that 10 moles of Fe^{2+} will lose 10 moles of electrons to produce a total of 10 moles of Fe^{3+}. Two moles of Mn^{7+} will gain 10 moles of electrons to produce 10 moles of Mn^{2+}:

$$\begin{array}{lrl} & 2Mn^{7+} \longrightarrow 2Mn^{2+} & - \cancel{10}e^- \\ \text{add} & 10Fe^{2+} \longrightarrow 5Fe_2{}^{3+} & + \cancel{10}e^- \\ \hline & \mathbf{2}Mn^{7+} + \mathbf{10}Fe^{2+} \longrightarrow \mathbf{2}Mn^{2+} & + \mathbf{5}Fe_2{}^{3+} \\ & 2 \text{ moles} + 10 \text{ moles} \longrightarrow 2 \text{ moles} & + 5 \text{ moles} \end{array}$$

VI. Replace the coefficients in the original equation:

$$\mathbf{2}KMnO_4 + \mathbf{10}FeSO_4 + H_2SO_4 \longrightarrow \mathbf{2}MnSO_4 + \mathbf{5}Fe_2(SO_4)_3 + K_2SO_4 + H_2O$$

VII. The completion of the balancing is done by inspection. The following order of inspection specified in Chapter 8 is again recommended:

a. Balance the metals.
b. Balance the nonmetals, including the polyatomic ions.
c. Balance the hydrogen.
d. Balance the oxygen.

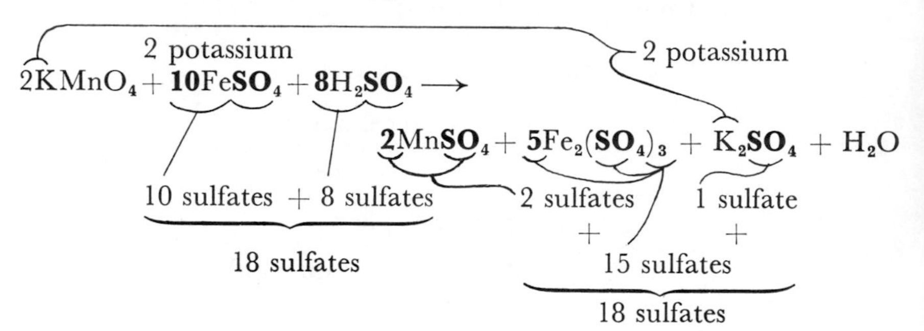

a. $8H_2SO_4$ gives a total of 16 hydrogens on the left.
b. The final balancing requires $8H_2O$ to produce 16 hydrogens on the right.
c. The total number of oxygens is also balanced at this point.
d. The balanced equation is

$$2KMnO_4 + 10FeSO_4 + 8H_2SO_4 \longrightarrow 2MnSO_4 + 5Fe_2(SO_4)_3 + K_2SO_4 + 8H_2O$$

Another example, more concisely developed, should reinforce the previous rules and principles.

EXAMPLE 12.5

Balance the following equation:

$$MnO_2 + HCl \longrightarrow MnCl_2 + Cl_2 + H_2O$$

1. The redox participants are identified.
   a. Chloride ion changes to free chlorine gas (clue #2).
   b. Manganese changes from Mn^{4+} in MnO_2 (manganese (IV) oxide) to Mn^{2+} in $MnCl_2$ (manganese (II) chloride) (clue #1).

2. Write the half-reactions.
   a. Balance the chlorine:

$$2Cl^- \longrightarrow Cl_2^0 + 2e^-$$

   Going from 1 – *up* to zero is oxidation. One mole of chloride ion loses one mole of electrons. Two moles lose two moles of electrons.

   b. One mole of manganese (IV) ion gains two moles of electrons in decreasing its oxidation number to manganese (II):

$$Mn^{4+} \longrightarrow Mn^{2+} - 2e^-$$

3. Combine and add the half-reactions, since the loss and gain of electrons are equal.

$$\begin{array}{lr} & 2Cl^- \longrightarrow Cl_2^0 + \cancel{2e^-} \quad \text{(oxidation)} \\ \text{add} & Mn^{4+} \longrightarrow Mn^{2+} - \cancel{2e^-} \quad \text{(reduction)} \\ \hline & Mn^{4+} + \mathbf{2}Cl^- \longrightarrow Cl_2^0 + Mn^{2+} \end{array}$$

4. Replace the coefficients in the original equation:

$$MnO_2 + \mathbf{2}HCl \longrightarrow MnCl_2 + Cl_2 + H_2O$$

5. Balance the equation by adjusting the number of chlorine units:

$$MnO_2 + \underbrace{4HCl}_{\text{4 chlorines}} \longrightarrow \underbrace{\underbrace{MnCl_2}_{\text{2 chlorines}} + \underbrace{Cl_2}_{\text{2 chlorines}}}_{\text{4 chlorines}} + H_2O$$

6. Balance the hydrogen and check the total number of oxygens:

$$\overbrace{MnO_2}^{\text{2 oxygen}} + \underbrace{4HCl}_{\text{4 hydrogen}} \longrightarrow MnCl_2 + Cl_2 + \overbrace{\underbrace{2H_2O}_{\text{4 hydrogen}}}^{\text{2 oxygen}}$$

   The equation is balanced.

EXAMPLE 12.6

Balance the following equation:

$$BiCl_3 + Na_2SnO_2 + NaOH \longrightarrow Bi + Na_2SnO_3 + NaCl + H_2O$$

1. Identify the redox participants.
   a. Bismuth (III) ion changes to free bismuth (clue #2).
   b. Tin changes from $Sn^{2+}(Na_2SnO_2)$ to $Sn^{4+}(Na_2SnO_3)$ (clue #3).
2. Write the half-reactions.
   a. Balance the bismuth:

$$Bi^{3+} \longrightarrow Bi^0$$

The oxidation number decreases from 3+ down to 0. Three moles of electrons are gained in the reduction:

$$Bi^{3+} \longrightarrow Bi^0 - 3e^-$$

   b. One mole of Sn^{2+} loses 2 moles of electrons in being oxidized to 1 mole of Sn^{4+}:

$$Sn^{2+} \longrightarrow Sn^{4+} + 2e^-$$

3. Combine the half-reactions, balance the loss and gain of electrons, and add.
   a. $2(Bi^{3+} \rightarrow Bi^0 - 3e^-)$
      $3(Sn^{2+} \rightarrow Sn^{4+} + 2e^-)$
   b. $2Bi^{3+} \rightarrow 2Bi^0 - \cancel{6e^-}$
      $3Sn^{2+} \rightarrow 3Sn^{4+} + \cancel{6e^-}$

$$\mathbf{2}Bi^{3+} + \mathbf{3}Sn^{2+} \rightarrow \mathbf{2}Bi^0 + \mathbf{3}Sn^{4+}$$

4. Replace the coefficients in the original equation:

$$2BiCl_3 + 3Na_2SnO_2 + NaOH \longrightarrow 2Bi + 3Na_2SnO_3 + NaCl + H_2O$$

5. Check the sodium and observe that there are *initially* 7 units on each side.
6. The 6 chlorines on the left must be balanced by 6 NaCl's on the right. This requires a readjustment of the sodium:

$$2BiCl_3 + 3Na_2SnO_2 + \mathbf{6}NaOH \longrightarrow 2Bi + 3Na_2SnO_3 + \mathbf{6}NaCl + H_2O$$

7. Six units of hydrogen on the left (6NaO**H**) require 3 moles of H_2O on the right:

$$2BiCl_3 + 3Na_2SnO_2 + 6NaOH \longrightarrow 2Bi + 3Na_2SnO_3 + 6NaCl + 3H_2O$$

8. Check the oxygen and find that there are 12 units on each side. The equation is complete.

Summary of Steps in the Oxidation Number Method of Balancing Molecular Equations

1. Identify the elements involved in the redox activity by using the three clues.
2. Write the half-reactions, showing the loss and gain of electrons.
3. Balance the number of units undergoing oxidation or reduction if subscripts are in the formulas.
4. Add the balanced half-reactions and transfer the coefficients to the original equation.
5. Balance the metals, nonmetals, and polyatomic ions, and then hydrogen, in that order.
6. Check the oxygens for final balancing.

Exercise 12.5

Balance the following equations by the oxidation number method:

a. $Bi_2S_3 + HNO_3 \rightarrow Bi(NO_3)_3 + S + NO + H_2O$
b. $MnO + PbO_2 + HNO_3 \rightarrow HMnO_4 + Pb(NO_3)_2 + H_2O$
c. $K_2Cr_2O_7 + HCl + FeCl_2 \rightarrow CrCl_3 + FeCl_3 + KCl + H_2O$
d. $PbCrO_4 + KI + HCl \rightarrow PbCl_2 + CrCl_3 + KCl + I_2 + H_2O$
e. $AuCl_3 + H_2C_2O_4 \rightarrow Au_2 + HCl + CO_2$
f. $KMnO_4 + HBr \rightarrow MnBr_2 + KBr + Br_2 + H_2O$
g. $(NH_4)_2Cr_2O_7 \rightarrow N_2 + Cr_2O_3 + H_2O$
h. $KIO_3 + HCl + H_2SO_3 \rightarrow ICl + KCl + H_2SO_4 + H_2O$
i. $Cu_2S + HNO_3 \rightarrow Cu(NO_3)_2 + NO_2 + S + H_2O$
j. $H_2S + Br_2 + H_2O \rightarrow HBr + H_2SO_4$
k. $HgS + HNO_3 + HCl \rightarrow HgCl_2 + S + NO + H_2O$
l. $KMnO_4 + HCl \rightarrow MnCl_2 + KCl + Cl_2 + H_2O$
m. $NaI + H_2SO_4 \rightarrow Na_2SO_4 + I_2 + H_2S + H_2O$
n. $K_2Cr_2O_7 + H_2S + HCl \rightarrow CrCl_3 + S + KCl + H_2O$
o. $I_2 + Cl_2 + H_2O \rightarrow HCl + HIO_3$

12.7 THE ION-ELECTRON METHOD

The ion-electron method is concerned with conservation of charge rather than with individual oxidation numbers of elements. The same reaction previously investigated looks different when stripped of "spectator" ions:

$$KMnO_4 + FeSO_4 + H_2SO_4 \longrightarrow MnSO_4 + Fe_2(SO_4)_3 + K_2SO_4 + H_2O$$

becomes

$$MnO_4^-(aq) + Fe^{2+}(aq) + H^+(aq) \longrightarrow Mn^{2+}(aq) + Fe^{3+}(aq) + H_2O$$

The oxidizing and reducing agents and their products often can be immediately identified. As familiarity is gained, an equation can be completed when only the reactants are given.

1. MnO_4^- (permanganate ion) in acid medium ($H^+(aq)$) is identified as an oxidizing agent that becomes reduced to manganese (II) ion (Table 12–2):

$$MnO_4^- \longrightarrow Mn^{2+}$$

2. The 4 units of oxygen on the left are balanced by the oxygen contained in 4 units of water on the right:

$$MnO_4^- \longrightarrow Mn^{2+} + 4H_2O$$

3. Since the reaction occurs in acid medium, the 8 hydrogens on the right are balanced by $8H^+(aq)$ (hydrogen ion in water solution) on the left:

$$MnO_4^- + 8H^+ \longrightarrow Mn^{2+} + 4H_2O$$

4. The total ionic charge on the left is 7+ (one negative permanganate unit added to 8 positive hydrogen ion units). The total ionic charge on the right is 2+, from the single unit of Mn^{2+}:

$$7+ \longrightarrow 2+$$

This imbalance is adjusted by adding 5 electrons to the left or, by convention, subtracting 5 electrons from the right:

$$MnO_4^- + 8H^+ \longrightarrow Mn^{2+} + 4H_2O - 5e^-$$

5. Fe^{2+} (iron (II) ion) is a reducing agent that becomes oxidized to Fe^{3+} (iron (III) ion):

$$Fe^{2+} \longrightarrow Fe^{3+}$$

6. Balancing the ionic charge is accomplished by adding one electron to the right side:

$$Fe^{2+} \longrightarrow Fe^{3+} + e^-$$

7. Add the two half-reactions after balancing the loss and gain of electrons by multiplying the iron half-reaction by 5:

$$\begin{array}{l} MnO_4^- + 8H^+ \longrightarrow Mn^{2+} + 4H_2O - 5e^- \\ 5Fe^{2+} \longrightarrow 5Fe^{3+} + 5e^- \\ \hline MnO_4^- + 5Fe^{2+} + 8H^+ \longrightarrow Mn^{2+} + 5Fe^{3+} + 4H_2O \end{array}$$

In conclusion, it can be seen that the ratio of 5 moles of Fe^{2+} reacting with 1 mole of MnO_4^- is exactly the same ratio that the stoichiometry of the molecular equation exhibits:

$$\mathbf{2}KMnO_4 + \mathbf{10}FeSO_4$$

$$2 \quad : \quad 10 \quad = 1:5$$

For future convenience Table 12–2 lists some common oxidizing and reducing agents.

Another example of this method may be seen in the reaction where ethanol (ethyl alcohol) is oxidized to acetic acid (ethanoic acid). In more familiar terms, this is what happens when wine turns to vinegar.

TABLE 12–2

Common Oxidizing Agents

Oxidizing Agents	Number of Electrons Gained	Common Products
$F_2(g)$	2	$2F^-(aq)$
$MnO_4^-(aq)$ (in acid)	5	$Mn^{2+}(aq)$, $MnO_2(s)$
$ClO_4^-(aq)$ (in acid)	8	$Cl^-(aq)$, $Cl_2(g)$
$Cl_2(g)$	2	$2Cl^-(aq)$
$Cr_2O_7^{2-}(aq)$ (in acid)	6	$2Cr^{2+}(aq)$
$O_2(g)$ (in acid)	4	$2H_2O(\ell)$
$Br_2(\ell)$	2	$2Br^-(aq)$
$NO_3^-(aq)$ (in acid)	3	$NO(g)$, $NO_2(g)$

Common Reducing Agents

Reducing Agent	Number of Electrons Lost	Product
$Li(s)$	1	$Li^+(aq)$
$Ca(s)$	2	$Ca^{2+}(aq)$
$Na(s)$	1	$Na^+(aq)$
$Zn(s)$	2	$Zn^{2+}(aq)$
$H_2(s)$	2	$2H^+(aq)$
$Cu(s)$	2	$Cu^{2+}(aq)$

The laboratory reaction is observed when potassium dichromate (oxidizing agent) is added to ethanol in an acidic medium provided by sulfuric acid.

EXAMPLE 12.7

Balance the following ionic equation:

$$\underset{\text{ethanol}}{C_2H_5OH} + Cr_2O_7^{2-} + H^+ \longrightarrow \underset{\text{acetic acid}}{C_2H_4O_2} + Cr^{3+} + H_2O$$

1. The dichromate ion is the oxidizing agent which becomes reduced to chromium (III) ion:

$$Cr_2O_7^{2-} \longrightarrow Cr^{3+}$$

2. The 7 units of oxygen on the left are balanced by 7 units of water on the right:

$$Cr_2O_7^{2-} \longrightarrow Cr^{3+} + 7H_2O$$

3. Since the reaction occurs in acid medium, the 14 hydrogens on the right are balanced by $14H^+(aq)$ on the left:

$$Cr_2O_7^{2-} + 14H^+ \longrightarrow Cr^{3+} + 7H_2O$$

4. The 2 chromium units on the left are balanced by placing the coefficient, 2, in front of Cr^{3+}:

$$Cr_2O_7^{2-} + 14H^+ \longrightarrow 2Cr^{3+} + 7H_2O$$

5. The total ionic charge on the left is 12+ (2 negative dichromate and 14+ hydrogen ion units). The total ionic charge on the right is 6+, from the two units of chromium (III) ion:

$$12+ \longrightarrow 6+$$

The imbalance is adjusted by subtracting 6 electrons from the right side:

$$Cr_2O_7^{2-} + 14H^+ \longrightarrow 2Cr^{3+} + 7H_2O - 6e^-$$

6. The oxidation of ethanol to acetic acid half-reaction is balanced by adding water to the left to obtain the extra oxygen atom needed, and by adding hydrogen ion on the right:

$$C_2H_5OH + H_2O \longrightarrow C_2H_4O_2 + 4H^+$$

7. The total ionic charge on the left is zero and the total of 4+ (due to $4H^+$) is balanced by adding 4 electrons to the right side:

$$0 \longrightarrow 4+ \;+ 4e^-$$
$$0 \longrightarrow 0$$

8. Balance the loss and gain of electrons and add the two half-reactions:

$$\text{multiply by } 2(Cr_2O_7^{2-} + 14H^+ \longrightarrow 2Cr^{3+} + 7H_2O - 6e^-)$$
$$\text{multiply by } 3(C_2H_5OH + H_2O \longrightarrow C_2H_4O_2 + 4H^+ + 4e^-)$$
$$2Cr_2O_7^{2-} + 28H^+ \longrightarrow 4Cr^{3+} + 14H_2O - \cancel{12e^-}$$
$$3C_2H_5OH + 3H_2O \longrightarrow 3C_2H_4O_2 + 12H^+ + \cancel{12e^-}$$
$$2Cr_2O_7^{2-} + 3C_2H_5OH + 28H^+ + 3H_2O \longrightarrow 4Cr^{3+} + 3C_2H_4O_2 + 12H^+ + 14H_2O$$

9. The equation should be simplified by subtracting $3H_2O$ and $12H^+$ from both sides:

$$2Cr_2O_7^{2-} + 3C_2H_5OH + 28H^+ + 3H_2\cancel{O}$$
$$- 12H^+ - \cancel{3}H_2O$$
$$\longrightarrow 4Cr^{3+} + 3C_2H_4O_2 + 12\cancel{H^+} + 14H_2O$$
$$-\cancel{12}H^+ - 3H_2O$$
$$2Cr_2O_7^{2-} + 3C_2H_5OH + 16H^+ \longrightarrow 4Cr^{3+} + 3C_2H_4O_2 + 11H_2O$$

The previous examples of oxidation-reduction were in acidic media. Redox in a basic medium involves some modification.

EXAMPLE 12.8

Permanganate ion reacts with cyanide ion (a strong base, since it is a vigorous proton acceptor) to produce manganese (IV) oxide and cyanate ion. Because of the CN^-, the reaction occurs in a basic environment (otherwise, it would evolve dangerously poisonous HCN gas):

$$MnO_4^- + CN^- \longrightarrow MnO_2 + OCN^-$$

1. The MnO_4^- undergoes reduction to MnO_2:

$$MnO_4^- \longrightarrow MnO_2$$

2. The 4 units of oxygen on the left compared to the 2 units on the right require water for balancing:

$$MnO_4^- \longrightarrow MnO_2 + 2H_2O$$

3. The 4 hydrogens on the right are balanced by $4H^+$. Just the same as an acidic medium reaction!

$$MnO_4^- + 4H^+ \longrightarrow MnO_2 + 2H_2O$$

4. *Now*, add $4OH^-$ units to both sides of the equation in order to convert the $4H^+$ water, since H^+ cannot remain as a reactant in a basic environment:

$$\begin{array}{lcl} MnO_4^- + 4H^+ & \longrightarrow & MnO_2 + 2H_2O \\ \quad\;\; + 4OH^- & & \qquad\qquad\;\; + 4OH^- \\ \hline MnO_4^- + 4H_2O & \longrightarrow & MnO_2 + 2H_2O + 4OH^- \end{array}$$

5. Simplify the half-reaction by subtracting $2H_2O$ from both sides:

$$\begin{array}{lcl} MnO_4^- + 4H_2O & \longrightarrow & MnO_2 + \cancel{2H_2O} + 4OH^- \\ \quad\;\; - 2H_2O & & \qquad\;\; - \cancel{2H_2O} \\ \hline MnO_4^- + 2H_2O & \longrightarrow & MnO_2 + 4OH^- \end{array}$$

6. Balance the ionic charge. Since the total on the left is 1 negative and the total on the right side is 4 negatives, subtract 3 electrons from the right side:

$$1- \longrightarrow (4-) - 3e^-$$
$$1- \longrightarrow 1-$$
$$MnO_4^- + 2H_2O \longrightarrow MnO_2 + 4OH^- - 3e^-$$

7. The oxidation of cyanide ion to cyanate ion follows the same pattern:

$$CN^- \longrightarrow OCN^-$$

8. Add water on the left to balance the oxygen and add H^+ on the right to balance the hydrogen in H_2O:

$$CN^- + H_2O \longrightarrow OCN^- + 2H^+$$

9. Add $2OH^-$ to both sides of the equation in order to convert the H^+ to water:

$$CN^- + H_2O + 2OH^- \longrightarrow OCN^- + 2H_2O$$

10. Simplify the half-reaction by subtracting one H_2O unit from both sides of the equation:

$$CN^- + 2OH^- \longrightarrow OCN^- + H_2O$$

11. Balance the ionic charge by adding 2 electrons to the right side:

$$\underbrace{CN^- + 2OH^-}_{\text{3 negatives}} \longrightarrow \underbrace{OCN^-}_{\text{1 negative}} + \underbrace{2e^-}_{\text{2 negatives}}$$

12. The two half-reactions are then balanced with regard to loss and gain of electrons. Finally, they are added:

$$\text{multiply by } 2(MnO_4^- + 2H_2O \longrightarrow MnO_2 + 4OH^- - 3e^-)$$
$$\text{multiply by } 3(CN^- + 2OH^- \longrightarrow OCN^- + H_2O + 2e^-)$$
$$2MnO_4^- + 4H_2O \longrightarrow 2MnO_2 + 8OH^- - \cancel{6e^-}$$
$$3CN^- + 6OH^- \longrightarrow 3OCN^- + 3H_2O + \cancel{6e^-}$$
$$2MnO_4^- + 3CN^- + 6OH^- + 4H_2O \longrightarrow 2MnO_2 + 3OCN^- + 8OH^- + 3H_2O$$

13. Simplify the final equation by subtracting $6OH^-$ units and $3H_2O$ units from both sides:

$$\begin{array}{lcl} 2MnO_4^- + 3CN^- + 6O\cancel{H}^- + 4H_2O & \longrightarrow & 2MnO_2 + 3OCN^- + 8OH^- + 3H_2\cancel{O} \\ \quad - \cancel{6}OH^- - 3H_2O & & \quad - 6OH^- - \cancel{3H}_2O \end{array}$$
$$2MnO_4^- + 3CN^- + H_2O \longrightarrow 2MnO_2 + 3OCN^- + 2OH^-$$

EXAMPLE 12.9

Balance the following reaction which occurs in *basic* medium:

$$CrO_4^{2-} + HSnO_2^- \longrightarrow HSnO_3^- + CrO_2^-$$

1. Balance the reduction half-reaction:

$$CrO_4^{2-} \longrightarrow CrO_2^-$$

a. Add $2H_2O$ to balance the oxygen:

$$CrO_4^{2-} \longrightarrow CrO_2^- + 2H_2O$$

b. Add $4H^+$ to balance the hydrogen:

$$4H^+ + CrO_4^{2-} \longrightarrow CrO_2^- + 2H_2O$$

c. Add $4OH^-$ to both sides of the equation in order to convert H^+ to H_2O:

$$\begin{array}{lcl} 4H^+ + CrO_4^{2-} & \longrightarrow & CrO_2^- + 2H_2O \\ + 4OH^- & & \qquad + 4OH^- \end{array}$$
$$4H_2O + CrO_4^{2-} \longrightarrow CrO_2^- + 2H_2O + 4OH^-$$

d. Simplify the half-reaction by subtracting $2H_2O$ from both sides:

$$2H_2O + CrO_4^{2-} \longrightarrow CrO_2^- + 4OH^-$$

e. Balance the ionic charge:

$$2- \longrightarrow 5-, \text{ subtract } 3e^-$$

$$2H_2O + CrO_4^{2-} \longrightarrow CrO_2^- + 4OH^- - 3e^-$$

2. Balance the oxidation half-reaction:

$$HSnO_2^- \longrightarrow HSnO_3^-$$

a. Add H_2O to balance the oxygen:

$$H_2O + HSnO_2^- \longrightarrow HSnO_3^-$$

b. Add $2H^+$ to the right side to balance the hydrogen:

$$H_2O + HSnO_2^- \longrightarrow HSnO_3^- + 2H^+$$

c. Add $2OH^-$ to the equation to convert H^+ to H_2O:

$$2OH^- + H_2O + HSnO_2^- \longrightarrow HSnO_3^- + 2H_2O$$

d. Subtract H_2O from the equation, then balance the ionic charge:

$$3- \longrightarrow 1-, \text{ add } 2e^-$$

$$2OH^- + HSnO_2^- \longrightarrow HSnO_3^- + H_2O + 2e^-$$

3. Add the two half-reactions after balancing the loss and gain of electrons:

$$\text{multiply by } 2(2H_2O + CrO_4^{2-} \longrightarrow CrO_2^- + 4OH^- - 3e^-)$$

$$\text{multiply by } 3(2OH^- + HSnO_2^- \longrightarrow HSnO_3^- + H_2O + 2e^-)$$

$$\text{(add)}\begin{cases} 4H_2O + 2CrO_4^{2-} \longrightarrow 2CrO_2^- + 8OH^- - \cancel{6e^-} \\ 6OH^- + 3HSnO_2^- \longrightarrow 3HSnO_3^- + 3H_2O + \cancel{6e^-} \end{cases}$$

$$4H_2O + 6OH^- + 2CrO_4^{2-} + 3HSnO_2^- \longrightarrow 2CrO_2^- + 3HSnO_3^- + 8OH^- + 3H_2O$$

4. Simplify the equation by reducing the number of H_2O and

OH^-. Subtract $3H_2O$ and $6OH^-$ from the equation. The final equation exhibits conservation of mass and charge:

$$2CrO_4^{2-} + 3HSnO_2^- + H_2O \longrightarrow 2CrO_2^- + 3HSnO_3^- + 2OH^-$$

Summary

The following checks should be applied to the balancing of redox equations when the ion-electron method is used:

1. The actual number of each species of atom must be balanced. H^+ and H_2O will have to be added in acidic medium, and OH^- and H_2O will have to be added in basic medium.
2. Electrons must be added or subtracted on the right side of the half-reactions to achieve electrical neutrality.
3. The loss and gain of electrons must be balanced before the half-reactions are added.
4. The final equation must be simplified by removing excess H^+, OH^-, and H_2O.

Exercise 12.6

Balance the following equations by the ion-electron method:

a. $Ag + NO_3^- \rightarrow Ag^+ + NO$ (acid solution)
b. $Mn^{2+} + Br_2 \rightarrow MnO_2 + Br^-$ (basic solution)
c. $Cr(OH)_4^- + H_2O_2 \rightarrow CrO_4^{2-}$ (basic solution)
d. $Cr_2O_7^{2-} + I^- \rightarrow Cr^{3+} + I_2$ (acid solution)
e. $PbO_2 + Cl^- \rightarrow Pb^{2+} + Cl_2$ (acid solution)
f. $MnO_4^- + H_2S \rightarrow Mn^{2+} + S$ (acid solution)
g. $Cr_2O_7^{2-} + HNO_2 \rightarrow Cr^{3+} + NO_3^-$ (acid solution)
h. $HS^- + MnO_4^- \rightarrow HSO^- + MnO_2$ (basic solution)
i. $MnO_4^- + NO_2^- \rightarrow MnO_2 + NO_3^-$ (basic solution)
j. $H_2C_2O_4 + MnO_2 \rightarrow Mn^{2+} + CO_2$ (acid solution)
k. $MnO_4^- + C_2H_4 \rightarrow Mn^{2+} + CO_2$ (acid solution)
l. $Br_2 \rightarrow BrO_3^- + Br^-$ (basic solution)
m. $MnO_2 + Cl^- \rightarrow Mn^{2+} + Cl_2$ (acid solution)
n. $Cr_2O_7^{2-} + Fe^{2+} \rightarrow Cr^{3+} + Fe^{3+}$ (acid solution)
o. $H_2S + I_2 \rightarrow S + I^-$ (acid solution)

12.8 REDOX STOICHIOMETRY

When an oxidation-reduction reaction is known to be a complete conversion of reactants to products (stoichiometric), the methods for performing useful calculations are essentially the same as those employed for nonredox reactions.

Gravimetric Problems

Whenever weights of solids are involved, the measurements required are described as gravimetric.

EXAMPLE 12.10

Calculate the weight of sodium iodide required for the production of 5.0 g of iodine in the following reaction:

$$NaI + H_2SO_4 \longrightarrow Na_2SO_4 + I_2 + H_2S + H_2O$$

1. Balance the equation to obtain the stoichiometry:

$$NaI + H_2SO_4 \longrightarrow Na_2SO_4 + I^2 + H_2S + H_2O$$

a. Sulfur and iodine fit the clues used to identify the redox participants:

$$2I^- \longrightarrow I_2^0 + 2e^- \qquad \text{(oxidation)}$$

$$H_2^+\ SO_4^{2-} = 2+ \ (?)\ 8- = 0 \qquad (6+)$$

$$H_2S_1O_4^{2-} = S$$

$$S^{6+} \longrightarrow S^{2-} - 8e^- \qquad \text{(reduction)}$$

b. Balance the half-reactions and replace the coefficients in the equation:

$$8I^- \longrightarrow 4I_2^0 + \cancel{8e^-}$$

$$S^{6+} \longrightarrow S^{2-} - \cancel{8e^-}$$

$$8I^- + S^{6+} \longrightarrow 4I_2^0 + S^{2-}$$

$$8NaI + H_2SO_4 \longrightarrow Na_2SO_4 + 4I_2 + H_2S + H_2O$$

c. Complete the balancing by inspection:

$$\mathbf{8}NaI + 5H_2SO_4 \longrightarrow 4Na_2SO_4 + \mathbf{4}I_2 + H_2S + 4H_2O$$

8 moles —produce→ 4 moles

or, more simply

2 moles of NaI will produce 1 mole of I_2

2. Organize the data:

$$2 \text{ moles NaI} = 2 \times \text{mole weight} = 2 \times 150 \text{ g} = 300 \text{ g}$$

$$\underset{23\,+\,127}{\text{NaI}} = 150 \text{ g mole}^{-1}$$

$$1 \text{ mole } I_2 = 2 \times \text{atomic wt} = 2 \times 127 \text{ g} = 254 \text{ g mole}^{-1}$$

3. Solve by the factor-label method:

$$\left(\frac{5.0 \text{ g } I_2}{254 \text{ g mol}^{-1}}\right) \times \left(\frac{2 \text{ mols NaI}}{1 \text{ mol } I_2}\right) \times 150 \text{ g mol}^{-1} = \chi$$

$$\chi = 5.9 \text{ g of NaI}$$

EXAMPLE 12.11

How many grams of solid potassium dichromate must be added to oxalic acid in acid medium to produce 500 ml of carbon dioxide measured at STP? The partial equation is

$$H_2C_2O_4 + Cr_2O_7^{2-} \text{ (acidic)} \longrightarrow$$

1. Complete the reaction with the knowledge (or reference to a table of redox half-reactions) that

   $Cr_2O_7^{2-}$ becomes reduced to Cr^{3+} in acidic medium
   $H_2C_2O_4$ becomes oxidized to CO_2

2. Balance the half-reactions:

$$Cr_2O_7^{2-} + 14H^+ \longrightarrow 2Cr^{3+} + 7H_2O - 6e^-$$

Multiply by three:

$$3(H_2C_2O_4 \longrightarrow 2CO_2 + 2H^+ + 2e^-)$$

or

$$3H_2C_2O_4 \longrightarrow 6CO_2 + 6H^+ + 6e^-$$

Add:

$$Cr_2O_7^{2-} + 14H^+ \longrightarrow 2Cr^{3+} + 7H_2O - \cancel{6e^-}$$

$$3H_2C_2O_4 \longrightarrow 6CO_2 + 6H^+ + \cancel{6e^-}$$

$$Cr_2O_7^{2-} + 3H_2C_2O_4 + 8H^+ \longrightarrow 2Cr^{3+} + 6CO_2 + 7H_2O$$

3. The pertinent stoichiometry is

$$1 \text{ mole } Cr_2O_7^{2-} \text{—produces} \longrightarrow 6 \text{ moles } CO_2$$

or

$$1 \text{ mole } K_2Cr_2O_7 \text{—produces} \longrightarrow 6 \text{ moles } CO_2$$

4. Organize the data:

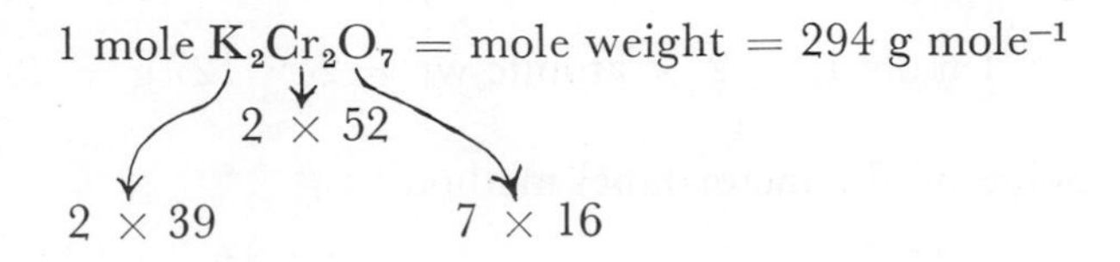

1 mole of gas occupies 22.4 liters at STP

500 ml of CO_2 = 0.5 liter

5. Solve by the factor-label method:

$$\left(\frac{0.5 \text{ liter } CO_2}{22.4 \text{ liter mol}^{-1}}\right) \times \left(\frac{1 \text{ mol } K_2Cr_2O_7}{6 \text{ mols } CO_2}\right) \times 294 \text{ g mol}^{-1} = \chi$$

$$\chi = 1.1 \text{ g of } K_2Cr_2O_7$$

Exercise 12.7

1. How many milligrams of sulfur can be obtained by reacting 50.0 mg of zinc with excess sulfuric acid? The skeleton equation is

$$Zn + H_2SO_4 \longrightarrow ZnSO_4 + S + H_2O$$

2. How many milliliters of NO_2 (*g*), at STP, will be liberated when 60.0 mg of copper is added to an excess of concentrated nitric acid? The skeleton equation is

$$Cu + HNO_3 \longrightarrow Cu(NO_3)_2 + NO_2 + H_2O$$

Volumetric Problems

In a manner analogous to acid-base reactions proceeding toward a completion called neutralization, redox reactions move toward an *end point* where electrons are no longer gained and lost. Two types of practical problems will be examined: **solution stoichiometry** and **redox titration.**

12.9 SOLUTION STOICHIOMETRY *

EXAMPLE 12.12

What is the molar concentration of nitric acid if 20 ml is required to completely oxidize 4.0 g of copper to copper (II) nitrate in the reaction

$$Cu + HNO_3 \longrightarrow Cu(NO_3)_2 + NO_2 + H_2O$$

1. The first and most critical step is to balance the equation:

$$Cu + HNO_3 \longrightarrow Cu(NO_3)_2 + NO_2 + H_2O$$

$$\frac{1+\ 5+\ 6- \ = 0}{\begin{matrix} 1+ & \boxed{5+} & 2- \\ H & N & O_3 \end{matrix}}$$

$$Cu^0 \longrightarrow Cu^{2+} + 2e^-$$
$$2N^{5+} \longrightarrow 2N^{4+} - 2e^-$$
$$Cu^0 + 2N^{5+} \longrightarrow Cu^{2+} + 2N^{4+}$$
$$Cu + 4HNO_3 \longrightarrow Cu(NO_3)_2 + 2NO_2 + 2H_2O$$
$$1 \text{ mole} + 4 \text{ moles} \longrightarrow \text{completion}$$

2. Organize the data:

$$1 \text{ mole Cu} = 63.5 \text{ g mole}^{-1}$$

$$\text{moles of } HNO_3(aq) = MV$$
$$(\text{molarity})(\text{volume in liters})$$

$$20 \text{ ml} = 0.02 \text{ liter}$$

3. Using the factor-label method

$$\left(\frac{4.0 \text{ g Cu}}{63.5 \text{ g mol}^{-1}}\right) \times \left(\frac{4 \text{ mols } HNO_3}{1 \text{ mol Cu}}\right) = \chi$$

$$\chi = 0.252 \text{ mole of } HNO_3$$

4. Finally, calculate the molarity:

$$M = \frac{n}{V} \quad \begin{matrix} \longleftarrow \text{moles} \\ \longleftarrow \text{volume in liters} \end{matrix}$$

$$M = \frac{0.252 \text{ mol}}{0.02 \text{ liter}} = 12.6 \text{ moles liter}^{-1}$$

Therefore, 20 ml of 12.6 M HNO_3 are needed.

EXAMPLE 12.13

What molar concentration of $MnO_4^-(aq)$ is used if 40.0 ml is needed to completely convert 25.0 ml of 0.2 M $Fe^{2+}(aq)$ to $Fe^{3+}(aq)$ in the reaction

$$MnO_4^- + 5Fe^{2+} + 8H^+ \longrightarrow Mn^{2+} + 5Fe^{3+} + 4H_2O$$

1. The stoichiometry states that 1 mole of MnO_4^- reacts with 5 moles of Fe^{2+}.
2. Calculate the actual number of moles of $Fe^{2+}(aq)$ used in the reaction:

$$n = MV$$

$$n = (0.2\ M)(0.025\ \ell)$$

$$n = 0.005 \text{ mole of } Fe^{2+}$$

3. Use the factor-label method to find the number of moles of $MnO_4^-(aq)$ that will combine with 0.005, or 5×10^{-3}, moles of Fe^{2+}:

$$(5 \times 10^{-2} \text{ mol } Fe^{2+}) \times \left(\frac{1 \text{ mol } MnO_4^-}{5 \text{ mols } Fe^{2+}}\right) = \chi$$

$$\chi = 1 \times 10^{-3} \text{ mole of } MnO_4^-(aq)$$

4. Find the molar concentration of $MnO_4^-(aq)$:

$$M = \frac{n}{v} = \frac{1 \times 10^{-3} \text{ mol}}{0.04 \text{ liter}} = 2.5 \times 10^{-2} \text{ mole } \ell^{-1}$$

$$MnO_4^-(aq) \text{ is } 0.025\ M$$

EXAMPLE 12.14

For the purpose of versatility, convert the 0.025 M concentration to normal concentration.

1. Recall that Chapter 9 (Solutions) developed the equation for interconverting molarity and normality:

$$N = (\text{total positive charge})(M)$$

or

$$N = (\text{number of replaceable } H^+ \text{ or } OH^-)(M)$$

2. Extending the original reasoning to the specific example of $MnO_4^-(aq)$, it is observed that 1 mole of MnO_4^- gains *five* moles of electrons in the redox reaction. Therefore, 1/5 mole of MnO_4^- gains one mole of electrons, which means that 1/5 mole is the equivalent weight; i.e., 1 equivalent = mass which reacts with 1 mole of electrons.
3. Since normality is based on equivalent weights, one molar MnO_4^- is equal to 5 normal MnO_4^-.
4. A new equation is obtained for interconverting molarity and normality in the case of oxidizing and reducing agents:

$$N = (\text{number of moles of electrons gained or lost per mole of reactant})(M)$$

5. The solution to the problem is

$$N = (0.025\ M)(5 \text{ moles of electrons per mole of } MnO_4^-)$$
$$\underline{N = 0.125}$$

Exercise 12.8

1. How many milliliters of 0.3 M nitric acid are required to produce 300.0 ml of NO gas (STP) according to the following skeleton equation?

$$PbS + HNO_3 \longrightarrow PbSO_4 + NO + H_2O$$

2. What is the molar concentration of $MnO_4^-(aq)$ if 75.0 ml are needed to completely oxidize 10.0 g of pure methanol (CH_3OH) to formaldehyde (CH_2O)? The skeleton equation is

$$MnO_4^- + CH_3OH + H^+ \longrightarrow MnO_2 + CH_2O + H_2O$$

3. Calculate the normal concentration of the MnO_4^- solution in the above problem.

12.10 REDOX TITRATION *

The principal advantage of using normal concentrations for oxidation-reduction titrations is that the stoichiometry is not required for determining unknown concentrations. Just as neutralization in acid-base titrations is described as the point when equivalent numbers of $H^+(aq)$ and $OH^-(aq)$ have formed water, redox titrations are concerned with balancing the number of equivalents of oxidizer and reducer. The titration equivalence point, in this case, occurs when there is no further electron transfer from the reducing agent to the oxidizing agent. The end point of a redox titration can sometimes be seen as a sudden and sharp color change.

Consider the $MnO_4^-(aq)$ and $Fe^{2+}(aq)$ redox system again. This time a laboratory titration will be described. A color change is employed in this system as an equivalence indicator. $MnO_4^-(aq)$ is a deep purple color. When $MnO_4^-(aq)$ has been completely reduced to $Mn^{2+}(aq)$, the purple color disappears. This marks the end point of the titration:

$$\underset{\text{purple}}{MnO_4^-(aq)} \longrightarrow \underset{\text{colorless}}{Mn^{2+}(aq)}$$

Example 12.15 describes a typical redox titration.

EXAMPLE 12.15

Pour 40.0 ml of a solution of an unknown concentration of $KMnO_4$ into a beaker, add a stirring bar and a few drops of concentrated H_2SO_4, and place the beaker on a magnetic stirrer. Fill a 50 ml buret with 0.20 N

$FeSO_4$ solution and add it carefully to the $KMnO_4$ solution. Stop immediately as the purple color disappears. Calculate the normality of the $KMnO_4$ from the data.

1. Set up the apparatus as shown in Figure 12–3.
2. Perform the experiment and record the results (Fig. 12–4).

$$\begin{aligned} \text{volume of } KMnO_4 &= 40.0 \text{ ml} \\ \text{volume of } FeSO_4 &= 25.0 \text{ ml} \\ \text{normality of } FeSO_4 &= 0.20 \text{ N} \\ \text{normality of } KMnO_4 &= ? \end{aligned}$$

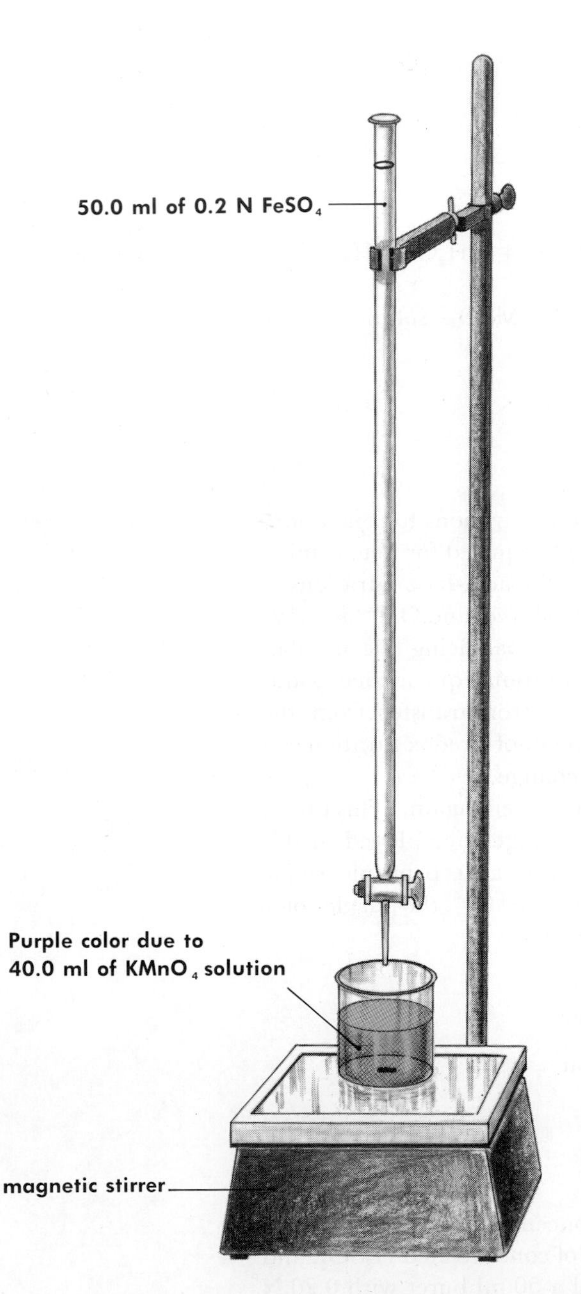

Figure 12–3 Redox titration assembly before starting.

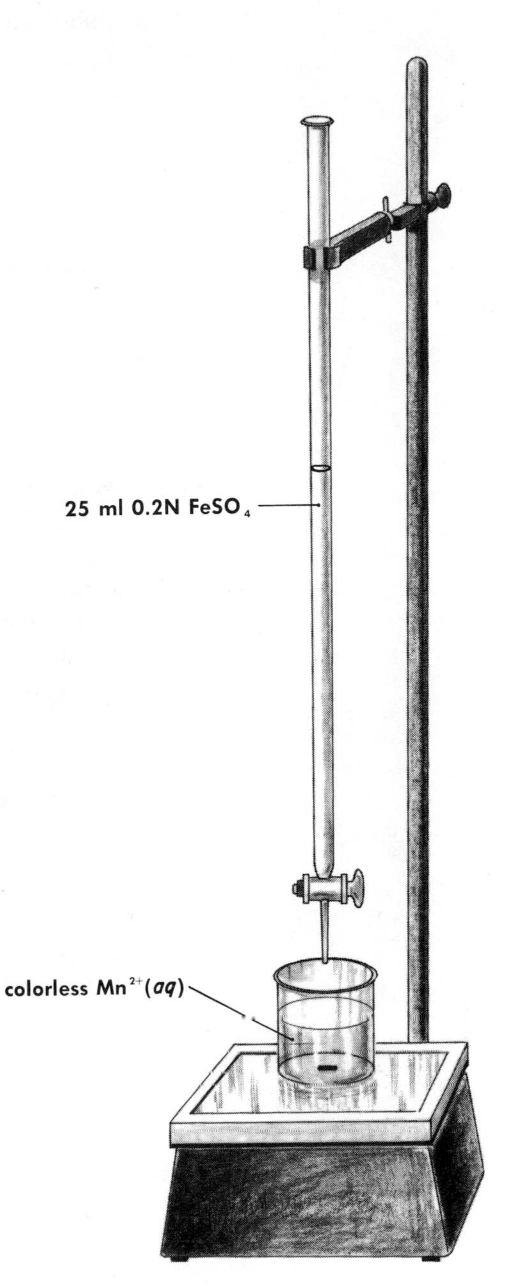

Figure 12–4 Redox tritration assembly at end point.

3. At the end point

the number of equivalents of oxidizing agent
= the number of equivalents of reducing agent

the number of equivalents = NV (liters)

the number of milliequivalents = NV (milliliters)

Therefore

$$NV_{(ox)} = NV_{(red)}$$

4. Substitute the data:

$$(40.0 \text{ ml}) \times (\text{N}) = (25.0 \text{ ml})(0.2 \text{ N})$$

$$\text{N} = \frac{25(0.2)}{40} = 0.125$$

$$\text{the } KMnO_4 \text{ is } \underline{0.125 \text{ N}}$$

This answer agrees with the result obtained in the solution stoichiometry approach.

EXAMPLE 12.16

A solution of oxalic acid contains 0.75 g of $H_2C_2O_4 \cdot 2H_2O$. In a titration, 20.0 ml of $KMnO_4(aq)$ was required to complete the redox reaction. Calculate the normal and molar concentration of the $KMnO_4$ solution.

1. Calculate the number of equivalents of oxalic acid from the data.

   a. $H_2C_2O_4 \cdot 2H_2O = 2 + 24 + 64 + 36 = 126 \text{ g mole}^{-1}$

   $2 \times 1 \quad 2 \times 12 \quad 4 \times 16 \quad 2 \times 18$

   b. The half-reaction

   $$H_2C_2O_4 \longrightarrow 2CO_2 + 2H^+ + 2e^-$$

   indicates 2 moles of electrons lost per mole of $H_2C_2O_4$ oxidized.

   c. Therefore, the equivalent weight is

   $$\text{g eq}^{-1} = \frac{\text{mole weight}}{\text{number of moles of electrons lost per mole of } H_2C_2O_4}$$

   $$\text{g eq}^{-1} = \frac{126 \text{ g } \cancel{\text{mol}^{-1}}}{2 \text{ eq } \cancel{\text{mol}^{-1}}} = 63 \text{ g eq}^{-1}$$

   d. The number of equivalents is calculated from the equation:

   $$\# \text{ eq} = \frac{\text{g}}{\text{g eq}^{-1}}$$

   $$\# \text{ eq} = \frac{0.75 \cancel{\text{g}}}{63 \cancel{\text{g}} \text{eq}^{-1}} = 0.012 \text{ equivalents}$$

2. In a titration, the number of equivalents of oxidizer equals the number of equivalents of reducer:

   $$\text{for } KMnO_4, \# \text{ eq} = \text{NV (liters)}$$

   $$\text{for } H_2C_2O_4 \cdot 2H_2O, \# \text{ eq} = 0.012$$

Therefore

$$NV_{(ox)} = 0.012_{(red)}$$

$$N = \frac{0.012\ \text{eq}}{0.020\ \text{liter}} = 0.60\ \text{eq liter}^{-1}$$

The $KMnO_4$ is 0.6 N.

3. The half-reaction for MnO_4^- reduction is

$$MnO_4^- + 8H^+ \longrightarrow Mn^{2+} + 4H_2O - 5e^-$$

Converting normality to molarity

$$M = \frac{N}{\text{(number of electrons gained or lost)}} = \frac{0.6}{5} = 0.12\ M$$

Exercise 12.9

1. What is the normality of a solution of $KMnO_4$ if 60.0 ml is titrated to the end point with 45.0 ml of 0.5 N $Na_2C_2O_4$ in acid medium?
2. What are the normality and molarity values of $KMnO_4$ if 40.0 ml is titrated to the end point with 0.4 g of KCN dissolved in water? The skeleton equation is

$$MnO_4^- + CN^- \longrightarrow Mn^{2+} + OCN^- \text{ (basic solution)}$$

3. How many grams of copper are needed to reduce completely all of the silver ion in 50.0 ml of 0.4 N $AgNO_3(aq)$?
4. What is the normality of an iodine solution if 2.5 ml are titrated to the end point (iodine color disappears) with 26.2 ml of 0.05 N $S_2O_3^{2-}(aq)$ (thiosulfate ion)?

QUESTIONS AND PROBLEMS

12.1 Define oxidation and reduction in terms of loss and gain of electrons and change in oxidation number.

12.2 What are oxidizing and reducing agents?

12.3 Label the following changes as oxidation or reduction:
a. $Al^{3+} \rightarrow Al^0$
b. $Au^0 \rightarrow AuCl_4^-$
c. $SO_3^{2-} \rightarrow SO_4^{2-}$
d. $Br^- \rightarrow Br_2^0$
e. $Cl_2^0 \rightarrow ClO_3^-$

12.4 Calculate the oxidation number of the underlined elements:

a. $K_2\underline{Cr}O_4$
b. $Ca\underline{S}_2O_3$
c. $\underline{Cl}O_2^-$
d. $\underline{As}_2S_5$
e. $H\underline{Sb}O_3^-$

12.5 Balance the following equations by the oxidation number method:

a. $HI + H_2SO_4 \rightarrow H_2S + I_2 + H_2O$
b. $H_2SO_4 + HBr \rightarrow SO_2 + Br_2 + H_2O$
c. $MnO_2 + HCl \rightarrow MnCl_2 + Cl_2 + H_2O$
d. $NaNO_3 + Fe \rightarrow Fe_2O_3 + NaNO_2$
e. $CuO + NH_3 \rightarrow Cu + N_2 + H_2O$
f. $As_4 + HNO_3 + H_2O \rightarrow H_3AsO_4 + NO$
g. $H_2O_2 + H_2S \rightarrow S + H_2O$
h. $Mo_2O_3 + KMnO_4 + H_2SO_4 \rightarrow MoO_3 + K_2SO_4 + MnSO_4 + H_2O$
i. $V_2O_2 + H_2SO_4 + KMnO_4 \rightarrow V_2O_5 + MnSO_4 + K_2SO_4 + H_2O$
j. $AuCl_3 + H_2S + H_2O \rightarrow Au_8 + HCl + H_2SO_4$

12.6 Balance the following equations by the ion-electron method:

a. $MnO_4^- + Br^- \rightarrow Br_2 + MnO_2$ (acid solution)
b. $Ag_2O + CH_2O \rightarrow Ag + HCO_2^-$ (basic solution)
c. $H_2SO_3 + Fe^{3+} \rightarrow Fe^{2+} + SO_4^{2-}$ (acid solution)
d. $Sn^{2+} + Cr_2O_7^{2-} \rightarrow Sn^{4+} + Cr^{3+}$ (acid solution)
e. $MnO_4^- + Sn^{2+} \rightarrow Mn^{2+} + Sn^{4+}$ (acid solution)
f. $CrO_4^{2-} + HSnO_2^- \rightarrow CrO_2^- + HSnO_3^-$ (basic solution)
g. $BaO_2 + Cl^- \rightarrow Ba^{2+} + Cl_2$ (acid solution)
h. $C_2H_4O + NO_3^- \rightarrow C_2H_4O_2 + NO$ (acid solution)
i. $CrO_2^- + ClO^- \rightarrow Cl^- + CrO_4^{2-}$ (basic solution)
j. $H_3AsO_4 + I^- \rightarrow H_3AsO_3 + I_2$ (acid solution)

12.7 How many grams of copper can be obtained when 1.6 g of CuO is allowed to react with an excess of ammonia? See the equation in Problem 12.5, part e.

12.8 What volume of SO_2 gas (STP) can be obtained when 0.03 mole of HBr reacts completely with H_2SO_4? See the equation in Problem 12.5, part b.

12.9 How many milliliters of 0.022 M HCl are required to produce 72.0 ml (STP) of chlorine gas when HCl reacts with MnO_2? See the equation in Problem 12.5, part c.

12.10 What is the molar concentration of dichromate ion if 34.0 ml is needed to oxidize completely the iron(II) ion in 8.0 g of $FeCl_2$? The skeleton equation is $Cr_2O_7^{2-} + Fe^{2+} \rightarrow Cr^{3+} + Fe^{3+}$ (acid solution).★

12.11 Find the normality of a solution of $K_2Cr_2O_7$ if 31.5 ml is titrated to the end point with 83.2 ml of 0.044 N Ni^{2+} in acid solution.★

12.12 If 17.0 ml of 0.26 N $S_2O_3^{2-}(aq)$ is used in reaching the titration end point with 38.3 ml of an iodine solution, find the number of grams of I_2 in solution. The skeleton equation is $S_2O_3^{2-} + I_2 \rightarrow S_4O_6^{2-} + I^-$ (neutral solution).★

APPENDICES

APPENDIX A

SCIENTIFIC MATHEMATICS

Man experiences the desire and almost the need for knowledge. From the distant origins of history we see him preoccupied with the study of the phenomena of the world which surrounds him, and with the endeavor to explain them.

Louis de Broglie, *Physics & Microphysics**

Descriptions of a physical nature are necessarily vague if the adjectives used are those like "much," "few," "heavy," "weak," "hot," and "dilute." The fact must be faced—and accepted with the knowledge that it is the scientist's means of escaping total chaos—that precise descriptions involve numerical values. The placement of a heavy chunk of metal in a hot solution of dilute acid may mean very different things to an analytical chemist and an industrial chemist. Using 5.0 grams of iron in 0.02 molar acid at 60 degrees Celsius is an altogether different kind of description.

The point is that when numbers are *used*—i.e., processed by appropriate mathematical operations that may require addition, multiplication, raising to a power, or conversion to a logarithmic value—the subject of *mathematics* is involved. The oft-stated axiom that mathematics is the language of science is a self-evident truth. The purpose of this appendix is to review the fundamental aspects of mathematics most relevant to the operations of chemistry.

SCIENTIFIC NOTATION

Scientific notation is a method of increasing efficiency in calculations by expressing cumbersome, many-digit numbers in a compact form. It is writing with exponents. The exponent indicates the number of times a value is multiplied by itself. Some examples are

$$2^2 = 2 \cdot 2 = 4$$

$$2^3 = 2 \cdot 2 \cdot 2 = 8$$

* New York, Pantheon Books, Inc., 1955.

$$N^4 = N \cdot N \cdot N \cdot N$$

$$10^3 = 10 \cdot 10 \cdot 10 = 1000$$

The last example provides a useful focal point, because the base number 10 is the heart of the international system (metric system) of weights and measures. The metric system is of tremendous importance and is discussed at length in Chapter 1. Note the relationship between the exponent and the number of digits (zeros in this case):

$$10^{③} = 1000 \qquad 3 \text{ digits}$$

The following examples will support the observation that the exponent of the base 10 is exactly the same as the number of places the decimal point is moved.

$$1.0 \times 10^{②} = 100$$

$$2.0 \times 10^{⑤} = 200{,}000$$

$$1.0 \times 10^{⑥} = 1{,}000{,}000$$

$$2.5 \times 10^{③} = 2500$$

This direct relationship between the exponent and the number of digits greatly simplifies the process of writing a large number in exponential form:

$$100{,}000 = 10^5 \quad \text{or} \quad 1.0 \times 10^5$$

$$3220 = 3.22 \times 10^3$$

$$96{,}500 = 9.65 \times 10^4$$

One exponential value of the base 10 is also known as an **order of magnitude** (a term that comes from astronomy). For example, 10^6 is four orders of magnitude larger than 10^2. When a chemist desires to decrease a quantity by two orders of magnitude, he would use 1/100 of the original amount. This last point raises the question of how some fraction, or any number less than *one* is written in the form of scientific notation. The method is to shift the decimal point in such a way as to express the number as a value between *one and ten,* and indicate this operation by use of a *negative* exponent. For example

$$0.1 = 1 \times 10^{-1}$$

$$0.003 = 3 \times 10^{-3}$$

$$0.000251 = 2.51 \times 10^{-4}$$

While the last example could be written

$$0.000251 = 25.1 \times 10^{-5}$$

or

$$0.000251 = 251 \times 10^{-6}$$

the conventional method is to shift the decimal far enough to make the number between one and ten, multiplied by the appropriate exponential value.

Two additional observations should be noted in connection with scientific notation. The first is the result of using *zero* as an exponent. This has the effect of converting any exponential value to unity, since

$$1 = \frac{a}{a} = \frac{10^n}{10^n} = 10^{n-n} = 10^0$$

For example

$$3^0 = 1$$

$$2.7 \times 10^0 = 2.7$$

$$10^0 = 1$$

$$\left(\frac{1}{3}\right)^0 = 1$$

$$Q^0 = 1$$

The second observation is concerned with any value to the *first* power. In this case, the exponent means no additional digits after the number. The number is unchanged. Some examples are

$$4^1 = 4$$

$$a^1 = a$$

$$\left(\frac{1}{5}\right)^1 = \frac{1}{5}$$

$$10^1 = 10$$

One other useful rule that can be derived from the previous examples is that *the process of shifting a decimal point corresponds to a change in the exponent.* Every digit included in a decimal shift is equivalent to a change of one order of magnitude (or one decimal place). Some examples are

$$100 \times 10^2 = 10 \times 10^3 = 1 \times 10^4$$

$$230 \times 10^2 = 23.0 \times 10^3 = 2.30 \times 10^4$$

$$0.05 \times 10^{-3} = 0.5 \times 10^{-4} = 5 \times 10^{-5}$$

$$2170 \times 10^{-2} = 2.17 \times 10^1 = 0.217 \times 10^2$$

Exercise A–1

1. Use scientific notation to express the following numerical values as numbers between one and ten, times the correct power of 10. For example, $7283 = 7.283 \times 10^3$.

a. 762
b. 538,000
c. 42,600,000
d. 0.038
e. 0.0000112

2. Increase the following values by two orders of magnitude:
   a. 3.35×10^{3}
   b. 7.6×10^{5}
   c. 2.8×10^{-7}
   d. 5.11×10^{-2}
3. Decrease the following values by four orders of magnitude:
   a. 6.2×10^{4}
   b. 7.18×10^{11}
   c. 3.08×10^{-3}
   d. 9.2×10^{-8}

CHANGING THE SIGN OF AN EXPONENT

It is often convenient in the process of performing a calculation to change the sign of an exponent. This may be done by writing the **reciprocal** value of a number. The reciprocal is really the result of inverting the numbers in a fraction. For example

$$\frac{100}{1} = \frac{1}{\frac{1}{100}} = \frac{1}{0.01}$$

In other words

$$\frac{10^{2}}{1} = \frac{1}{10^{-2}}$$

Other examples are

a. $\frac{1}{10^{-4}} = 10^{4}$

b. $2 \times 10^{-3} = \frac{2}{10^{3}}$

c. $\frac{6}{2 \times 10^{-2}} = \frac{6 \times 10^{2}}{2} = 3 \times 10^{2} = 300$

Proving the last example

$$\frac{6}{2 \times 10^{-2}} = \frac{6}{0.02} = 300$$

RULES FOR ADDITION AND SUBTRACTION PROBLEMS INVOLVING EXPONENTS

In a list of numbers having different exponents, the coefficients must be altered until their exponents are identical. For example

$$\begin{array}{r} 1.25 \times 10^4 \\ +30.60 \times 10^3 \\ 602.0 \times 10^2 \\ \hline \end{array}$$

must be changed to

$$\begin{array}{r} 1.25 \times 10^4 \\ +3.06 \times 10^4 \\ 6.02 \times 10^4 \\ \hline 10.33 \times 10^4\text{, or } 1.033 \times 10^5 \end{array}$$

Notice that the exponential values are *not* added. The "$\times 10^4$" is a unit, a tag, a designation, like apples or grams or meters. The error of adding the exponents in an addition operation may be illustrated by the following example

$$\begin{array}{rcl} 1 \times 10^3 & = & 1000 \\ +2 \times 10^3 & = & 2000 \\ 3 \times 10^3 & = & 3000 \\ \hline 6 \times \mathbf{10^3} & = & 6000 \end{array} \qquad \begin{array}{rcl} 1 \times 10^3 & = & 1000 \\ +2 \times 10^3 & = & 2000 \\ 3 \times 10^3 & = & 3000 \\ \hline 6 \times \mathbf{10^9} & = & 6{,}000{,}000{,}000 \end{array}$$

correct *wrong*

If the exponents were added, the answer would be 6×10^9, which is 6,000,000,000. This is obviously a gross error.

The same rule applies to subtraction. For example

$$\begin{array}{rcr} 2.76 \times 10^{-5} & = & 2.76 \times 10^{-5} \\ -4.3 \times 10^{-6} & = & -0.43 \times 10^{-5} \\ \hline & & 2.33 \times 10^{-5} \end{array}$$

Exercise A–2

1. Write the reciprocals of the following:

a. x^{-a}

b. $\dfrac{1}{10^{-7}}$

c. $3^{-1/2}$

d. 2^5

2. Add the following terms:
$(36.2 \times 10^4) + (1.12 \times 10^3) + (5.22 \times 10^5)$
3. Find the difference between
581.3×10^4 and 4.14×10^6

RULES FOR THE MULTIPLICATION AND DIVISION OF EXPONENTS

The rules governing the multiplication and division of exponential values may be deduced from examples. Consider $10^2 \times 10^4$, substituting a parenthesis for the "times" sign ($\times$):

$$(10^2)(10^4) = (10 \cdot 10)(10 \cdot 10 \cdot 10 \cdot 10)$$

Observe that the multiplication in its expanded form shows the number 10 multiplying itself six times. Therefore, the answer would be

$$(10^2)(10^4) = 10^6$$

This amounts to *adding* the exponents rather than multiplying them. If it were assumed that $(10^2)(10^4) = 10^8$, the answer would be incorrect. The derived rule is *when multiplying exponential values, add the exponents.* It must be understood that any coefficients of the exponential values are handled traditionally, and the exponential values themselves must have the *same* base, such as 10. Some examples are

$$(10^2)(10^5) = 10^7$$

$$(10^3)(10^4)(10^{-2}) = 10^5$$

$$(2 \times 10^3)(3 \times 10^7) = 6 \times 10^{10}$$

Consider the example 10^5 *divided* by 10^2:

$$\frac{10^5}{10^2} = \frac{\cancel{10} \cdot \cancel{10} \cdot 10 \cdot 10 \cdot 10}{\cancel{10} \cdot \cancel{10}} = 10^3$$

This is the same as *subtracting* the exponents. The rule is as follows: *When dividing exponential values, subtract the exponents.* The coefficients, once again, are handled normally. Some examples are

$$\frac{10^7}{10^5} = 10^2$$

$$\frac{10^3}{10^8} = 10^{-5}$$

$$\frac{8 \times 10^5}{2 \times 10^2} = 4 \times 10^3$$

An alternative method is to use the reciprocals, thereby converting division operations to multiplication. For example

$$\frac{10^7}{10^5} = \frac{(10^7)(10^{-5})}{1} = 10^2$$

$$\frac{6 \times 10^3}{4 \times 10^8} = \frac{(1.5 \times 10^3)(10^{-8})}{1} = 1.5 \times 10^{-5}$$

RULES FOR RAISING AN EXPONENT TO A POWER OR FINDING A ROOT OF AN EXPONENT

Analysis of the expression $(2^3)^2$ suggests a simplified form:

$$(2^3)(2^3) = (2 \cdot 2 \cdot 2)(2 \cdot 2 \cdot 2) = 2^6$$

The rule that may be quickly extracted from this observation is *when raising an exponent to a power, multiply the exponents.* For examples

$$(a^4)^3 = a^{12}$$

$$(2 \times 10^4)^2 = 4 \times 10^8$$

$$(180)^2 = (1.8 \times 10^2)^2 = 3.2 \times 10^4$$

$$(2 \times 10^{-3})^2 = 4 \times 10^{-6}$$

$$(3 \times 10^{-2})^{-3} = 3^{-3} \times 10^6 = \frac{1 \times 10^6}{3^3} = \frac{1 \times 10^6}{27}$$

$$= 0.37 \times 10^6 = 3.7 \times 10^5$$

Examine the problem $90000^{1/2}$. This value may be expressed as $(9 \times 10^4)^{1/2}$. The fact to be remembered is that any value to the 1/2 power may be described as the *square root* of that value. Therefore, $(9 \times 10^4)^{1/2} = \sqrt{9 \times 10^4}$. However, if $(10^4)^{1/2}$ is considered by itself, it can be observed that the previous rule indicates an answer of 10^2, since that is the result of the multiplication of the exponents. The rule deduced from this example is *when taking the square root of an exponential value, multiply the exponent by one-half.* This rule may be extrapolated to cube roots, in which case the exponential value would be multiplied by one-third. The answer to the example is

$$\sqrt{9 \times 10^4} = 3 \times 10^2$$

Some further examples are

$$\sqrt{8.1 \times 10^5} = \sqrt{81 \times 10^4} = 9 \times 10^2$$

$$\sqrt{640 \times 10^{-7}} = \sqrt{64.0 \times 10^{-6}} = 8 \times 10^{-3}$$

$$(2.5 \times 10^9)^{1/2} = \sqrt{25 \times 10^8} = 5 \times 10^4$$

$$(2.7 \times 10^{-5})^{1/3} = \sqrt[3]{27 \times 10^{-6}} = 3 \times 10^{-2}$$

Problems dealing with multiple operations involving exponents will be deferred until the slide rule is considered (Appendix B). The rules and the overall value of scientific notation are infinitely more meaningful when practically applied.

Exercise A–3

1. Multiply the following and express the coefficients as one digit numbers before the decimal point:
   a. $(10^3)(10^5) =$
   b. $(4 \times 10^2)(6 \times 10^6) =$
   c. $(2 \times 10^2)(1 \times 10^{-3})(3 \times 10^4) =$
   d. $(216.5 \times 10^2)(0.018 \times 10^2) =$
2. Divide the following and express the coefficients as one digit numbers before a decimal point:
   a. $\frac{10^4}{10^2} =$
   b. $\frac{1 \times 10^{-3}}{2 \times 10^2} =$
   c. $\frac{15 \times 10^7}{30 \times 10^{11}} =$
   d. $\frac{617 \times 10^3}{0.012 \times 10^8} =$
3. Raise the following values to the indicated power:
   a. $(x^2)^5 =$
   b. $(3 \times 10^3)^3 =$
   c. $(4 \times 10^2)^{-2} =$
   d. $(20000)^3 =$
4. Perform the indicated operation on the following values:
   a. $\sqrt{360 \times 10^3} =$
   b. $\sqrt{0.049 \times 10^{-3}} =$
   c. $(3.6 \times 10^5)^{1/2} =$
   d. $(80 \times 10^8)^{1/3} =$

SIGNIFICANT FIGURES

The term significant figures refers to the number of digits, in some numerical value, that has been obtained with a generally acceptable level of accuracy. The accuracy of the numerical value, and the number of digits representing the order of magnitude to which that accuracy extends, depend on the *nature of the object being measured, the technical skill of the measurer, and the* **precision** *allowed by the instrument used to perform the measurement.* It should be made clear that *precision* and *accuracy* are related but they are not synonymous. Precision refers to a degree of *reproducibility* of measurements, while accuracy is a matter of exactness as opposed to approximation. The distinction between these two terms may be illustrated by the analogy of shooting at a target (Fig. A–1).

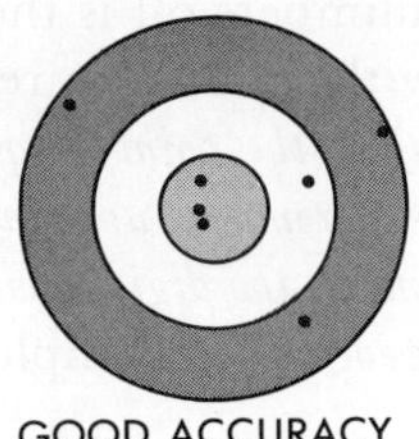

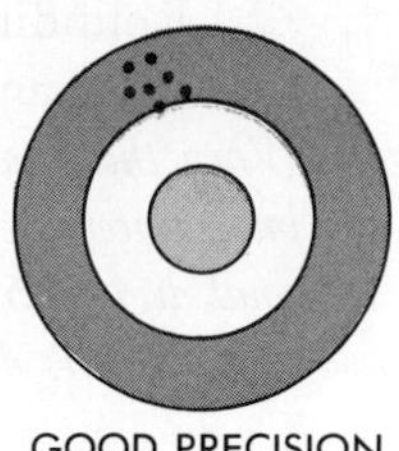

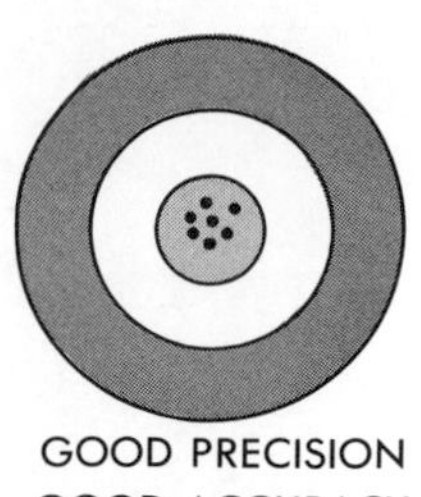

Figure A–1 Representation of the difference between accuracy and precision.

The rules governing the writing of significant figures (often abbreviated to "sig figs") are reasonably straightforward, and they will be explained by examples. However, a few common distortions of "sig fig" usage should be mentioned to justify the discussion. One type of error that clearly indicates a misunderstanding of "sig figs" is illustrated in Example A–1.

EXAMPLE A–1

A student calculates the area of a circle having a diameter of 6.30 centimeters.

1. The equation is, area of circle $= \pi r^2$.
2. If the diameter is 6.30 centimeters, the radius is 3.15 centimeters.
3. If π is 3.14

$$\text{area} = 3.14(3.15 \text{ centimeters})^2$$

4. Using a calculator

$$\text{area} = 30.842650 \text{ centimeters squared}$$

This is absurd! The accuracy of the answer is limited to *three* significant figures because the information given in the beginning has a certainty of only three figures. The answer cannot be more accurate than the measurement. The answer should read, area = 30.8 centimeters squared. The major misconception in the minds of many students is that the greater the number of digits in the answer, the more exquisite the accuracy. This is false unless the digits are *significant* figures.

Other common errors appear when values of many significant figures are mysteriously obtained when relatively crude, low-precision instruments are used. To suggest that a meter stick is an inappropriate instrument to measure the length of a flea would hardly be described as being unreasonably fussy. It is simply a matter of common sense. This common-sense approach to handling numbers and "rounding off" is the process of recognizing and using significant figures.

Rounding numbers off is the usual method of avoiding a pretense toward accuracy that does not really exist. The rule is simple enough: *When the final digit (the point of questionable accuracy) is 4 or less, the digit immediately before it remains unchanged while the last digit is dropped. If the final digit is 5 or more, the digit immediately before it is increased by one, and the final digit is dropped.* For example:

five digit number	*rounded off*
23.716	23.72
23.715	23.72
23.713	23.71

An application of the rule for rounding off in addition or subtraction is illustrated in Example A–2, which is based on the principle that says: *The accuracy indicated by the answer to an addition or subtraction problem is limited to the least number of decimal places found in any single value.*

EXAMPLE A–2

Calculate the average of the following gram weights:

12.22 g
11.50 g
11.6 g
12.272 g

1. Note the third number. It is 11.6, not 11.60.
2. Therefore, the number could be anywhere between 10.55 and 11.64.
3. Since the number 11.6 has only one decimal place, the answer to the problem is restricted to the same limit.
4. Round off all the numbers to three digits and calculate the answer:

12.2 g
11.5 g
11.6 g
12.3 g
47.6 g

$$\frac{47.6}{4} = 11.9$$

average = 11.9 g

A great deal of self-deception can be avoided when a slide rule is used to do calculations. Since a slide rule is largely restricted to representing only three digits of a number, it helps prevent the common error of confusing many digits with great accuracy.

Exercise A–4

1. How many significant figures are there in each of the following:
   a. 163.5 c. 0.0003
   b. 0.015 d. 5.17×10^3
2. Perform the following calculations within the limits of "sig figs" or decimal places.
   a. $2.02 \times 3.5 =$
   b. $\frac{4.15}{2.077} =$
   c. $(8.1 \times 10^2)(1.23 \times 10^3) =$
   d. $4.22 + 0.308 + 65.4 =$
   e. $2.13 \times 10^2 - 0.18 \times 10^3 =$
3. Round off the following values to three significant figures:
   a. 3.081 c. 20809
   b. 0.07635 d. 14.22

BASIC ALGEBRA

An algebraic equation is the most common method of expressing the majority of mathematical relationships in chemistry. The particular type of equation most frequently used is called a **first degree** (no exponents larger than 1) equation, which is limited to one unknown.

Consider Example A–3 in light of the central principle of algebra, which says: *An equation represents a balance, and whatever operation is performed on one side of that balance must also be performed on the other side if the equal sign is to remain a statement of truth.*

Example A–3, may strike the student as being remarkably simple-minded. If the solution to the actual problem were the point, it would be an insult to the intelligence. However, that is not the point. The object is to review terminology and present some recommended techniques in problem solving. There are numerous instances throughout this text in which simple numbers are used to explain the method of solving a problem. These methods continue to work when more formidable data are encountered.

EXAMPLE A–3

Solve for Q when 5Q = 3.

1. In order to "isolate" Q on one side of the equation, the 5 must be eliminated:

$$5Q = 3 \qquad \text{divide } \textit{both} \text{ sides of the equation by 5}$$

$$\frac{\not{5}Q}{\not{5}} = \frac{3}{5}$$

$$Q = \frac{3}{5} = 0.6$$

EXAMPLE A–4

If $PV = \frac{gRT}{M}$, write an equation for the solution of M.

1. $PV = \frac{gRT}{M}$ multiply *both* sides of the equation by M

$$MPV = \frac{gRT}{\cancel{M}}(\cancel{M})$$

2. $\frac{MP\cancel{V}}{\cancel{PV}} = \frac{gRT}{PV}$ divide *both* sides of the equation by PV

3. The result is $M = \frac{gRT}{PV}$

EXAMPLE A–5

Solve for the unknown in the equation $7x = (2/3) - x$.

1. $7x = \frac{2}{3} - \cancel{x}$ add x to *both* sides
   $+x \quad +\cancel{x}$

2. $(3)8x = \frac{2}{\cancel{3}}(\cancel{3})$ multiply *both* sides by 3

3. $24x = 2$ divide *both* sides by 24

4. $\frac{\cancel{24}x}{\cancel{24}} = \frac{2}{24}$

$$x = \frac{2}{24} = \frac{1}{12}$$

Exercise A–5

1. Solve the following equations for x:

   a. $2x = 5$

   b. $2x = 8 - x$

   c. $x^2 = 81$

d. $\frac{5}{x} = 17$

e. $\frac{8.30}{2.10} = \frac{12.2}{x}$

2. Solve for V_2 in the equation $\frac{P_1V_1}{T_1} = \frac{P_2V_2}{T_2}$.

3. Solve for g in the equation $PV = \frac{gRT}{M}$.

4. Solve for v in the equation $E_k = \frac{mv^2}{2}$.

5. Solve for h in the equation $E = \frac{hc}{\lambda}$.

6. Solve for r in the equation $V = \frac{4\pi r^3}{3}$.

The previous examples were intended to be a brief refresher of some basic operations in algebra. Throughout the text, abundant examples are presented as practical applications in chemical problem solving. However, the concept of proportionality and the graphical representation of proportionality are natural considerations at this time.

PROPORTIONAL RELATIONSHIPS

There are many measurements in chemistry in which simple mathematical relationships can be seen when the data are examined. For example, a look at the measurements of pieces of a particular metal would indicate that its mass (weight) has a simple and direct relationship to its volume (space occupied). In short, the larger the piece of metal, the more it weighs—doubling the weight doubles the size, or reducing the size by one-fourth will reduce the weight by the same fraction. In this type of relationship between two quantities, it could be said that the mass of the object is directly proportional to the volume. Symbolically expressed, this relationship is

$$M \propto V$$

where the symbol $\propto$, is a *sign of proportionality*. This expression is observed to be true as long as no other variable factor, such as temperature, is introduced.

Another way of stating the relationship is to say that the volume is dependent on the mass. Such a point of view establishes the mass as an **independent variable** and the volume as a **dependent variable.** The chemist may describe the relationship of the variables by saying that the volume of an object is a **function** of its mass. The mathematical expression is

$$V = f(M)$$

If the volume is always triple the value of the mass, the functional relationship may be expressed as an equation:

$$V = 3M$$

In this equation, the number 3 is a **constant**—i.e., it does not vary regardless of the value of the mass and the dependent variable, volume. Examine the data in Table A–1. This constant is appropriately called a **proportionality constant.** Changing the expression

$$M \propto V$$

to a mathematical expression incorporates the constant, which may be symbolized as k:

$$M = kV$$

When the relationship between variables is inversely proportional, the equation takes a different form. For example, if the volume occupied by a gas *decreases* proportionately as the pressure on the gas *increases*, it can be said that the volume is **inversely proportional** to the pressure. This means, in effect, that the dependent variable, volume, is directly proportional to the reciprocal of the pressure:

$$V \propto \frac{1}{P}$$

Once again, the relationship may be stated functionally, where the volume is a function of the reciprocal of the pressure:

$$V = f(1/P)$$

TABLE A–1

Constant		Mass (grams)		Volume (milliliters)	
3	×	2	=	6	The volume is observed
3	×	4	=	12	to be triple the mass in
3	×	5	=	15	every case.
3	×	0.7	=	2.1	

TABLE A–2

Constant (k)		Pressure (P) (atmospheres)		Volume (V) (liters)
2	=	1	×	2
2	=	2	×	1
2	=	4	×	1/2
2	=	1/2	×	4

Starting with the observation that doubling the pressure on a gas reduces the volume to one-half its original value, examine the data in Table A–2.

Since the constant, k, equals the product of P times V, the first degree equation may be rewritten:

$$k = PV \quad \text{divide the equation by P}$$

$$\frac{k}{P} = \frac{\not{P}V}{\not{P}}$$

or

$$V = k \times \frac{1}{P} = \frac{k}{P}$$

The equation says that the volume is directly proportional to the reciprocal of the pressure. If the volume were assumed to be the independent variable, the possible error in that description would not necessarily lead to a mathematical error. There may be, however, some inconvenience when a graphical representation of the data is made. Nevertheless, the equation could be expressed:

$$P = \frac{k}{V}$$

Once a proportionality constant has been obtained for a functional relationship, the solution to many related problems may be obtained. A common method of finding a proportionality constant is by the graphical representation of data.

GRAPHING PROPORTIONAL RELATIONSHIPS—BASIC RULES OF GRAPHING

Before getting into the analysis and methods of graphing, a few ground rules and graph labels ought to be explained:

1. The labels applied to a graph are illustrated in Figure A–2.
2. The independent variable data are usually distributed along

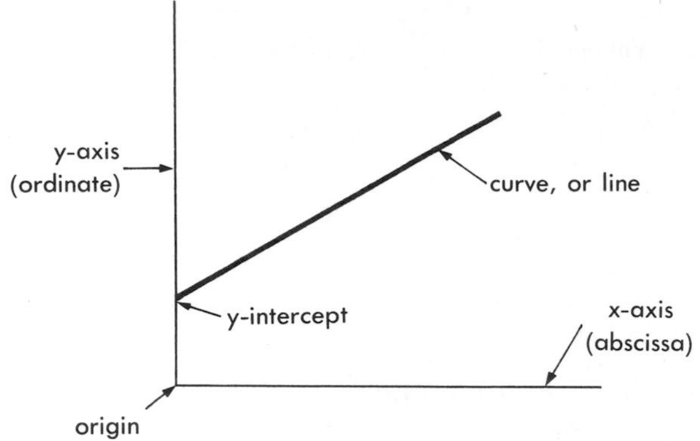

Figure A–2 The common labels applied to a graph.

the x-axis, while the dependent variable data are distributed along the y-axis. The units of measure are included in the labels, as shown in Figure A–3.

3. The numerical data should be *uniformly* distributed over a maximum area (Fig. A–4).
4. The data should be distributed along the axes in the form of scientific notation. Compare the two examples in Figure A–5.
5. A *smooth line* that represents an average value should be drawn through the points if a simple relationship is indicated. Figure A–6 illustrates the correct form and a faulty form. Notice that "error halos" are drawn around each point. The "halo" represents the uncertainty of the measurement due to imperfect experimental methods and apparatus. The lines do not have to pass through the points, just through the halos.

THE SLOPE OF THE LINE AND PROPORTIONALITY

It was mentioned before that the volume of an object is a function of its mass. The proportionality of these physical measurements was symbolically written as $M \propto V$, and mathematically expressed as $M = kV$. When the **slope** of a straight line is defined as the ratio of the

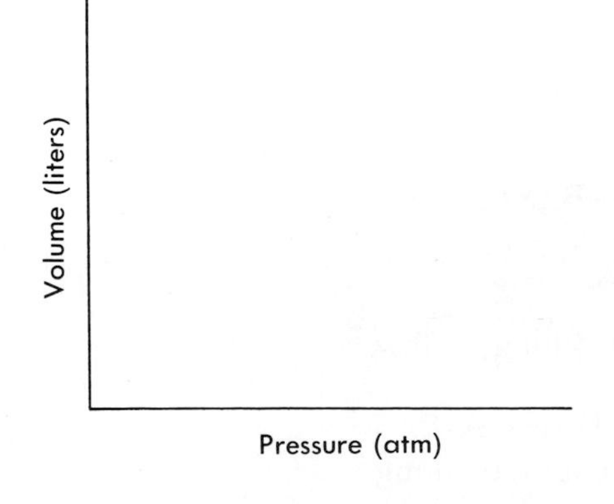

Figure A–3 Labeling the axes of a graph.

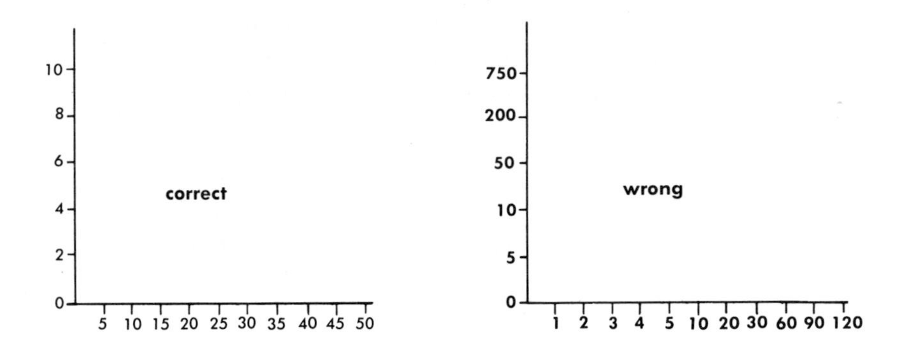

Figure A-4 Uniform distribution of data.

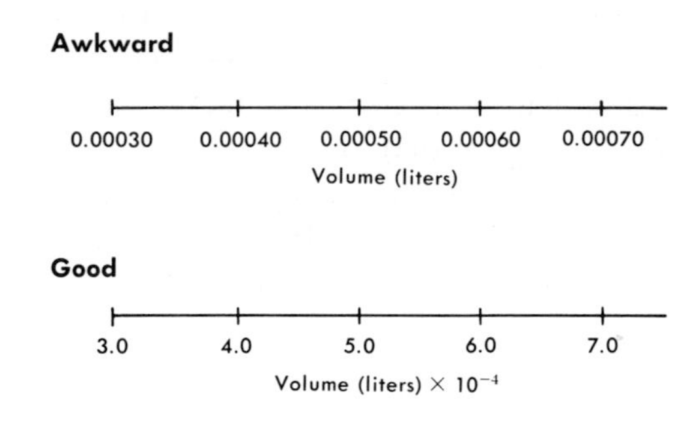

Figure A-5

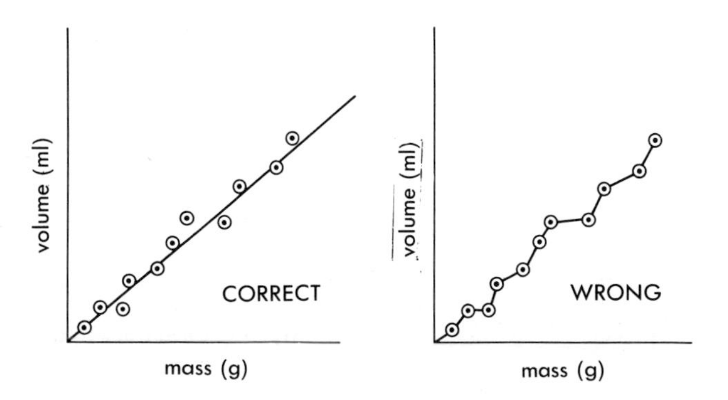

Figure A-6 Mass-volume relationship of an alloy.

change on the y-axis (Δy) to the corresponding change on the x-axis (Δx), the quotient is the same as the proportionality constant. The significance of this relationship is that an alternate method for obtaining a proportionality constant is available. Having a proportionality constant in hand may permit solutions to many practical problems.

Considering the direct proportionality of mass and volume, observe how the solution for k in the equation $M = kV$ agrees with the value for k as the slope of the line. Use the data given in Table A–3.

TABLE A–3

Mass (grams)	Volume (milliliters)
2.2	1.0
4.4	2.0
11.0	5.0
15.4	7.0
26.4	12.0
30.8	14.0

The graphical representation of these data is shown in Figure A–7. Notice that the slope in Figure A–7 is a positive number. This is always the case when the angle of the slope to the x-axis is less than 90 degrees. An angle of more than 90 degrees indicates a negative proportionality constant. The slope will then have a negative value. The value of the slope in Figure A–7, 2.2, does equal the solution for k in the equation $M = kV$, since

$$\frac{M}{V} = \frac{k\cancel{V}}{\cancel{V}} \quad \text{divide the equation by V}$$

$$\frac{M}{V} = k \quad \text{substitute data}$$

$$\frac{2.2 \text{ grams}}{1.0 \text{ milliliter}} = 2.2 \text{ (grams per milliliter)}$$

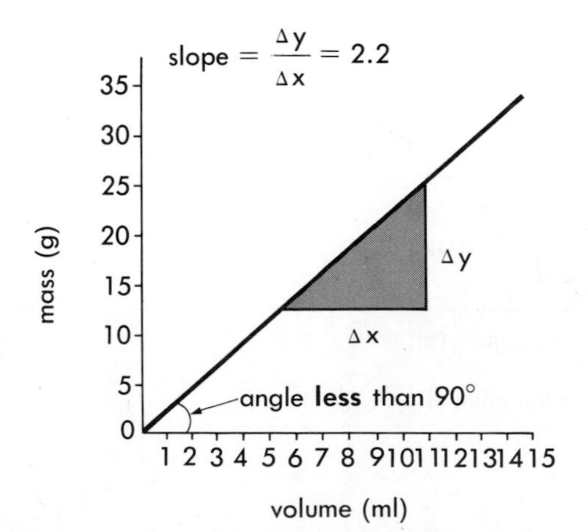

Figure A–7 The assignment of the labels as independent and dependent variables is arbitrary in this case.

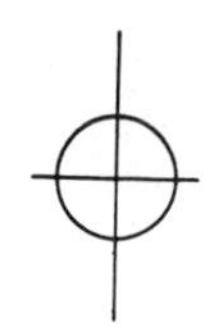

TABLE A-4

Volume (liters)	Pressure (atmospheres)
10.00	0.50
5.00	1.00
2.50	2.00
1.25	4.00
1.00	5.00

Now, compare the inverse relationship between volume and pressure. Use the data in Table A–4.

The graphical representation of these data is shown in Figure A–8. The slope of the curve in Figure A–8 can be determined most accurately

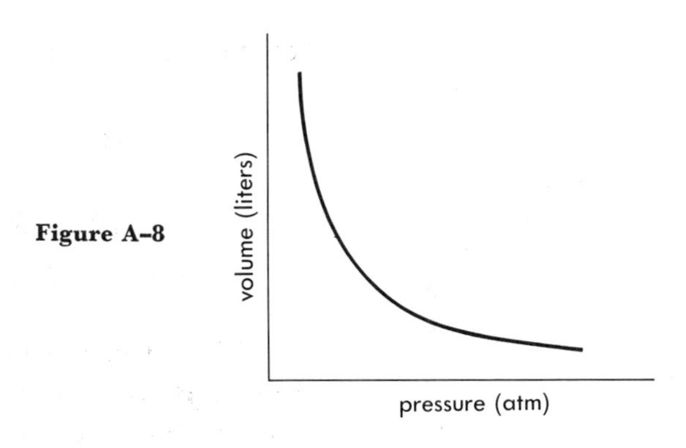

Figure A–8

if the curve is "straightened out" as shown in Figure A–9. This may be done by plotting the volume against the *reciprocal* of the pressure, since the expression $V = k \times \frac{1}{P}$ may be written $V = kP^{-1}$. The term P^{-1} may be called "reciprocal pressure."

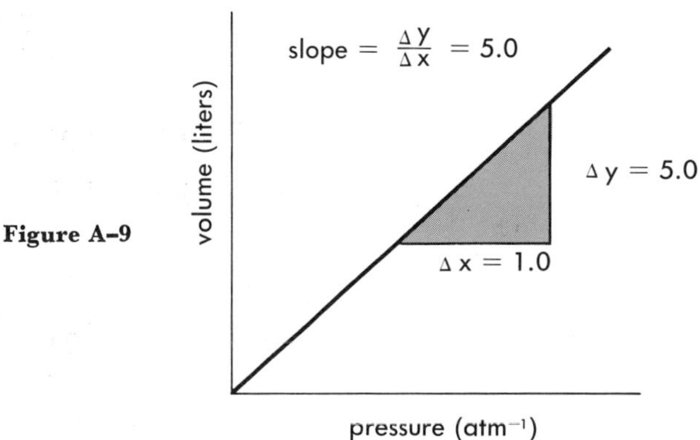

Figure A–9

TABLE A–5

Volume	Pressure	Reciprocal of Pressure (1/P)
10.00	0.50	2.00
5.00	1.00	1.00
2.50	2.00	0.50
1.25	4.00	0.25
1.00	5.00	0.20

Once again, the slope of the line equals the proportionality constant as it is obtained from the solution of the equation

$$V = k\frac{1}{P}$$

Solving for the constant, k

$$P(V) = \frac{k}{\cancel{P}}(\cancel{P}) \quad \text{multiply the equation by P}$$

$$PV = k \quad \text{substitute data}$$

$$(1.00 \text{ atm})(5.00 \text{ liters}) = 5.00 \text{ liter-atm}$$

The interpretation of the slope as a **conversion factor** and the **intercept** as a **correction factor** is discussed in Chapter 1. Also reserved for Chapter 1 is the consideration of dimensional analysis, which is a technique highly recommended in performing calculations.

Exercise A–6

1. Draw a graph representing y as a function of x, given the following data:

x	y	x	y
0.0	0	0.3	15
0.1	5	0.4	20
0.2	10	0.5	25

Find the slope of the line from the graph.

2. Graph the following data where A is a function of the reciprocal of B:

A	B	A	B
20	1.00	80	0.25
40	0.50	10	2.00
60	0.33	5	4.00

Find the slope of the line from the graph.

3. If the disintegration of substance Q is a linear function of time, calculate the rate of breakdown from the slope, given the following data:

Micrograms of Q	Time (seconds)
80	0.0
65	0.5
50	1.0
35	1.5
20	2.0
5	2.5

APPENDIX B

LOGARITHMS AND THE SLIDE RULE

Just as to Columbus long philosophic meditation led him to the fixed belief of the existence of a yet untrodden world beyond that waste of Atlantic waters, so to our most keen-eyed chemists, physicists, and philosophers a variety of phenomena suggest the conviction that the elements of ordinary assumption are not the ultimate boundary in this direction of the knowledge which man may hope to attain.

William Crookes, *Presidential Address, 1886*

A slide rule is one of the most useful instruments a chemist can possess. It is one of the least expensive, too. A perfectly satisfactory plastic slide rule can be purchased for a dollar or two.

Although circular slide rules have their own particular virtues, the discussion in this chapter will be restricted to linear models. The principles are the same; only the mechanics of operation differ.

A whole appendix devoted to presenting the slide rule as an invaluable instrument for arithmetical calculations is not meant to suggest its superiority to a desk calculator. A calculator is unquestionably faster and more accurate. It is a well-used fixture in many laboratories. However, desk calculators are notoriously expensive and a student will certainly find it much more practical to carry a slide rule.

The question of why a slide rule has such a peculiar array of scales, and how it works, is explained in terms of **logarithms**. The slide rule is, in fact, a table of logarithms distributed over lengths of wood, metal, or plastic. What logarithms are, and what they can do, is an essential introductory topic in this appendix.

THE BASIS OF THE SLIDE RULE

Given the two numbers 30,815,724 and 6,153,809 and the choice of adding, or subtracting, or multiplying, or dividing, the average person would probably elect to add or subtract. Multiplying or dividing so many digits is a long, dull job. The natural response to a dull job is to select the shortest method of getting it done. While these two large numbers could be added or subtracted in seconds, multiplication or division would be quite time consuming without the aid of some type of automatic calculator. The beauty of the slide rule, in addition to its handy size, is that it is based on a number system that permits addition to substitute for multiplication, and subtraction to substitute for division.

The explanation for this ability to alter a mathematical operation becomes apparent when some of the rules for the multiplication and division of exponents are reviewed. Remember that the sign of *multiplication* means the *addition* of exponents (see Appendix A):

$$(2^2)(2^4)(2^3) = 2^9$$

$$(10^4)(10^{0.301}) = 10^{4.301}$$

$$(10^a)(10^x) = 10^{a+x}$$

The sign of *division* means the *subtraction* of exponents:

$$\frac{2^7}{2^4} = 2^3$$

$$\frac{10^{0.778}}{10^{0.301}} = 10^{0.477}$$

$$\frac{10^a}{10^{-x}} = 10^{a+x}$$

These examples suggest the path to be followed. If any number can be converted to exponential form, the operations of addition and subtraction can be substituted for multiplication and division. For the sake of convenience and ease of handling, the exponents of the base 10 are commonly used. Furthermore, there is a special name given to the exponent of the base 10 — **logarithm**, abbreviated to **log**. While this is not a very sophisticated definition, it is a very useful one. Some examples are listed in Table B–1.

TABLE B–1

10,000	$= 10^4$ and the log is 4
1000	$= 10^3$ and the log is 3
100	$= 10^2$ and the log is 2
10	$= 10^1$ and the log is 1
1	$= 10^0$ and the log is 0
0.1	$= 10^{-1}$ and the log is -1
0.01	$= 10^{-2}$ and the log is -2
0.001	$= 10^{-3}$ and the log is -3

From the examples in Table B–1, it can be seen that $100 \times 1000 = 10^2 \times 10^3 = 10^5$. Using the logarithms, $2 + 3 = 5$. This is quite

TABLE B–2

Number		Exponent		Log
1	=	10^0	=	0
2	=	10?	=	?
3	=	10?	=	?
↓		↓		↓
10	=	10^1	=	1

convenient when the steps are whole orders of magnitude. But what about the numbers in between? What are the logarithms of 2, 3, 11, 19, 53, and so on (Table B–2)? Observe that the log of 1 equals zero and the log of 10 equals 1. The conclusion is that the exponents of the numbers between 1 and 10 must have *fractional* values. An easy way of approximating these numbers is to represent graphically the relationship of numbers to known logarithms and pick the answers off the line (Fig. B–1). Information extracted from calculations that are far more accurate than points on the exponential curve is summarized in Table B–3.

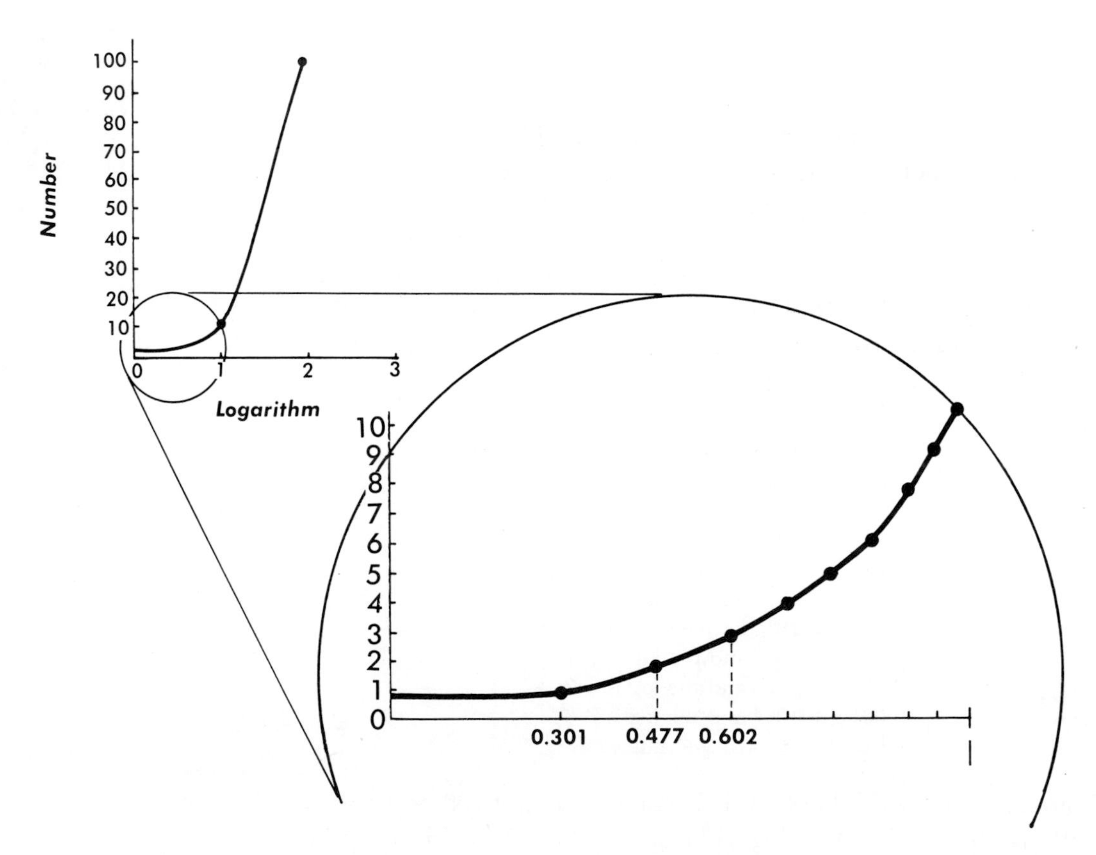

Figure B–1

TABLE B–3

Number	Exponent	Log
1	$10^{0.000}$	0.000
2	$10^{0.301}$	0.301
3	$10^{0.477}$	0.477
4	$10^{0.602}$	0.602
5	$10^{0.699}$	0.699
6	$10^{0.778}$	0.778
7	$10^{0.843}$	0.843
8	$10^{0.903}$	0.903
9	$10^{0.954}$	0.954
10	$10^{1.000}$	1.000

The logarithms may be used as shown by the following examples:

a. $(2)(3) = (10^{0.301})(10^{0.477}) = 10^{0.778}$

multiplication becomes addition

Logarithmically:

$0.301 + 0.477 = 0.778$. The **antilog** of $0.778 = 6$.

The concept of antilog means using the exponent of base 10 to find the number represented by the exponential form.

b. $\frac{8}{4} = \frac{10^{0.903}}{10^{0.602}} = 10^{0.301}$

division becomes subtraction
Logarithmically

$0.903 - 0.602 = 0.301$. The antilog of $0.301 = 2$.

Of course, the purpose of the slide rule is to solve problems more difficult than 2 times 3 or 8 divided by 4. The important thing, however, is that the theory and method are the same, regardless of the complexity of the calculation.

Exercise B–1

1. Using a table of logarithms (base_{10}), find the answers to the following problems by the log-antilog method:
   a. $(4)(7) =$
   b. $(2)(5)(3) =$
   c. $(3)(6)^2 =$
   d. $\frac{8}{2} =$
   e. $\frac{(4)(7)(9)}{(3)(2)} =$

2. Find the approximate logarithms of the following numbers on the exponential curve (Fig. B–1):
   a. 2.5
   b. 4.2
   c. 6.9
   d. 7.5

The relationship between logarithms and the slide rule is due to the fact that the slide rule—those pieces of wood, plastic, or metal—gives substance to the fractional exponents. For example, $10^{0.301}$ is equivalent to the number 2. The fraction 0.301 (approximately 3/10) is called the logarithm. But 3/10, simply in the form of an unrelated fraction, has limited value. A person might be asked if he would like 3/10. Well, 3/10 of *what*? If it's 3/10 of a million dollars—fine! If its 3/10 of the national debt—no thanks! The slide rule gives substance to this fraction by using a length of some real material, such as wood or plastic, and then measuring off 3/10 of whatever its length is. In other words, the fraction is translated into a distinct length of material. Take a 10 cm line and mark off 3/10:
Now, instead of simply adding fractions (logarithms of the numbers between 1 and 10) on paper, lengths of slide rule material can be physically added. Look again at (3)(2) = 6. Since $3 = 10^{0.477}$ and $2 = 10^{0.301}$, these two lengths of slide rule material can be added (Fig. B–2). The antilog of 0.778 is 6, and 6 is the number seen on the slide rule.

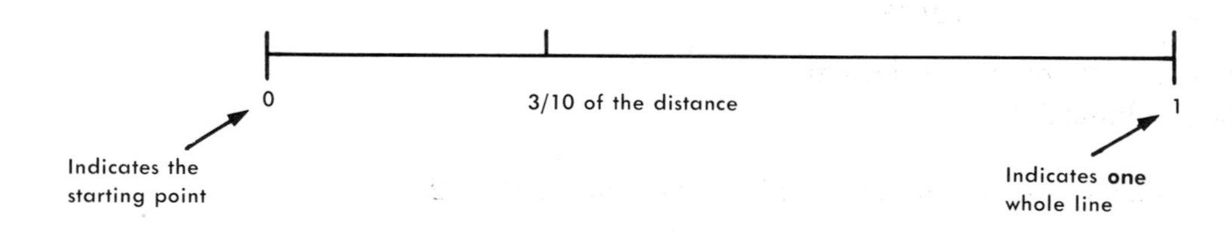

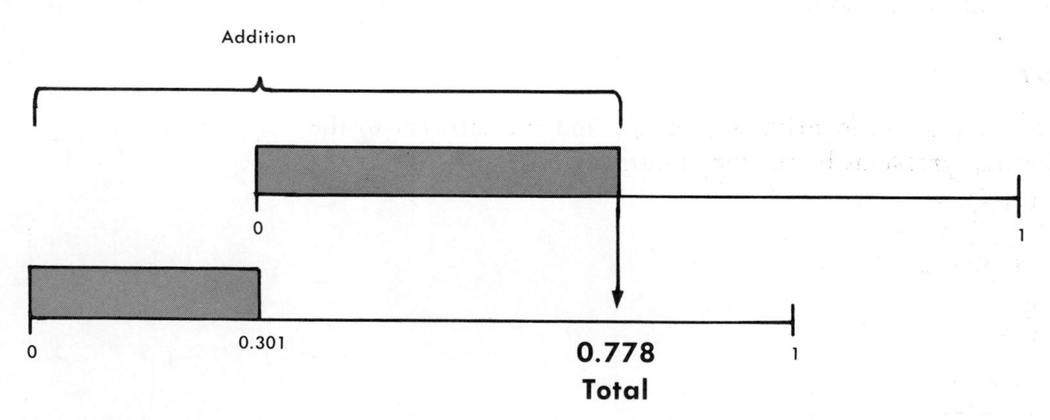

Figure B–2

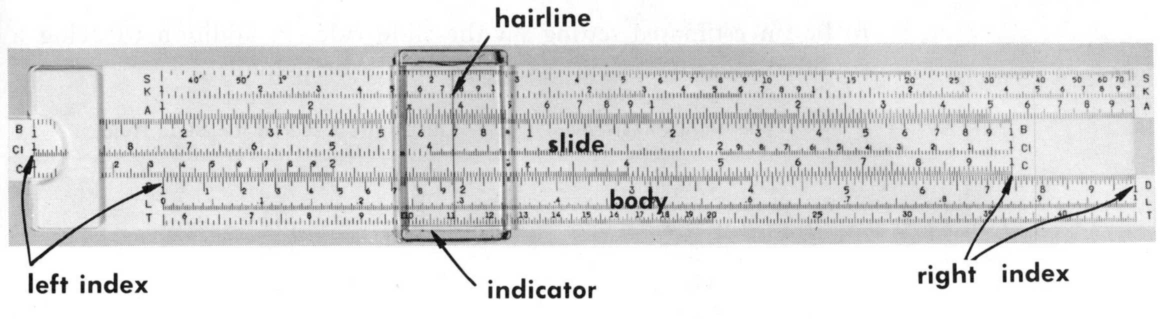

Figure B–3

THE STRUCTURE OF THE SLIDE RULE

The typical slide rule is composed of three principal parts: the *body*, the *slide*, and the *indicator* with a hairline etched down its center (Fig. B–3). The operations of multiplication and division require the use of the body and slide sections, while squares, cubes (raising a number to the third power), square roots, cube roots, logarithms, and antilogs use only the body on most slide rules. Before any mathematical operations can be discussed, it is necessary to learn how to read the scales and set the numbers which are involved in the calculation.

The scales most commonly used in chemistry are the K, A, C, CI, D, and L scales. A common arrangement and the related operations are shown in Figure B–4.

SETTING NUMBERS ON THE SLIDE RULE

Since the D scale is used in the greatest variety of calculations, it is appropriate to use it as a focal point in an attempt to understand the graduations on the slide rule. *The reliability of the numbers set or read on any scale is usually limited to three digits.* Most numbers having four or more digits should be rounded off. The slide rule operator ought to allow some flexibility for the value of the third digit, because it usually has

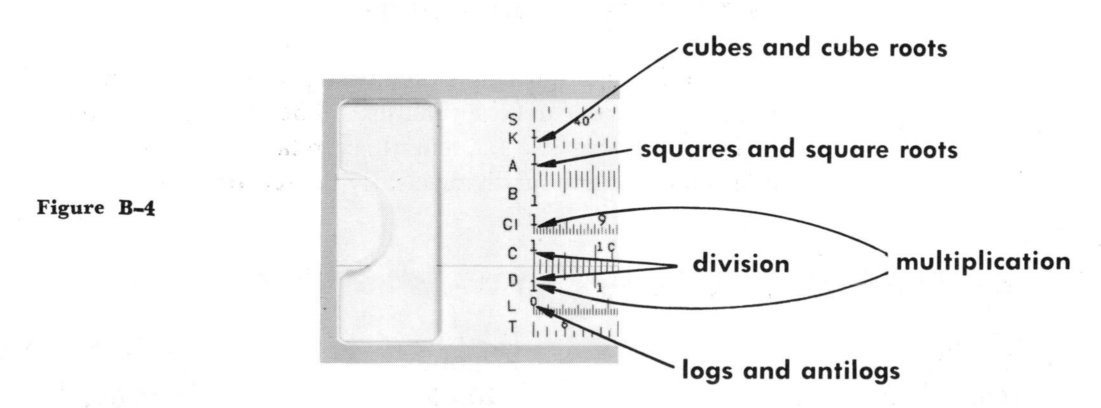

Figure B–4

to be an estimated setting on the slide rule, in addition to being a rounded-off value. Some examples might be helpful:

The Number Given	The Digits Used on the Slide Rule
3.05	305
47,135	471
22.581	226
0.00053	530
6.017	602

It should also be mentioned that *the slide rule gives no information about the decimal point placement* in a number. This is a separate operation handled by use of scientific notation.

SETTING THE FIRST DIGIT

The first digit of any numerical value is set on one of the large, boldly printed numbers that extend from 1 to 10 across the entire length of the slide rule. The zeros before a digit are not counted. For example, the first set digit in the number 0.002 is 2. The first digit is known as a *primary graduation.* The primary graduation, 2, represents *any* value having the digit 2 as its *first* digit, regardless of whether it is 0.002 or 20,000. Examples are presented in Figure B–5.

SETTING THE SECOND DIGIT

The second digit is related to the *secondary graduations* on a scale. The space between any two primary graduations is divided into ten parts. Each part represents the second digit of a number. For example, Figure B–6 shows the secondary graduations between the bold numbers, and includes sample values for illustration.

SETTING THE THIRD DIGIT

The last digit, which is the finest graduation (often called *tertiary*) usually requires careful estimation. The third digit is read as a subdivision between two sequential secondary divisions. Since the slide rule is too small to subdivide clearly the secondary graduations into ten

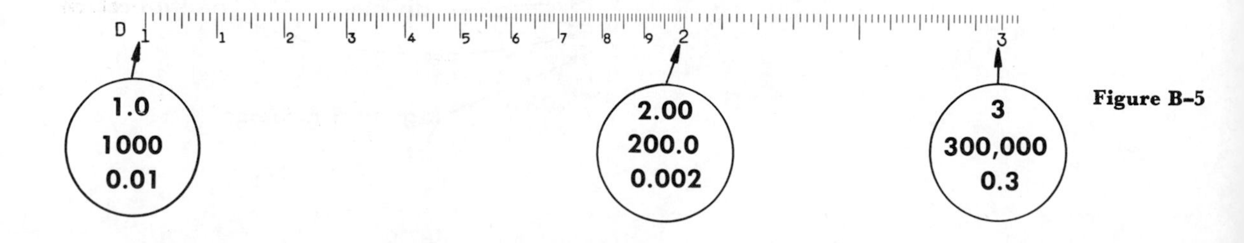

Figure B-5

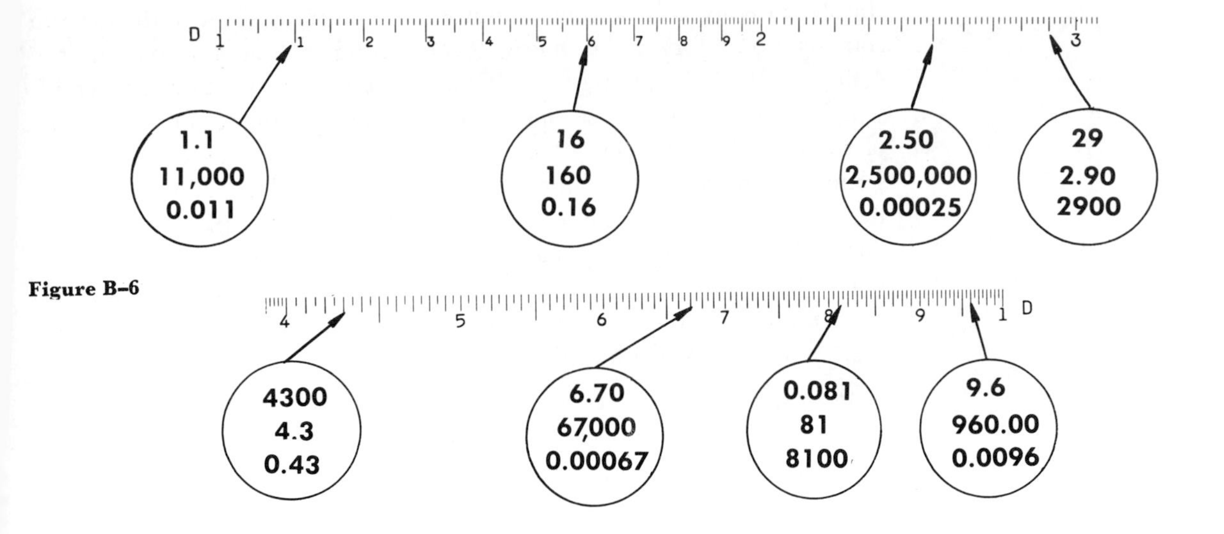

Figure B–6

parts, except between 1.0 and 2.0, the operator must judge by eye as carefully as possible. A magnifying glass type of indicator is helpful, although not essential. Examine the following examples in Figure B–7.

Consider the number 433 in Figure B–7 for purposes of reviewing

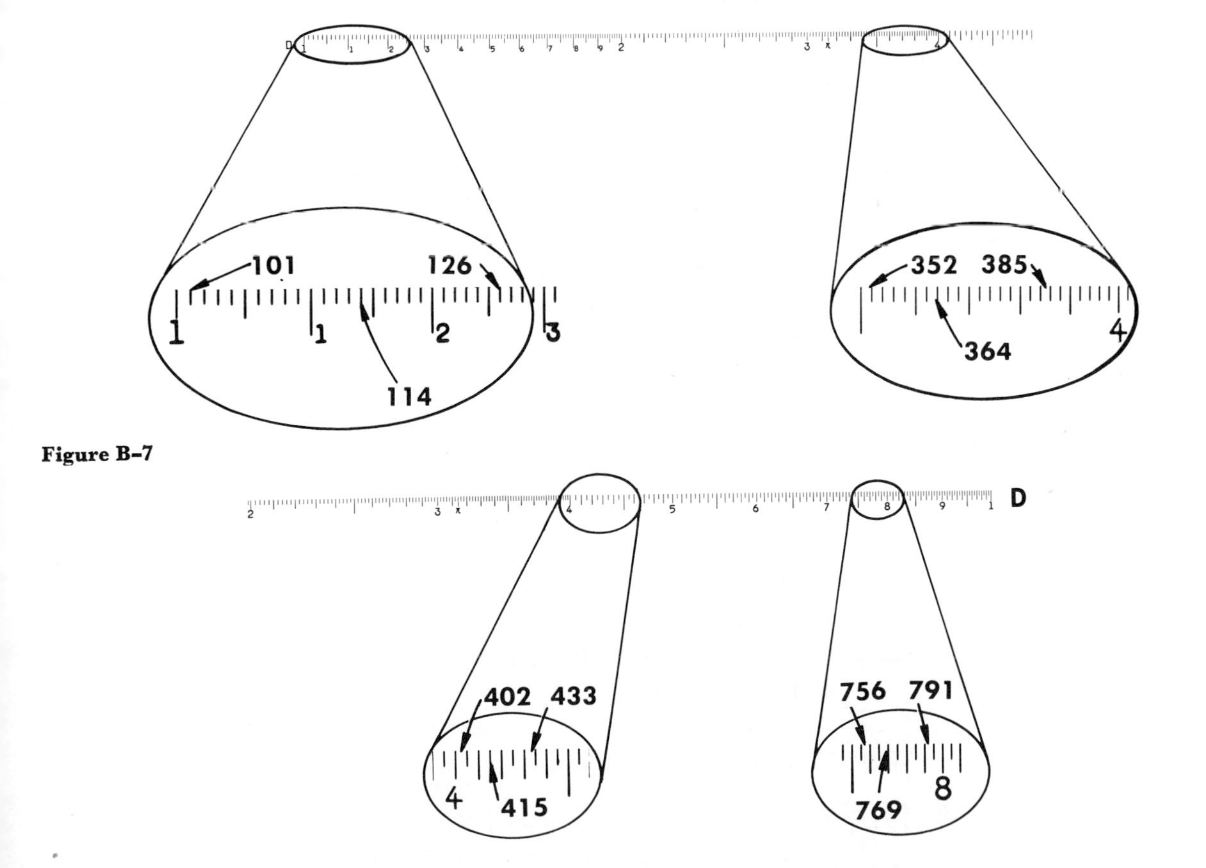

Figure B–7

the steps involved in setting a number. The first digit, 4, indicates the starting point for any numerical value having 4 as the first digit—4, 40, 40,000, 0.004, and so on. The second digit makes the number 43, which means using the secondary graduations to place the value 3/10 of the distance between 40 and 50. If the number were 43,000, the same placement of the hairline would indicate 3/10 of the distance between 40,000 and 50,000. Remember that the final placement of the decimal point is determined by scientific notation. The third digit makes the number 433—a value which is 3/10 of the distance between 430 and 440. In summary: first digit is primary graduation, second digit is secondary graduation, and third digit is the actual or estimated tertiary graduation.

EXAMPLE B–1

Set the number 14.730 (Fig. B–8) on the D scale.

1. Round the number off to three digits:

   147

2. Find the digits on the D scale, and set the hairline:

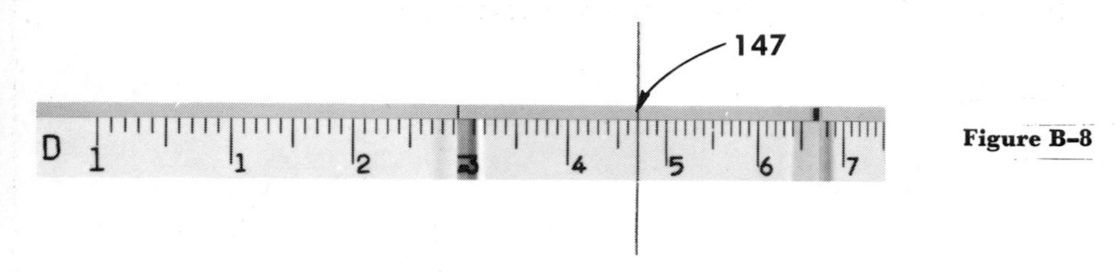

Figure B–8

EXAMPLE B–2

Set the number 0.03637 (Fig. B–9) on the D scale.

1. Round the number off to three digits:

   364

2. Find 364 on the D scale, and set the hairline.

EXAMPLE B–3

Set the number 8.23×10^3 (Fig. B–10) on the CI scale.

1. Use the three digits 823.
2. Find 823 on the CI scale.

   Note: The CI scale runs in the reverse direction when compared to the D scale.

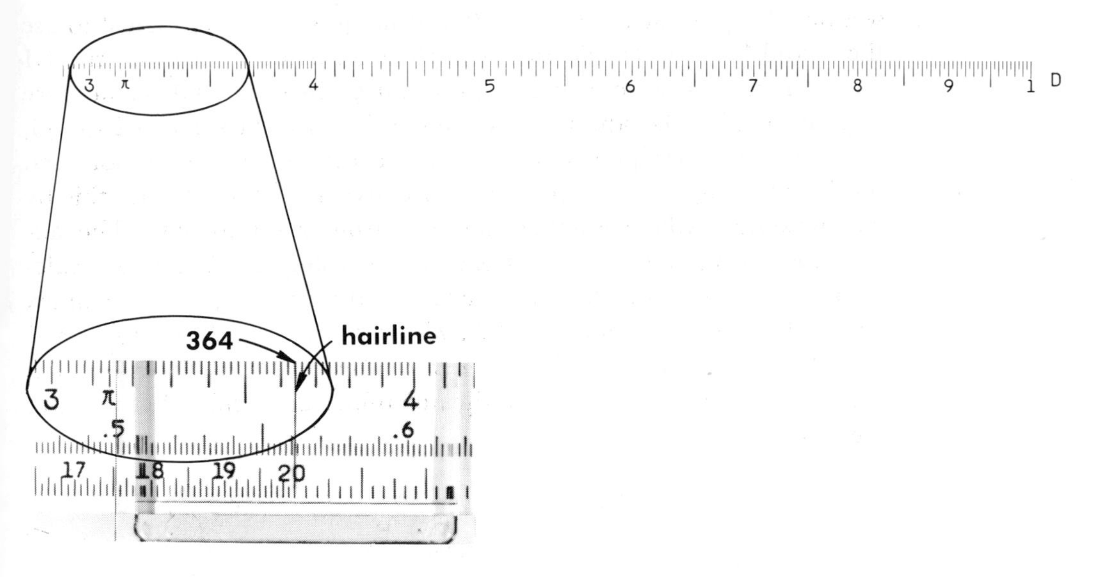

Figure B–9

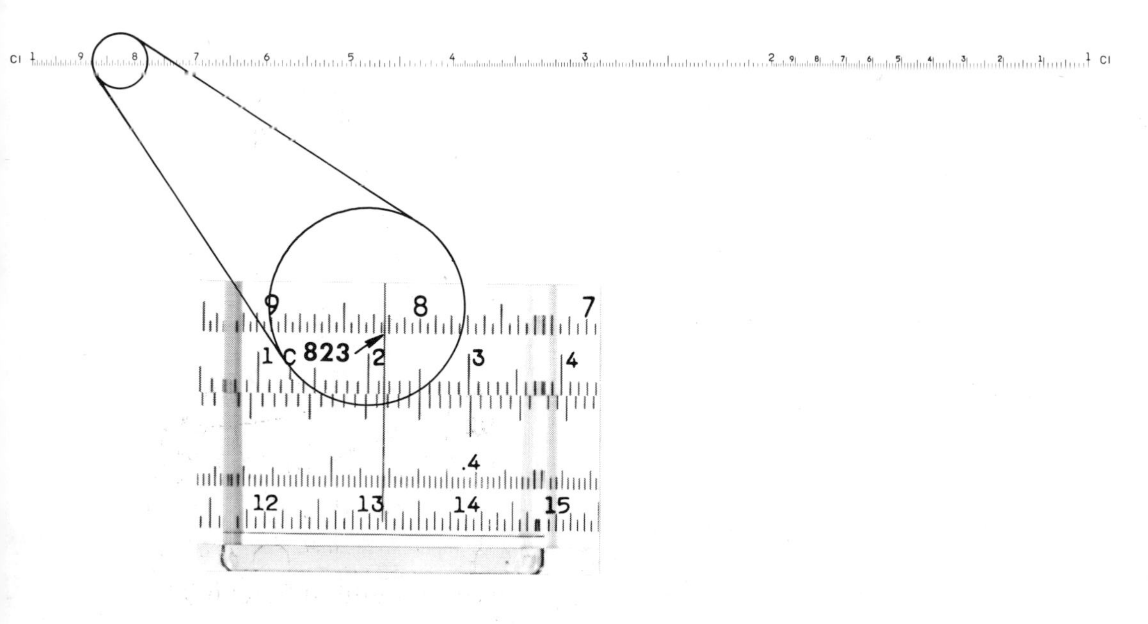

Figure B–10

THE MULTIPLICATION PROCESS

The most efficient method of multiplication is to use the CI and D scales. The CI scale is often printed in red near the center of the slide section. Despite the observation that it might seem more logical to use the C and D scales to add the logarithms for a multiplication, the CI scale substitutes a subtraction procedure, since the CI values are reciprocals of the D values. For example, 2×3 is the same as $2 \div 1/3$, where 1/3 is the reciprocal of 3. However, the slide rule operator need not be concerned about reciprocals because the CI scale does this by having simple whole numbers increase from right to left. The advantage is simplicity of operation. While using the C and D scales involves some guesswork with regard to setting the right or left index on the D scale multiplier, *the CI scale simply requires the setting of both multipliers on the hairline. The answer is found on the D scale under the index.* There will be only one answer indicated under the right *or* left index of the C scale.

EXAMPLE B–4

Multiply (602) (31.87) (Fig. B–11).

1. Select either number (make it 602) and set it on the D scale with the hairline.
2. Move the 319 (31.87 rounded off) on the CI scale to the hairline.

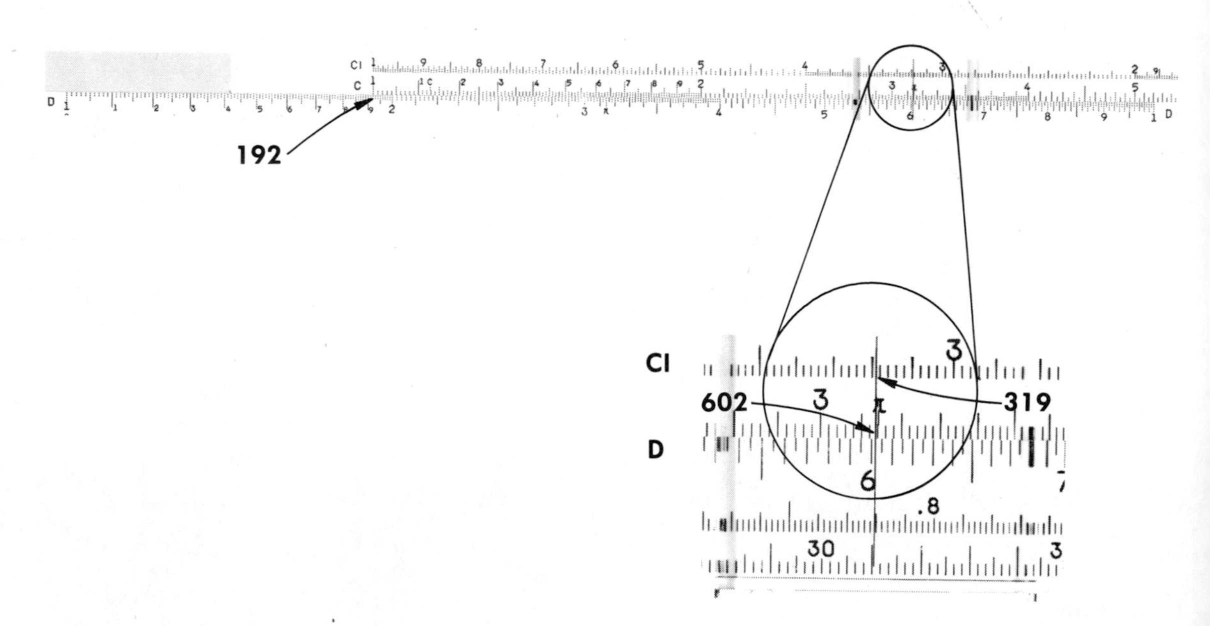

Figure B–11

Caution: Remember that the numbering of the CI scale runs in reverse—right to left.

3. Read the answer to three digits under the C scale index. It is 192.
4. Use scientific notation with numbers drastically rounded off to find the true answer:

$$\text{let } 602 \text{ be } 600 = 6 \times 10^2$$

$$\text{let } 31.87 \text{ be } 30 = 3 \times 10^1$$

Multiply as indicated:

$$(6 \times 10^2)(3 \times 10^1) = 18 \times 10^3$$

5. The answer is a two digit number $\times$ 10^3 or, preferably, a one digit number $\times$ 10^4:

$$\mathbf{19}.2 \times 10^3 \quad \text{or} \quad \mathbf{1}.92 \times 10^4 \text{ (preferred)}$$

2 digits 1 digit

Exercise B–2

1. Use a slide rule to perform the following multiplications. Find the answers to three significant figures by scientific notation. Express the answers with one digit before the decimal point:
   a. $(1.71)(2.04) =$
   b. $(82.37)(406) -$
   c. $(36710)(5322) =$
   d. $(6.84 \times 10^3)(7.23 \times 10^4) =$
   e. $(2.95 \times 10^{-4})(9.05 \times 10^{-2}) =$
   f. $(0.01278)(0.000312) =$
   g. $(0.00511)(6239130) =$
   h. $(7.18 \times 10^4)\left(\frac{2.24}{10^2}\right) =$

THE DIVISION PROCESS

Division uses the C and D scales. In this case the *dividend* must be set on the D scale and the *divisor* on the C scale. The answer is found on the D scale under the left or right index. For example, $6 \div 2$ (or, $6/2 = 2\overline{)6}$) indicates that the dividend, 6, is placed on the D scale,

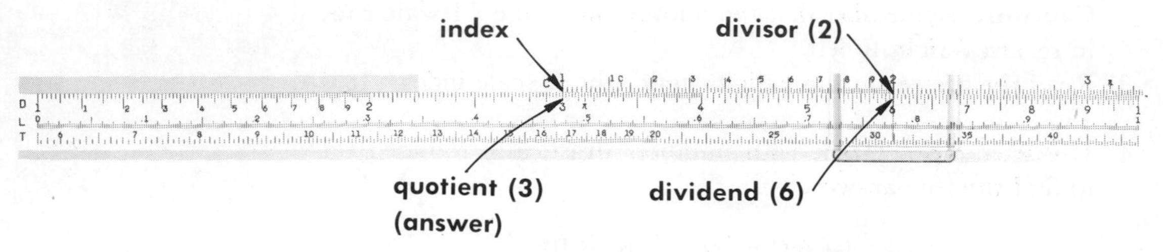

Figure B–12

and the divisor, 2, is placed above it on the hairline. The answer, 3, will appear under the left index (Fig. B–12).

EXAMPLE B–5

Divide 471.6 by 23,041.0 (Fig. B–13).

1. The dividend, rounded off to 472, is placed on the D scale and marked with the hairline.
2. The divisor, rounded off to 230, is located on the C scale and set on the hairline above 472.
3. The three digit answer, 205, is located under the left index.
4. Use scientific notation to determine the actual answer:

$$\text{let } 471.6 \text{ be } 500 = 5 \times 10^2$$

$$\text{let } 23041.0 \text{ be } 20000 = 2 \times 10^4$$

Divide as indicated:

$$\frac{5 \times 10^2}{2 \times 10^4} = 2.5 \times 10^{-2}$$

5. The answer is a one digit number $\times\ 10^{-2}$:

$$\mathbf{2}.05 \times 10^{-2}$$

one digit

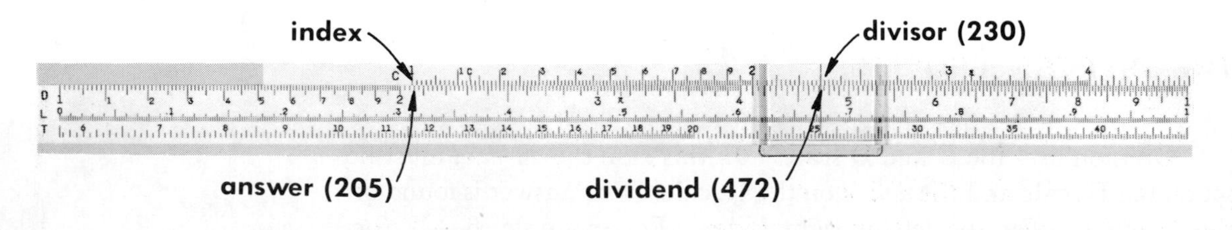

Figure B–13

Exercise B–3

1. Use a slide rule to perform the following divisions. Find the answers to three significant figures with one digit before the decimal point:

   a. $\dfrac{5.08}{2.71} =$

   b. $\dfrac{43.17}{618} =$

   c. $\dfrac{525}{0.0266} =$

   d. $\dfrac{0.0004827}{0.00741} =$

   e. $\dfrac{5.18 \times 10^{3}}{2.455 \times 10^{-3}} =$

   f. $\dfrac{8.27 \times 10^{-2}}{0.0511 \times 10^{-4}} =$

   g. $\dfrac{4.38}{7.18 \times 10^{2}} =$

   h. $\dfrac{2.316 \times 10^{5}}{\dfrac{1}{4.12 \times 10^{-3}}} =$

THE SQUARE AND SQUARE ROOT PROCESS

The finding of the square root of a number or raising a number to the second power is accomplished by using the A and D scales.

It is important to note that the A scale is divided into right and left sections in which both sections cover the distance between 1 and 10 (Fig. B–14). The significance of this arrangement is related to the

Figure B–14

number of digits in the answer when a number on the D scale is squared, and to the number of digits in the value which is the subject of the square root calculation. The rule is that *numerical values having an* **odd** *number of digits to the left of the decimal point are related to the* **left**-*hand side of the A scale, while values having an* **even** *number of digits are found on the* **right**-*hand side* (Fig. B–15). An example of number placement on the A scale is illustrated in Figure B–16.

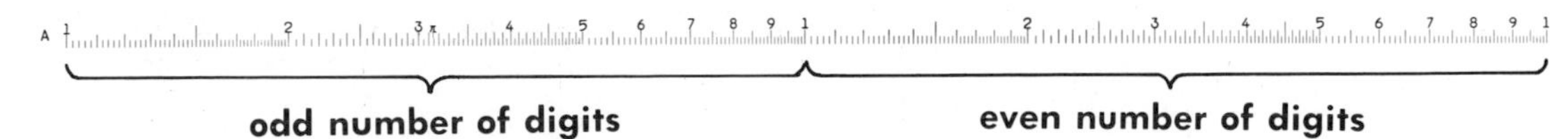

Figure B–15

Figure B–16 The square root of 13.5. (An even number of digits before the decimal point.)

The square root of 13.5 is found on the D scale under the hairline (Fig. B–17).

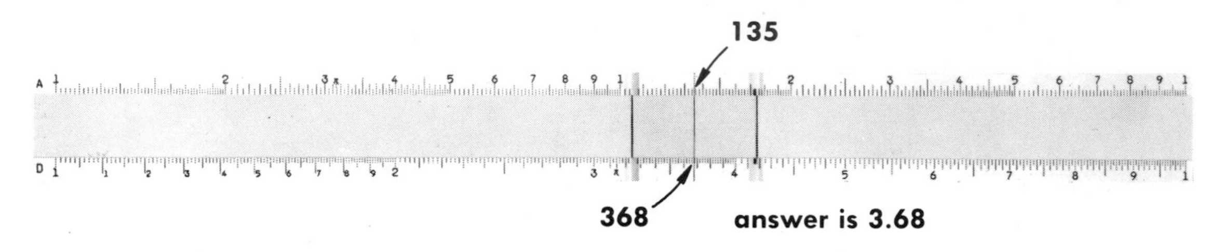

Figure B–17

EXAMPLE B–6

Find the square root of 240.50.

1. The value 240.50 is a three digit number, which is, of course, an odd number.
2. Set the hairline on the rounded off value, 241, on the *left* hand side of the A scale and read the answer (temporarily ignoring the placement of the decimal point), 155, under the hairline on the D scale (Fig. B–18).
3. A quick estimation provides the answer. The square root of 240.50 is 15.5 (to three significant figures).

   Note: A method of estimating the decimal placement in the answer is to bracket the value in question in *pairs*, starting from its decimal point. There is one digit in the answer for each pair. Some examples are

$$\sqrt{3.457} \quad \text{expanded to} \quad \overset{1.\ 8\ 6}{\sqrt{\underbrace{03}.\underbrace{45}\underbrace{70}}}$$

$$\sqrt{127.3} \quad \text{expanded to} \quad \overset{1\ 1.\ 3}{\sqrt{\underbrace{01}\underbrace{27}.\underbrace{30}}}$$

$$\sqrt{0.00361} \quad \text{expanded to} \quad \overset{0.\ 0\ 6\ 0\ 1}{\sqrt{\underbrace{00}.\underbrace{00}\underbrace{36}\underbrace{10}\underbrace{00}}}$$

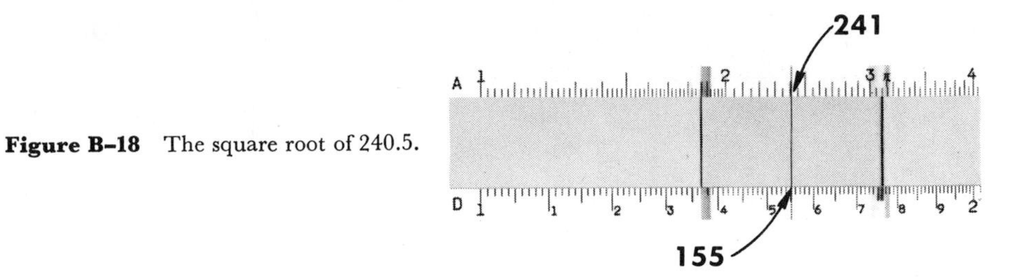

Figure B–18 The square root of 240.5.

A preferable alternative method is to express this value in the form of scientific notation:

$$\sqrt{0.00361} = \sqrt{36.1 \times 10^{-4}} = \overset{6.\ 0\ 1}{\sqrt{\underbrace{36}.\underbrace{10}\underbrace{00}}} \times 10^{-2}$$

EXAMPLE B–7

What is $\sqrt{7.22 \times 10^{-5}}$?

1. Adjust the number so that its exponent can be easily multiplied by 1/2. Remember that an exponential value raised to the 1/2 power is the same as taking the square root:

$$\sqrt{7.22 \times 10^{-5}} = \sqrt{72.2 \times 10^{-6}}$$

Note: Never forget that the changing of numerical value by a factor of 10 is a change of one order of magnitude.

2. $\sqrt{72.2 \times 10^{-6}} = \overset{?.\ ?}{\sqrt{\underbrace{72}.\underbrace{20}}} \times 10^{-3}$

3. Set the digits 722 on the *right* hand side of the A scale, since the value 72.2 has two (even number) digits before the decimal point (Fig. B–19). The answer is 8.50×10^{-3}.

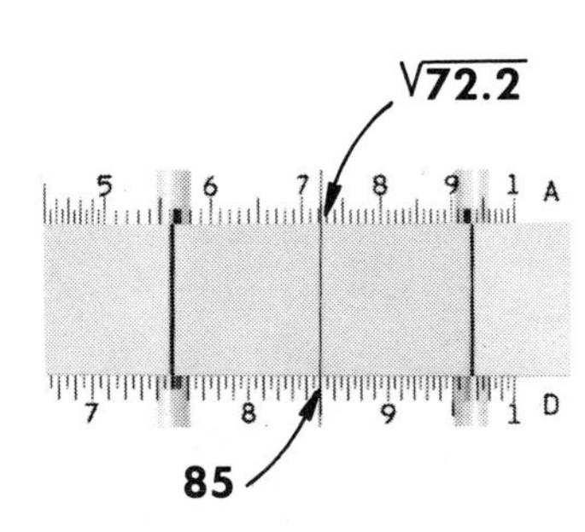

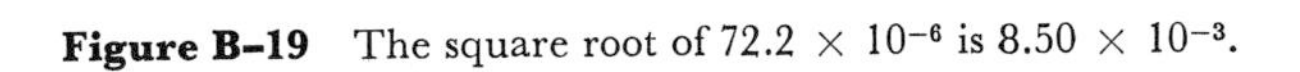

Figure B–19 The square root of 72.2×10^{-6} is 8.50×10^{-3}.

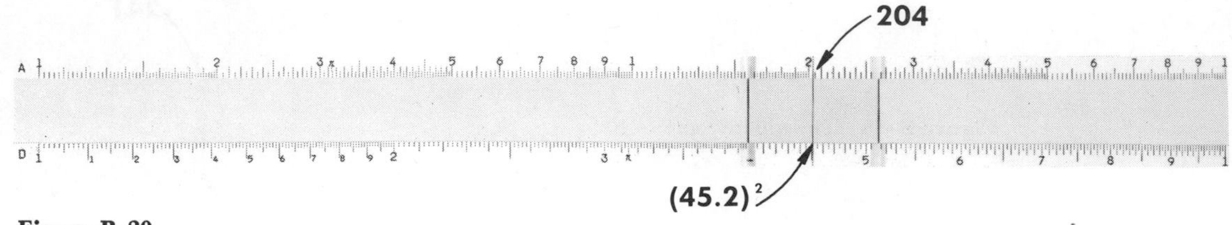

Figure B–20

Raising a number to the second power is simply the inverse operation of finding the square root. It is actually easier because the setting on the D scale of the number to be squared does not involve a right or left hand scale decision. The operator merely sets the hairline on the number on the D scale and reads the squared value under the hairline on the A scale. The answer clearly emerges as a numerical value having an odd or even number of digits, depending on which section of the A scale is indicated. The use of scientific notation may make the placement of the decimal point even easier. For example, $(45.2)^2$, set on the D scale, is read as the digits 204 on the A scale (Fig. B–20).

The answer, to three significant digits, may be estimated by scientific notation. For example, let 45.2 be 4.5×10^1. Then $(4.5 \times 10^1)^2 = \sim 20 \times 10^2 = \sim 2 \times 10^3$. The "squiggle" ($\sim$) means "approximately." The answer, $\sim 2 \times 10^3$, indicates an answer of one digit $\times 10^3$. The actual answer, then, is 2.04×10^3. If 2.04×10^3 is written as 2040, it can be seen that it is a four (even number) digit number that properly falls on the right-hand section of the A scale.

EXAMPLE B–8

What is 397.20 squared?

1. Set the three digits, 397, on the D scale and mark it with the hairline.
2. Find the three digit answer on the A scale. It is 158 (Fig. B–21).

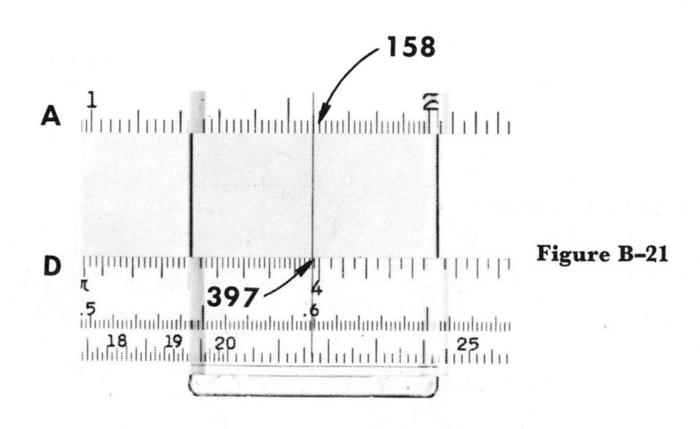

Figure B–21

3. The actual answer is obtained by scientific notation:

$$\text{let } 397.20 \text{ be } 400 = 4 \times 10^2$$

$$\text{then, } (4 \times 10^2)^2 = 16 \times 10^4$$

$$= 1.6 \times 10^5$$

Notice how the reduction of the number 16 (by a factor of ten) to 1.6 necessitates an increase of the exponent 10^4 (by one order of magnitude) to 10^5.
The answer is a one digit number before the decimal point $\times$ 10^5:

$$1.58 \times 10^5$$

Exercise B–4

1. Find the answers on the slide rule to the following questions. Express the answers with one digit before the decimal point and with three significant figures.
   a. $\sqrt{2.805}$
   b. $\sqrt{371.6 \times 10^2} =$
   c. $\sqrt{0.000324} =$
   d. $(8.92)^{1/2} =$
2. Raise the following values to the indicated power. Express the answers as described above.
   a. $(16.4)^2 =$
   b. $(7292000)^2 =$
   c. $(0.005238)^2 =$
   d. $(31.3 \times 10^6)^2 =$

THE CUBE AND CUBE ROOT PROCESS

Raising a number to the third power or finding its cube root is accomplished by using the D and K scales. A number set on the D scale can be read to the third power on the K scale, under the hairline. Conversely, the D scale reveals the cube root of a number set on the K scale.

Observe the K scale. Notice that it is marked off in three distinct sections, each of which goes from 1 to 10. Instead of referring to these sections as left, middle, and right, it is more convenient to refer to them as *first cycle* (left-hand section), *second cycle* (middle section), and *third cycle* (right-hand section). See Figure B–22.

Raising a number to the third power is simply a matter of setting it on the D scale and reading the answer on the K scale. The placement

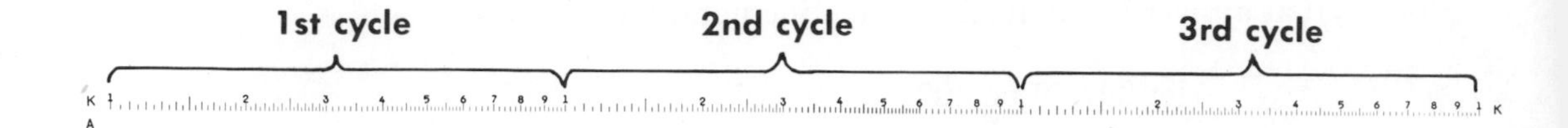

Figure B–22 The three cycles of the K scale.

of the decimal point can be determined by scientific notation as in the case of squares of numbers. A decision with regard to cycle is required for the extraction of cube roots. The rules may be summarized:

K Scale.

$\sqrt[3]{\text{one}}$ digit number = first cycle
$\sqrt[3]{\text{two}}$ digit number = second cycle
$\sqrt[3]{\text{three}}$ digit number = third cycle
$\sqrt[3]{\text{four}}$ digit number = first cycle
$\sqrt[3]{\text{five}}$ digit number = second cycle
$\sqrt[3]{\text{six}}$ digit number = third cycle

A sample cube root extraction is illustrated in Figure B–23, where the one, two, and three digit numbers, 8, 27, 125, are shown to have cube roots of 2, 3, and 5, respectively.

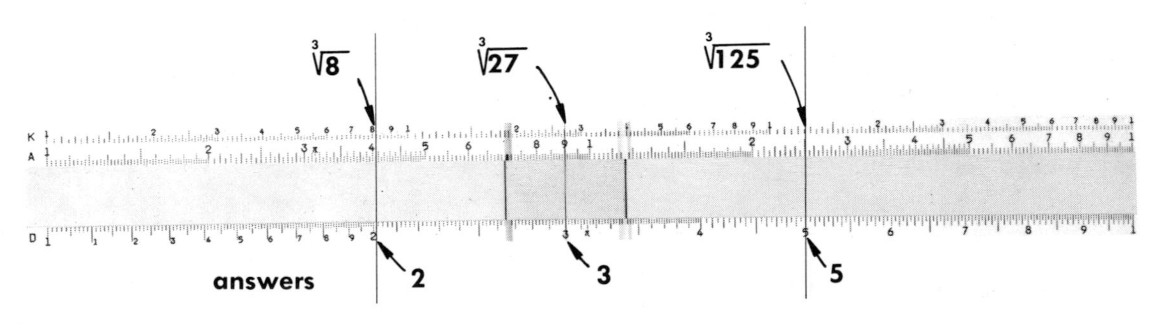

Figure B–23 The cube roots of 8, 27, and 125.

EXAMPLE B–9

Find the cube root of 84.00.

1. Since 84 is a two digit number, it is set under the hairline on the second cycle of the K scale.
2. The digits, 438, are read under the hairline on the D scale.
3. The actual value is quickly estimated as 4.38 (Fig. B–24).

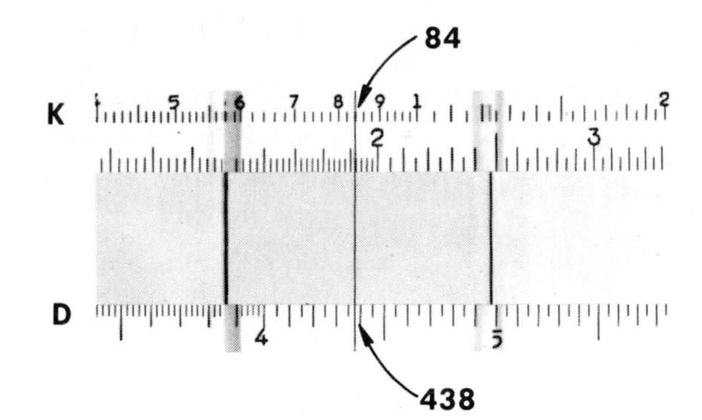

Figure B–24 The cube root of 84.

EXAMPLE B–10

What is $\sqrt[3]{72.1 \times 10^{-8}}$?

1. Alter the numerical value so that the exponential term yields a small whole number when multiplied by 1/3. Remember that the cube root of an exponent is the same as multiplying that exponent by 1/3. For example, $(2^6)^{1/3} = 2^2$, since 1/3 of 6 is 2.
2. Now find the cube root of the coefficient on the slide rule:

$$\sqrt[3]{72.1 \times 10^{-8}} = \sqrt[3]{721 \times 10^{-9}}$$
$$\sqrt[3]{721 \times 10^{-9}} = \sqrt[3]{721} \times 10^{-3}$$

$\sqrt[3]{721}$ is obtained by setting the digits 721 on the *third* cycle of the K scale. The three digit answer is 897 (Fig. B–25).

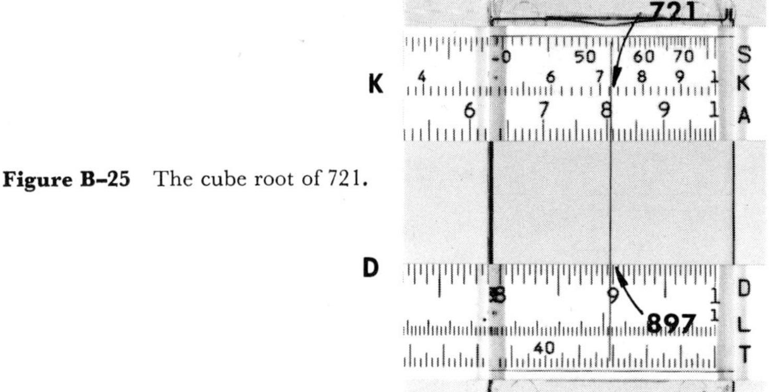

Figure B–25 The cube root of 721.

3. The actual cube root of 721 is easily estimated to 8.97.
4. The final answer is 8.97×10^{-3}

THE LOG AND ANTILOG PROCESS

The slide rule provides a much faster method for finding logarithms and antilogarithms than do standard tables. It also eliminates the time-consuming process of interpolating. Interpolation is a method of estimating a value that is not exactly determined on a log table.

The log-antilog relationship is exhibited directly between the D and L scales. The D scale is the number and the L scale is its logarithm.* For example, the log of 7.2 is found by setting 72 under the hairline on the D scale and reading its logarithm on the L scale as it is indicated by the hairline (Fig. B–26).

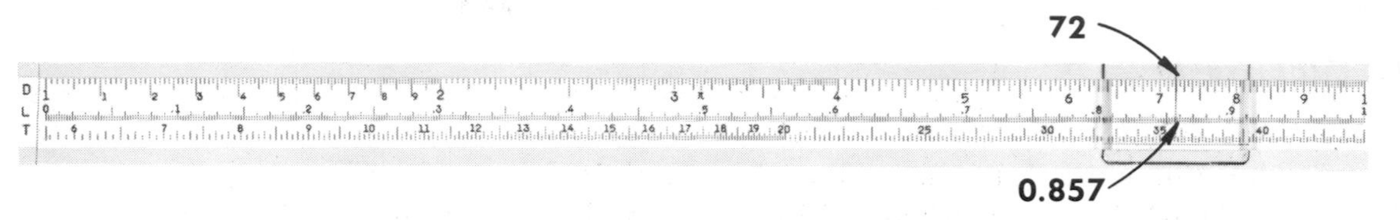

Figure B–26 The logarithm of 7.2.

If the number is larger than one digit, scientific notation is used. For example, the log of 200 is determined by writing 200 as 2×10^2. Since the log of a number is the same as the exponent of 10, the log of 10^2 is 2. This is added to the log of 2, which is $10^{0.301}$. Remember that $(10^{0.301})(10^2) = 10^{2.301}$. The log of 200, therefore, is 2.301.

EXAMPLE B–11

Find the log of 6722.

1. Change 6722 to 6.722×10^3.
2. Find the log of 6.72 (the first three significant figures).
3. The answer is 0.828, as illustrated in Figure B–27.
4. Since the log of 10^3 is 3, the answer is 3.828.

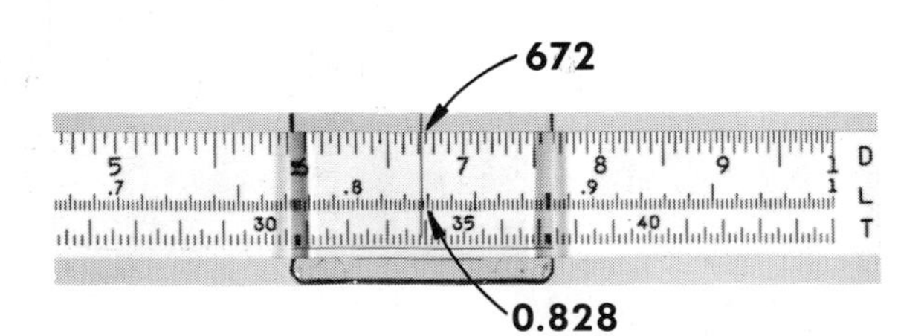

Figure B–27 The logarithm of 6.722.

* On some slide rules, the L scale is on the reverse side of the slide. However, these slide rules are usually equipped with a special index so that the slide does not have to be flipped over.

EXAMPLE B–12

Find the antilog of 2.65.

1. The antilog of 2 is 10^2.
2. The antilog of 0.65 is 4.47 (Fig. B–28).

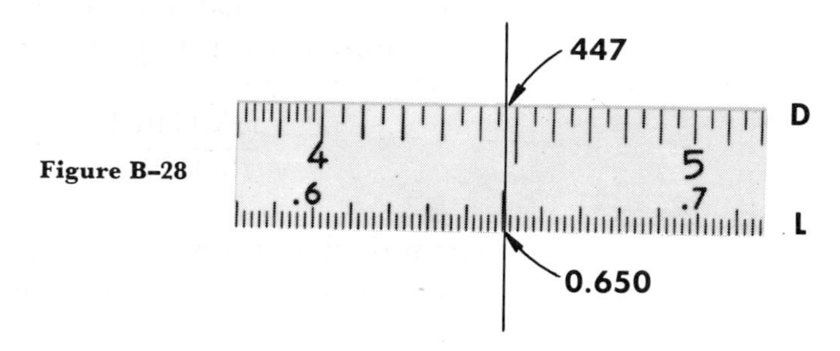

Figure B–28

3. The answer is the product of the two logs, 4.47×10^2.

Exercise B–5

1. Find the answers to the following problems on the slide rule. Express the answers to three significant figures with the decimal point after the first digit.
   a. $\sqrt[3]{18} =$
   b. $\sqrt[3]{47280} =$
   c. $\sqrt[3]{0.00752} =$
   d. $(6.15 \times 10^{-6})^{1/3} =$
2. Find the answers to the following in the same manner as described above:
   a. $(3.41)^3 =$
   b. $(724000)^3 =$
   c. $(0.00254)^3 =$
   d. $(8.26 \times 10^4)^3 =$
3. What are the logarithms of the following numbers?
   a. 3.88 =
   b. 247 =
   c. $2.73 \times 10^4 =$
   d. $6.05 \times 10^5 =$
4. Find the antilogs of the following:
   a. 0.622 =
   b. 4.38 =
   c. 3.041 =
   d. 6.0827 =

MULTIPLE OPERATIONS

Calculations in chemistry problems often involve more than a simple multiplication or division of two numbers. It is in the case of a more complex calculation that the great utility of the slide rule becomes most obvious. For example, if (23.17) (8400) is divided by (0.062) (190), the long-hand operation requires that the product of the numerator be divided by the product of the denominator:

$$\frac{(23.17)(8400)}{(0.062)(190)} = \frac{\text{numerator product}}{\text{denominator product}}$$

The procedure necessitates three separate and time-consuming operations. However, since the slide rule is a physical representation of logarithms, the four numbers are handled by addition-subtraction operations. Given a list of numbers to be added and subtracted, the operator can take them in *any* order, without concern for the proper order of operations that must be observed in multiplication and division.

The example above might be tackled as described in Example B–13.

EXAMPLE B–13

$$\frac{(23.17)(8400)}{(0.062)(190)} = ?$$

1. Round off the numbers to the three digits that are set on the slide rule:

$$\frac{(232)(84)}{(62)(19)}$$

2. Use the CI and D scales to multiply (in reality to add the logarithms) 232 and 84.
3. Set the hairline on the answer found on the D scale under the left index. The answer is 195.
4. Find the first divisor, 62, on the C scale and move it to the hairline. This is dividing 62 into the product of the numerator (in reality, subtracting the log of 62 from the sum of the logs of 232 and 84). The new answer is 314, found on the D scale under the right index.
5. Move the hairline to the indicated 314, and divide by 19 on the C scale (this is the subtraction of the last logarithm). The final three digit answer read on the D scale under the left index is estimated as 165 or 166. Both sets of digits are acceptable because of normal visual error.

6. Use scientific notation to determine the placement of the decimal point. Round off the numbers with reckless abandon, since the only purpose at this point is to find the decimal place:

a. $$\frac{(23.17)(8400)}{(0.062)(190)} = \frac{(20)(8000)}{(0.06)(200)} = \frac{(2 \times 10^1)(8 \times 10^3)}{(6 \times 10^{-2})(2 \times 10^2)}$$

b. Use a crude canceling approach to the reduction of the coefficients. Speed is the key!

$$\frac{(\not{2} \times 10^1)(8 \times 10^3)}{(6 \times 10^{-2})(\not{2} \times 10^2)} = \sim 1.5 \times 10^?$$

The coefficients reduce to a one digit number that is somewhere between 1 and 2.

c. Collect all the exponents, while being careful to observe the rules governing their multiplication and division (multiplication = add exponents; division = subtract exponents):

$$\frac{(10^1)(10^3)}{(10^{-2})(10^2)} = \frac{10^4}{10^0} = 10^4$$

d. The approximate answer is

1.5 × 10^4
one digit

The actual answer is the slide rule value as it is described by the approximation:

1.65 × 10^4
one digit

Exercise B–6

1. Solve the following problems by slide rule and scientific notation. Express the answers with one digit before the decimal point.

a. $$\frac{(273)(1.2)}{(14.6)} =$$

b. $$\frac{(821.3)(16.5)}{(72.4)(0.0812)} =$$

c. $$\frac{(3.29 \times 10^5)(22.4)}{(4.2 \times 10^4)(8255)} =$$

d. $\dfrac{17.62}{(0.00342)(1.082)} =$

e. $\dfrac{(7514)(30.16)(0.0445)}{(0.0000219)(72300)} =$

A typical slide rule has scales that have not been discussed. The varieties of operations that can be performed depend on the scales present. Expensive slide rules are most often made for greater versatility rather than for greater accuracy. However, the scales and operations discussed in this chapter have been limited to those that are relevant to ordinary chemistry calculations. An attempt to answer the needs of an engineer would be beyond the scope of this text.

It must be emphasized that the considerable gap between knowing how to use a slide rule and actually using it with speed, accuracy, and confidence can be bridged only by regular use. The situation is analogous to using a typewriter—an instrument that could be quite useless if the aspiring typist did not put the learning of the "touch system" into practice.

APPENDIX C

REFERENCE TABLES

A SUMMARY OF COMMON OXIDATION NUMBERS (OR, IONIC CHARGES)

Cations

1+	2+		3+	4+
H^+	Mg^{2+}	Fe^{2+}	Al^{3+}	C^{4+}
Li^+	Ca^{2+}	Cu^{2+}	Co^{3+}	Si^{4+}
Na^+	Ba^{2+}	Pb^{2+}	Ni^{3+}	Pb^{4+}
K^+	Sr^{2+}	Mn^{2+}	Fe^{3+}	Mn^{4+}
NH_4^+	Cd^{2+}	Sn^{2+}	Cr^{3+}	Sn^{4+}
Hg_2^{2+}	Zn^{2+}	Cr^{2+}	Bi^{3+}	
Cu^+	Co^{2+}		Sb^{3+}	
Ag^+	Hg^{2+}		As^{3+}	
Rb^+	Ni^{2+}			
Cs^+				

Anions

3−	2−		1−		
N^{3-}	O^{2-}	CrO_3^{2-}	H^-	OCN^-	HCO_3^-
P^{3-}	S^{2-}	CrO_4^{2-}	OH^-	SCN^-	MnO_4^-
PO_3^{3-}	CO_3^{2-}	$Cr_2O_7^{2-}$	F^-	ClO^-	$H_2PO_4^-$
PO_4^{3-}	SO_3^{2-}	HPO_4^{2-}	Cl^-	ClO_2^-	HSO_3^-
AsO_3^{3-}	SO_4^{2-}	$C_2O_4^{2-}$	Br^-	ClO_3^-	HSO_4^-
AsO_4^{3-}	$S_2O_3^{2-}$		I^-	ClO_4^-	O_2^{2-}
	$S_2O_4^{2-}$		CN^-	NO_2^-	OAc^-
				NO_3^-	HS^-

VAPOR PRESSURE OF WATER

Temp. °C	Torr	Temp. °C	Torr
0	4.6	28	28.3
5	6.5	29	30.0
10	9.2	30	31.8
15	12.8	31	33.7
16	13.6	32	35.7
17	14.5	33	37.7
18	15.5	34	39.9
19	16.5	35	42.2
20	17.5	40	55.3
21	18.6	50	92.5
22	19.8	60	149.3
23	21.0	70	233.7
24	22.4	80	355.1
25	23.8	90	525.8
26	25.2	100	760.0
27	26.7		

BOILING AND FREEZING POINT DATA FOR SELECTED SOLVENTS

Solvent	Formula	T_f(°C)	K_f(°C)	T_b(°C)	K_b(°C)
benzene	C_6H_6	5.48	5.12	80.15	2.53
carbon disulfide	CS_2	—	—	46.25	2.34
carbon tetrachloride	CCl_4	−22.8	29.8	76.8	5.02
acetic acid	HOAc	16.6	3.90	118.1	3.07
chloroform	$CHCl_3$	−63.5	4.68	61.2	3.63
water	H_2O	0	1.86	100	0.52
naphthalene	$C_{10}H_8$	80.2	6.80	—	—
camphor	$C_{10}H_{16}O$	178.4	37.7	208.3	5.95
cyclohexane	C_6H_{12}	6.5	20.0	80.9	2.79

SIMPLIFIED TABLE OF THE SOLUBILITY OF COMMON SALTS IN WATER AT 20°C

Anion	Cation	Solubility
acetate, chlorate, nitrate	nearly all	soluble
chloride, bromide, iodide	lead(II), silver, mercury(I)	insoluble
	all others	soluble
hydroxide	group I metals, barium strontium	soluble
	all others*	insoluble
sulfate	mercury(I) and (II), calcium, barium, strontium, silver, lead(II)	insoluble
	all others	soluble
carbonate, phosphate, chromate	group I metals,** ammonium	soluble
	all others	insoluble
sulfide	group I metals, ammonium, magnesium, calcium, barium	soluble
	all others	insoluble

* Li_3PO_4 insoluble.
** $Ca(OH)_2$ slightly soluble.

BRØNSTED-LOWRY ACID-BASE CONJUGATE PAIRS

(Acids: INCREASING STRENGTH ↑ ; Bases: INCREASING STRENGTH ↓)

Acid	Base
$HClO_4$	ClO_4^-
HI	I^-
H_2SO_4	HSO_4^-
HNO_3	NO_3^-
HCl	Cl^-
H_3O^+	H_2O
H_2SO_3	HSO_3^-
HSO_4^-	SO_4^{2-}
H_3PO_4	$H_2PO_4^-$
HF	F^-
HNO_2	NO_2^-
$HOAc$	OAc^-
$Al(H_2O)_6^{3+}$	$Al(H_2O)_5(OH)^{2+}$
H_2S	HS^-
HSO_3	SO_3^{2-}
NH_4^+	NH_3
HCN	CN^-
HCO_3^-	CO_3^{2-}
HPO_4^{2-}	PO_4^{3-}
HS^-	S^{2-}
H_2O	OH^-
CH_3OH	CH_3O^-
NH_3	NH_2^-
OH^-	O^{2-}
H_2	H^-

INCREASING STRENGTH

INCREASING STRENGTH

EQUILIBRIUM CONSTANTS OF WEAK ACIDS AT 25°C

(INCREASINGLY WEAK ACIDS ↓)

Name	First Ionization Reaction	K_a
oxalic	$H_2C_2O_4 \rightleftharpoons H^+ + HC_2O_4^-$	5.6×10^{-2}
sulfurous	$H_2SO_3 \rightleftharpoons H^+ + HSO_3^-$	1.7×10^{-2}
phosphoric	$H_3PO_4 \rightleftharpoons H^+ + H_2PO_4^-$	5.9×10^{-3}
hydrofluoric	$HF \rightleftharpoons H^+ + F^-$	6.7×10^{-4}
nitrous	$HNO_2 \rightleftharpoons H^+ + NO_2^-$	5.1×10^{-4}
acetic	$HOAc \rightleftharpoons H^+ + OAc^-$	1.8×10^{-5}
hydrocyanic	$HCN \rightleftharpoons H^+ + CN^-$	4.8×10^{-10}

TABLE OF ATOMIC WEIGHTS

(Based on Carbon-12)

	Symbol	*Atomic No.*	*Atomic Weight*
Actinium	Ac	89	227
Aluminum	Al	13	26.9815
Americium	Am	95	[243]*
Antimony	Sb	51	121.75
Argon	Ar	18	39.948
Arsenic	As	33	74.9216
Astatine	At	85	[210]
Barium	Ba	56	137.34
Berkelium	Bk	97	[249]
Beryllium	Be	4	9.0122
Bismuth	Bi	83	208.980
Boron	B	5	10.811
Bromine	Br	35	79.909
Cadmium	Cd	48	112.40
Calcium	Ca	20	40.08
Californium	Cf	98	[251]
Carbon	C	6	12.01115
Cerium	Ce	58	140.12
Cesium	Cs	55	132.905
Chlorine	Cl	17	35.453
Chromium	Cr	24	51.996
Cobalt	Co	27	58.9332
Copper	Cu	29	63.54
Curium	Cm	96	[247]
Dysprosium	Dy	66	162.50
Einsteinium	Es	99	[254]
Erbium	Er	68	167.26
Europium	Eu	63	151.96
Fermium	Fm	100	[253]
Fluorine	F	9	18.9984
Francium	Fr	87	[223]
Gadolinium	Gd	64	157.25
Gallium	Ga	31	69.72
Germanium	Ge	32	72.59
Gold	Au	79	196.967
Hafnium	Hf	72	178.49
Hahnium	Ha	105	[260]
Helium	He	2	4.0026
Holmium	Ho	67	164.930
Hydrogen	H	1	1.00797
Indium	In	49	114.82
Iodine	I	53	126.9044
Iridium	Ir	77	192.2
Iron	Fe	26	55,847
Krypton	Kr	36	83.80
Kurchatovium	Ku	104	[257]
Lanthanum	La	57	138.91
Lawrencium	Lw	103	[257]
Lead	Pb	82	207.19
Lithium	Li	3	6.939
Lutetium	Lu	71	174.97
Magnesium	Mg	12	24.312
Manganese	Mn	25	54.9380
Mendelevium	Md	101	[256]
Mercury	Hg	80	200.59
Molybdenum	Mo	42	95.94
Neodymium	Nd	60	144.24
Neon	Ne	10	20.183
Neptunium	Np	93	[237]
Nickel	Ni	28	58.71
Niobium	Nb	41	92.906
Nitrogen	N	7	14.0067
Nobelium	No	102	[253]
Osmium	Os	76	190.2
Oxygen	O	8	15.9994
Palladium	Pd	46	106.4
Phosphorus	P	15	30.9738
Platinum	Pt	78	195.09
Plutonium	Pu	94	[242]
Polonium	Po	84	210
Potassium	K	19	39.102
Praseodymium	Pr	59	140.907
Promethium	Pm	61	[145]
Protactinium	Pa	91	231
Radium	Ra	88	226.05
Radon	Rn	86	222
Rhenium	Re	75	186.2
Rhodium	Rh	45	102.905
Rubidium	Rb	37	85.47
Ruthenium	Ru	44	101.07
Samarium	Sm	62	150.35
Scandium	Sc	21	44.956
Selenium	Se	34	78.96
Silicon	Si	14	28.086
Silver	Ag	47	107.870
Sodium	Na	11	22.9898
Strontium	Sr	38	87.62
Sulfur	S	16	32.064
Tantalum	Ta	73	180.948
Technetium	Tc	43	[99]
Tellurium	Te	52	127.60
Terbium	Tb	65	158.924
Thallium	Tl	81	204.37
Thorium	Th	90	232.038
Thulium	Tm	69	168.934
Tin	Sn	50	118.69
Titanium	Ti	22	47.90
Tungsten	W	74	183.85
Uranium	U	92	238.03
Vanadium	V	23	50.942
Xenon	Xe	54	131.30
Ytterbium	Yb	70	173.04
Yttrium	Y	39	88.905
Zinc	Zn	30	65.37
Zirconium	Zr	40	91.22

*A value given in brackets denotes the mass number of the longest-lived or best-known isotope.

PERIODIC CHART

	IA	IIA	IIIB	IVB	VB	VIB	VIIB	VIII	VIII	VIII	IB	IIB	IIIA	IVA	VA	VIA	VIIA	O
1	1 1.00797 H Hydrogen																	2 4.0026 He Helium
2	3 6.939 Li Lithium	4 9.0122 Be Beryllium											5 10.811 B Boron	6 12.01115 C Carbon	7 14.0067 N Nitrogen	8 15.9994 O Oxygen	9 18.9984 F Fluorine	10 20.183 Ne Neon
3	11 22.9898 Na Sodium	12 24.312 Mg Magnesium											13 26.9815 Al Aluminum	14 28.086 Si Silicon	15 30.9738 P Phosphorus	16 32.064 S Sulfur	17 35.453 Cl Chlorine	18 39.948 Ar Argon
4	19 39.102 K Potassium	20 40.08 Ca Calcium	21 44.956 Sc Scandium	22 47.90 Ti Titanium	23 50.942 V Vanadium	24 51.996 Cr Chromium	25 54.938 Mn Manganese	26 55.847 Fe Iron	27 58.933 Co Cobalt	28 58.71 Ni Nickel	29 63.54 Cu Copper	30 65.37 Zn Zinc	31 69.72 Ga Gallium	32 72.59 Ge Germanium	33 74.922 As Arsenic	34 78.96 Se Selenium	35 79.909 Br Bromine	36 83.80 Kr Krypton
5	37 85.47 Rb Rubidium	38 87.62 Sr Strontium	39 88.905 Y Yttrium	40 91.22 Zr Zirconium	41 92.906 Nb Niobium	42 95.94 Mo Molybdenum	43 (98) Tc Technetium	44 101.07 Ru Ruthenium	45 102.905 Rh Rhodium	46 106.4 Pd Palladium	47 107.870 Ag Silver	48 112.40 Cd Cadmium	49 114.82 In Indium	50 118.69 Sn Tin	51 121.75 Sb Antimony	52 127.60 Te Tellurium	53 126.904 I Iodine	54 131.30 Xe Xenon
6	55 132.905 Cs Cesium	56 137.34 Ba Barium	57 138.91 ★ La Lanthanum	72 178.49 Hf Hafnium	73 180.948 Ta Tantalum	74 183.85 W Wolfram	75 186.2 Re Rhenium	76 190.2 Os Osmium	77 192.2 Ir Iridium	78 195.09 Pt Platinum	79 196.967 Au Gold	80 200.59 Hg Mercury	81 204.37 Tl Thallium	82 207.19 Pb Lead	83 208.980 Bi Bismuth	84 (210) Po Polonium	85 (210) At Astatine	86 (222) Rn Radon
7	87 (223) Fr Francium	88 (226) Ra Radium	89 (227) ★★ Ac Actinium															

★ 6	58 140.12 Ce Cerium	59 140.907 Pr Praseodymium	60 144.24 Nd Neodymium	61 (147) Pm Promethium	62 150.35 Sm Samarium	63 151.96 Eu Europium	64 157.25 Gd Gadolinium	65 158.924 Tb Terbium	66 162.50 Dy Dysprosium	67 164.930 Ho Holmium	68 167.26 Er Erbium	69 168.934 Tm Thulium	70 173.04 Yb Ytterbium	71 174.97 Lu Lutetium
★★ 7	90 232.038 Th Thorium	91 (231) Pa Protactinium	92 238.03 U Uranium	93 (237) Np Neptunium	94 (242) Pu Plutonium	95 (243) Am Americium	96 (245) Cm Curium	97 (249) Bk Berkelium	98 (251) Cf Californium	99 (254) Es Einsteinium	100 (255) Fm Fermium	101 (256) Md Mendelevium	102 (254) No Nobelium	103 (257) Lw Lawrencium

APPENDIX D

ANSWERS TO NUMERICAL PROBLEMS

CHAPTER 1

Exercise 1.1

a. 1.53×10^{-2} m
b. 27 m
c. 1.53×10^{-5} cm
d. 2.41×10^{-5} cm
e. 1.2×10^{-6} m
f. 8.1×10^{4} nm
g. 2.3×10^{-5} km
h. 15 nm
i. 4.4×10^{-5} μm
j. 3.5 mm

Exercise 1.2

a. 2.72×10^{-2} ℓ
b. 84 ml
c. 0.176 ml
d. 2.0×10^{4} μl
e. 56 ℓ
f. $2 \times 10^{3} \lambda$
g. 5.4 ml
h. 0.260 ℓ
i. 31 cm^3
j. 49 μl

Exercise 1.3

a. 0.325 g
b. 1.4×10^{-3} kg
c. 3.3×10^{3} mg
d. 30 μg
e. 0.15 μg
f. 4.28×10^{-3} g
g. 9×10^{-5} g
h. 0.6 g
i. 2.3×10^{-5} mg
j. 7.2×10^{4} μg

Exercise 1.4

a. −8.3 °C
b. 392 °F
c. −14 °C
d. −364 °F
e. 283 K
f. −158 °C
g. 242.5 K
h. 277.4 K

Exercise 1.5

1. 0.64 g ml^{-1}
2. 0.19 ml
3. 4.83 cm^3
4. 1.84
5. 1.43 g ℓ^{-1}

Problems and Questions

1.1 a. 0.351 km
b. 1.4 cm
c. 6.50×10^{-5} cm
d. 520 Å
e. 2×10^{-5} μm
f. 13 nm

1.2 a. 4×10^{2} g
b. 0.512 g
c. 7×10^{-2} g
d. 4.5×10^{3} μg
e. 2 mg
f. 6.7×10^{2} mg

1.3 a. 120 ml
b. 3.7×10^{-2} μl
c. 0.185 ℓ
d. 28 cm^3
e. 8×10^{-4} ml
f. 5 ml

1.6 a. 41 °F
b. −94 °F
c. 71.1 °C
d. 193 K
e. −43 °C
f. 263.6 K

1.7 4.2×10^7 erg cal^{-1}
1.8 33.4 ml
1.9 0.96 g cm^{-3}
1.10 924 g
1.11 50.2 ml
1.12 18.31 g ml^{-1}

CHAPTER 2

Exercise 2.1

1. 1.25×10^5 ergs
2. 1.25×10^{-2} J
3. 1.0×10^7 cm s^{-1}
4. 639 cal
5. 0.583 cal g^{-1} $°C^{-1}$

Exercise 2.2

1. 4.8×10^{-4} cal
2. 4.5×10^2 ergs
3. 2.0×10^{-6} kcal
4. 3.4×10^3 ℓ-atm
5. 15.0 eV
6. 1.21 cal
7. 4.2×10^4 eV
8. 2.1×10^6 cal

Exercise 2.3

1. 1.5×10^{15} Hz
2. 7.5×10^{-3} cm
3. 331 m s^{-1}
4. 1.4×10^{16} Hz

Exercise 2.4

1. 9.93×10^{-12} erg
2. 1.1×10^{-5} cm
3. 1.22×10^{15} Hz
4. 230 kcal

Exercise 2.5

1. 4×10^5 Ω
2. 24.0 Ω
3. 600 volts

Problems and Questions

2.7 2.8×10^3 cm s^{-1}
2.8 3×10^2 ergs
2.9 715 cal
2.10 6.15 liter-atm, 3.9×10^{21} eV
2.11 9.4×10^{14} Hz
2.12 3.3×10^{-11} erg

CHAPTER 3

Exercise 3.2

1. 28.09 amu
2. 3.35 g
3. 2.22×10^{-3} mole
4. 3.44×10^{-22} g
5. 7.2×10^{21} atoms

Exercise 3.3

1. a. 199.9 g $mole^{-1}$
   b. 68.0 g $mole^{-1}$
   c. 60.0 g $mole^{-1}$
   d. 342.0 g $mole^{-1}$
   e. 249.5 g $mole^{-1}$
2. 4.5×10^{-2} g
3. 3.5×10^{-3} mole
4. 7.3×10^{-23} g
5. 2.45×10^{23} molecules

Exercise 3.4

1. 1.3×10^{-2} mole
2. 7.2×10^{-2} liter
3. 5.4×10^{22} molecules
4. 1.96 g $liter^{-1}$
5. 43.7 g $mole^{-1}$

Problems and Questions

3.7 0.25 mole
3.8 6.3×10^{-3} mole
3.9 7.3×10^{-2} g
3.10 3.27×10^{-22} g
3.11 3.26 g
3.12 5.6 liters
3.13 0.057 cal g^{-1} $°C^{-1}$
3.14 42.8 g eq^{-1}
3.15 84.9 g $mole^{-1}$

CHAPTER 4

Exercise 4.1

1. 2.15 liters
2. 135 ml
3. 3.33 atm
4. 701.4 torr

Exercise 4.2

1. 1.56 liters
2. 7.11×10^{3} ml
3. 60.9 ml
4. 2.98 g $liter^{-1}$, 66.8 g $mole^{-1}$

Exercise 4.3

1. 1.3×10^{-2} mole
2. 0.798 atm
3. 50.8 g $mole^{-1}$
4. 0.79 g
5. 0.46 g $liter^{-1}$

Problems and Questions

4.8 231 torr
4.11 252 ml
4.12 32.8 g $mole^{-1}$
4.13 0.73 liter
4.14 64.8 g $mole^{-1}$
4.15 17.3 g $liter^{-1}$

CHAPTER 5

Exercise 5.1

1. 7.2×10^{14} Hz
2. 278.7 kcal $mole^{-1}$
3. 1.2×10^{-5} cm, 2.46×10^{15} Hz
4. 4.8×10^{-19} J, 3.0 eV

Problems and Questions

5.5 9.9×10^{-3} cm
5.8 73.5 kcal $mole^{-1}$
5.9 62 eV

CHAPTER 7

Exercise 7.7

1. a. 32.85% K, 67.15% Br
   b. 54.1% Ca, 43.2% O, 2.7% H
   c. 2.0% H, 32.7% S, 65.3% O
   d. 29.1% Na, 40.5% S, 30.4% O
   e. 27.9% Fe, 0.8% H, 23.3% P, 48.0% O
2. 51.2% water
3. 51.0% In, 49.0% Cl

Exercise 7.8

1. a. CH_4
   b. Fe_3O_4
   c. $Mg_2P_2O_7$
   d. Fe_2O_3
   e. $C_{12}H_{11}N_4O_2$
2. Na_3PO_4
3. C_6H_{12}

Problems and Questions

7.12 a. 48.2% K, 19.7% O, 14.8% C, 17.3% N
b. 31.0% Fe, 15.6% N, 53.4% O
c. 72.4% Fe, 27.6% O
d. 17.7% N, 6.3% H, 15.2% C, 60.8% O
7.13 1.41 g
7.14 As_2O_3
7.15 C_6H_6

CHAPTER 8

Exercise 8.3

1. a. 1/16 mole
   b. 3.6 g
   c. 1.7 g
2. b. 0.2 mole
   c. 1.8×10^{-2} g
   d. 6.7×10^{-3} mole

Exercise 8.4

1. 0.397 liter
2. 3.45×10^{-2} g
3. 59.7 g
4. 0.23 liter

Problems and Questions

8.6 1.6 mole
8.7 26.0 g
8.8 0.49 liter
8.9 1.45 g
8.10 1.96 g
8.11 0.52 g
8.12 0.114 g
8.13 1.12 liters
8.14 909 ml
8.15 38.5%
8.16 H_2 is reaction limiting factor.

CHAPTER 9

Exercise 9.2

1. a. 11.9 g
   b. 25.8 g
2. 0.89 M
3. 10.3 M
4. 6.2×10^{21} molecules

Exercise 9.3

1. 27.3 ml
2. 8.75 ml
3. 7.5×10^{-2} liter
4. 29.2 ml
5. 5.83 ml

Exercise 9.4

1. a. 2.66 g
   b. 61.5 g
2. 6.67 ml
3. 44.75 N
4. 14.92 M

Exercise 9.5

1. 10.5 g
2. 1.43×10^{-4} ppm
3. 0.36 torr
4. 0.37 m
5. 5.7 g

Exercise 9.6

1. 76.0 g $mole^{-1}$
2. 0.48 g
3. 2.30 mole
4. $C_5H_{10}O_5$

Problems and Questions

9.7 1.49 g
9.8 0.45 g
9.9 0.18 M
9.10 18.1 M
9.11 0.83 ml
9.12 3.2 ml
9.13 12.3 g
9.14 0.58 N
9.15 1.0 ml
9.16 a. 4.0 M
b. 0.66 N
c. 0.16 M
d. 0.16 N
e. 2.5 F
9.17 4.0 g
9.18 0.073
9.19 1.52 torr
9.20 2.88 g
9.21 1.02 °C
9.22 $C_{12}H_{15}$

CHAPTER 10

Exercise 10.4

1. $[OH^-] = 3.1 \times 10^{-12}$ M
2. a. pH = 0.7
   b. pH = 1.1
   c. pH = 2.36
   d. pH = 8.77
   e. pH = 10.91
3. a. pOH = 3.7
   b. pOH = 1.0
   c. pOH = 7.43

d. pOH = 1.28
e. pOH = 3.86

4. a. 5×10^{-10} M
b. 2.5×10^{-12} M
c. 0.4 M
d. 6.3×10^{-13} M
e. 6.3×10^{-7} M

Exercise 10.5

1. 1.28 g
2. 1.17 ml
3. 0.34 g
4. 1.4 liters
5. 236 ml

Exercise 10.6

1. 0.49 N
2. 0.043 M
3. 0.14 N, 0.07 M

Problems and Questions

10.11 pH = 1.17, pOH = 12.83
10.12 a. pH = 1.4
b. pH = 3.89
c. pH = 2.49
d. pH = 7.21
e. pH = 9.65
10.13 a. 6.3×10^{-5} M
b. 5.0×10^{-13} M
c. 4.0×10^{-3} M
d. 5.0×10^{-9} M
e. 0.16 M
10.14 3.09 g
10.15 6.14 ml
10.16 4.49 ml
10.17 2.92 N
10.18 0.446 N, 0.223 M

CHAPTER 12

Exercise 12.7

1. 8.15 mg
2. 42.3 ml

Exercise 12.8

1. 44.6 ml
2. 2.78 M
3. 8.33 N

Exercise 12.9

1. 0.375 N
2. 0.31 N, 0.06 M
3. 0.635 g
4. 0.52 N

Problems and Questions

12.7 1.28 g
12.8 0.34 liter
12.9 584 ml
12.10 0.31 M
12.11 0.12 N
12.12 0.56 g

APPENDIX A

Exercise A–1

1. a. 7.62×10^2
   b. 5.38×10^5
   c. 4.26×10^7
   d. 3.8×10^{-2}
   e. 1.12×10^{-5}
2. a. 3.35×10^5
   b. 7.6×10^7
   c. 2.8×10^{-5}
   d. 5.11
3. a. 6.2
   b. 7.18×10^7
   c. 3.08×10^{-7}
   d. 9.2×10^{-12}

Exercise A–2

1. a. $\frac{1}{x^a}$
   b. 10^7
   c. $\frac{1}{3^{1/2}}$
   d. $\frac{1}{2^{-5}}$
2. 8.85×10^5
3. 1.67×10^6

Exercise A–3

1. a. 10^8
   b. 2.4×10^9
   c. 6×10^3
   d. 3.90×10^4
2. a. 10^2
   b. 5×10^{-6}
   c. 5×10^{-5}
   d. 5.14×10^{-1}
3. a. x^{10}
   b. 2.7×10^{10}
   c. 6.25×10^{-6}
   d. 8.0×10^{12}
4. a. 6×10^2
   b. 7×10^{-3}
   c. 6×10^2
   d. 2×10^3

Exercise A–4

1. a. 4
   b. 2
   c. 1
   d. 3
2. a. 7.1
   b. 2.00
   c. 9.96×10^5
   d. 69.9
   e. 3.3×10^1
3. a. 3.08
   b. 0.0764
   c. 208
   d. 14.2

Exercise A–5

1. a. 2.5
   b. 2.67
   c. 9
   d. 0.29
   e. 3.09

APPENDIX B

Exercise B–1

1. a. 28
   b. 30
   c. 108
   d. 4
   e. 42
2. a. 0.40
   b. 0.62
   c. 0.84
   d. 0.88

Exercise B–2

1. a. 3.49
   b. 3.34×10^4
   c. 1.95×10^8
   d. 4.95×10^8
   e. 2.67×10^{-5}
   f. 3.99×10^{-6}
   g. 3.19×10^4
   h. 1.61×10^3

Exercise B–3

1. a. 1.87
   b. 6.98×10^{-2}
   c. 1.97×10^4
   d. 5.39×10^{-2}
   e. 2.11×10^6
   f. 1.62×10^4
   g. 6.10×10^{-3}
   h. 9.54×10^2

Exercise B–4

1. a. 1.68
   b. 1.93×10^{2}
   c. 1.80×10^{-2}
   d. 2.99
2. a. 2.69×10^{2}
   b. 5.32×10^{13}
   c. 2.74×10^{-5}
   d. 9.80×10^{14}

Exercise B–5

1. a. 2.62
   b. 3.61×10^{1}
   c. 1.95×10^{-1}
   d. 1.83×10^{-2}
2. a. 3.97×10^{1}
   b. 3.80×10^{17}
   c. 1.64×10^{-8}
   d. 5.64×10^{14}
3. a. 0.589
   b. 2.393
   c. 4.436
   d. 5.782
4. a. 4.18
   b. 2.40×10^{4}
   c. 1.10×10^{3}
   d. 1.21×10^{6}

Exercise B–6

1. a. 2.24×10^{1}
   b. 2.31×10^{3}
   c. 2.13×10^{-2}
   d. 4.76×10^{3}
   c. 6.37×10^{3}

INDEX

Note: Page numbers followed by (t) indicate tables.

1969 INTERNATIONAL ATOMIC WEIGHTS

Based on the assigned relative atomic mass of $^{12}C = 12$

The following values apply to elements as they exist in materials of terrestrial origin and to certain artificial elements. When used with footnotes, they are reliable to ±1 in the last digit, or ±3 if that digit is in small type. (A value in parentheses denotes mass number of most stable known isotope. These have been added to the IUPAC table.)

	SYMBOL	ATOMIC NUMBER	ATOMIC WEIGHT		SYMBOL	ATOMIC NUMBER	ATOMIC WEIGHT		SYMBOL	ATOMIC NUMBER	ATOMIC WEIGHT
actinium	Ac	89	(227)	hahnium[h]	Ha	105	(260)	promethium	Pm	61	(147)
aluminum	Al	13	26.9815[a]	helium	He	2	4.00260[b,c]	protactinium	Pa	91	231.0359[a]
americium	Am	95	(243)	holmium	Ho	67	164.9303[a]	radium	Ra	88	226.0254[a,f,g]
antimony	Sb	51	121.75	hydrogen	H	1	1.0080[b,d]	radon	Rn	86	(222)
argon	Ar	18	39.948[b,c,d,g]	indium	In	49	114.82	rhenium	Re	75	186.2
arsenic	As	33	74.9216[a]	iodine	I	53	126.9045[a]	rhodium	Rh	45	102.9055[a]
astatine	At	85	(210)	iridium	Ir	77	192.22	rubidium	Rb	37	85.4678[c]
barium	Ba	56	137.34	iron	Fe	26	55.847	ruthenium	Ru	44	101.07
berkelium	Bk	97	(247)	krypton	Kr	36	83.80	rutherfordium[h]	Rf	104	(261)
beryllium	Be	4	9.01218[a]	kurchatovium[h]	Ku	104	(261)	samarium	Sm	62	150.4
bismuth	Bi	83	208.9806[a]	lanthanum	La	57	138.9055[b]	scandium	Sc	21	44.9559[a]
boron	B	5	10.81[c,d,e]	lawrencium	Lr	103	(257)	selenium	Se	34	78.96
bromine	Br	35	79.904[c]	lead	Pb	82	207.2[d,g]	silicon	Si	14	28.086[d]
cadmium	Cd	48	112.40	lithium	Li	3	6.941[c,d,e]	silver	Ag	47	107.868[c]
calcium	Ca	20	40.08	lutetium	Lu	71	174.97	sodium	Na	11	22.9898[a]
californium	Cf	98	(249)	magnesium	Mg	12	24.305[c]	strontium	Sr	38	87.62[g]
carbon	C	6	12.011[b,d]	manganese	Mn	25	54.9380[a]	sulfur	S	16	32.06[d]
cerium	Ce	58	140.12	mendelevium	Md	101	(256)	tantalum	Ta	73	180.9479[b]
cesium	Cs	55	132.9055[a]	mercury	Hg	80	200.59	technetium	Tc	43	98.9062[f]
chlorine	Cl	17	35.453[c]	molybdenum	Mo	42	95.94	tellurium	Te	52	127.60
chromium	Cr	24	51.996[c]	neodymium	Nd	60	144.24	terbium	Tb	65	158.9254[a]
cobalt	Co	27	58.9332[a]	neon	Ne	10	20.179[c]	thallium	Tl	81	204.37
copper	Cu	29	63.546[c,d]	neptunium	Np	93	237.0482[b]	thorium	Th	90	232.0381[a]
curium	Cm	96	(245)	nickel	Ni	28	58.71	thulium	Tm	69	168.9342[a]
dysprosium	Dy	66	162.50	niobium	Nb	41	92.9064[a]	tin	Sn	50	118.69
einsteinium	Es	99	(254)	nitrogen	N	7	14.0067[b,c]	titanium	Ti	22	47.90
erbium	Er	68	167.26	nobelium	No	102	(254)	tungsten	W	74	183.85
europium	Eu	63	151.96	osmium	Os	76	190.2	uranium	U	92	238.029[b,c,e]
fermium	Fm	100	(255)	oxygen	O	8	15.9994[b,c,d]	vanadium	V	23	50.9414[b,c]
fluorine	F	9	18.9984[a]	palladium	Pd	46	106.4	wolfram	W	74	183.85
francium	Fr	87	(223)	phosphorus	P	15	30.9738[a]	xenon	Xe	54	131.30
gadolinium	Gd	64	157.25	platinum	Pt	78	195.09	ytterbium	Yb	70	173.04
gallium	Ga	31	69.72	plutonium	Pu	94	(244)	yttrium	Y	39	88.9059[a]
germanium	Ge	32	72.59	polonium	Po	84	(210)	zinc	Zn	30	65.37
gold	Au	79	196.9665[a]	potassium	K	19	39.102	zirconium	Zr	40	91.22
hafnium	Hf	72	178.49	praseodymium	Pr	59	140.9077[a]				

Reproduced by permission of the International Union of Pure and Applied Chemistry. From *Pure and Applied Chemistry*. **21**(1), (1970).

[a] Mononuclidic element.
[b] Element with one predominant isotope (about 99 to 100% abundance).
[c] Element for which the atomic weight is based on calibrated measurements.
[d] Element for which variation in isotopic abundance in terrestrial samples limits the precision of the atomic weight given.
[e] Element for which users are cautioned against the possibility of large variations in atomic weight due to inadvertent or undisclosed artificial isotopic separation in commercially available materials.
[f] Most commonly available long-lived isotope.
[g] In some geological specimens this element has a highly anomalous isotopic composition, corresponding to an atomic weight significantly different from that given.
[h] Name and symbol not officially accepted. See page 337.